Single Variable Differential and Integral Calculus

Mathematical Analysis

Elimhan Mahmudov

Istanbul Technical University
Macka, 34367 Istanbul
Turkey

Azerbaijan National Academy of Sciences
Institute Cybernetics
Azerbaijan

ATLANTIS PRESS

AMSTERDAM – PARIS – BEIJING

Atlantis Press

8, square des Bouleaux
75019 Paris, France

For information on all Atlantis Press publications, visit our website at: *www.atlantis-press.com*

All Atlantis Press books are published in collaboration with Springer.

Copyright

ISBNs
Print: 978-94-6239-040-9
E-Book: 978-94-91216-86-2
ISSN: 1875-7634

About the author

Elimhan N. Mahmudov was born in Kosali, and after leaving school with golden medal, attended Azerbaijan State University, Mechanical – Mathematical Faculty. Between 1973 and 1977 he was post- graduate student at the Academy of Sciences of Ukraine, where he was guided by advisor Professor B. N. Pshenichnyi, Academician of the Academy of Sciences of Ukraine – one of the most prominent authorities in the field of theory of extremal problems, numerical methods and game theory. E. N. Mahmudov defended his thesis in Ukraine in 1980 and since that time as a Ph. Doctor and senior scientific associate worked at the Academy of Sciences of Azerbaijan until 1995. Besides he taught part time at the Azerbaijan State University and was a permanent member of the Physical-Mathematical Doctoral Sciences Committee in Azerbaijan. On the recommendation of the Physical-Mathematical Doctoral Committee of the Taras Shevchenko's Kiev State University, since 1992 he became a Doctor of Physical-Mathematical Sciences. Both Ph. D and Doctor of Sciences Diplomas are given by Moscow. In 1992, he received a GRANT Support for Mathematics by National Science Foundation in Washington, DC, and became a member of the American Mathematical Society.

Professor E. N. Mahmudov have numerous research papers devoted to Advances in Pure Mathematics, Convex Analysis, Approximation Theory, Mathematical Methods of Optimal Control Theory and Dual problems, Theory of Ordinary and Partial Differential Equations which have been published in the SCI journals with high levels in the USSR, USA and the European countries. He has devoted the majority of his time to textbooks entitled *"Mathematical Analysis and Its Applications"* (Papatya, 2002) and *"Approximation and Optimization of Discrete and Differential Inclusions"* (Elsevier, 2011; ISBN:978-0-12-388428-2). He was always invited for lectures on the problems of Mathematical Methods Optimal Control at the International Conferences in United Kingdom, France, Switzerland, Italy,

Germany, Poland, Russia, Turkey, etc. and in the World Mathematical Banach Center in Warsaw, Poland. He is an Editor of the Journal "*Advances in Pure Mathematics*" (APM).

Professor E. N. Mahmudov currently works in Istanbul Technical University, where he is teaching lectures on the problems of Nonlinear Mathematical Programming, Mathematical Analysis, Differential Equations, Game Theory and its applications, Nonlinear optimization and Numerical Analysis. He enjoys playing chess, traditional Azerbaijanian music in a national stringed instruments, traveling and is an avid painter and sculptor.

Preface

I do not know what I may appear to the world, but to myself I seem to have been only like a boy playing on the sea-shore, and diverting myself in now and then finding a smoother pebble or a prettier shell than ordinary, whilst the great ocean of truth lay all undiscovered before me

Isaac Newton (1643–1727)

Nature and Nature's laws lay hid in night;
God said, Let Newton be! and all was light.

Alexander Pope (1688–1744)

In an era of improving science and technology, the training of competent experts becomes increasingly important for a rising standard of living. Therefore, undergraduate and post-graduate students as well as scientists and engineers should improve their mastery of the various subsections of mathematics, especially of mathematical analysis, the base of mathematics.

In general, mathematical analysis consists of the function theory of real and complex variables, the theory of sequences and series, differential or integral equations, and the variational calculus. It provides general methods for the solution of various problems of both pure and applied mathematics.

Briefly, this book covers real and complex numbers, vector spaces, sets and some of their topological properties, series and sequences of functions (including complex-valued functions and functions of a complex variable), polynomials and interpolation, differential and integral calculus, extrema of functions, and applications of these topics.

The book differs from conventional studies in its content and style. Special attention is given to the right placement of subjects and problems. For instance, the properties of the

Newton, Riemann, Stieltjes, and Lebesgue integrals are handled and explained together. As is well known, mathematical analysis is also referred to as infinitesimal (incalculably small) analysis; but statements involving phrases such as "infinite numbers" may lead to serious misunderstandings. To prevent such ambiguity, we extend the idea of the number of elements in a set from finite sets to infinite sets.

In this book, approximate calculation (which plays an ever-increasing role in science) is mentioned and some care is devoted to interpolation methods and the calculation of integrals, elementary functions, and the other calculations of mathematical analysis.

Complex numbers and the corresponding formulas of De Moivre, logarithms of complex numbers, antilogarithms of complex variables, etc., are ordinary tasks for any teacher of school-mathematics. On the other hand, especially for the theory of analytic functions, there are important practical applications of the theory of complex variables to hydro- and aero-mechanics, integral calculus, algebra, and stability problems. Therefore, the arguments in this book are elaborated for complex variables.

One of the eventual objectives of this book is the application of these methods of mathematical analysis. Applied problems are presented in almost every chapter after the third, and they consist in finding the constrained or unconstrained extreme value of a function of a single variable. For instance, one of the historically important problems is the problem of Euclid, where in a given triangle it is required to inscribe the parallelogram with greatest area. Another is Johannes Kepler's problem: in a ball with radius R, inscribe the cylinder with the greatest volume. These are both examples of extremum problems of a function of one variable. It should be noted that the required parallelogram in Euclid's problem is the parallelogram with the midpoints of the triangle's sides as three of its vertices. And the volume in Kepler's problem is $4\pi R^3/3\sqrt{3}$.

The famous brachistochrone problem was posed by the illustrious Swiss mathematician Johann Bernoulli in 1696: this problem consists in finding the shape of the curve down which a bead sliding from rest and accelerated by gravity will slip (without friction) from one point to another in the least time. The solution of this problem is a cycloid: this requires finding an extremum of a function of infinitely many variables. The history of mathematics made a sudden jump: from dimension one it passed at once to infinite dimensions, or from the theory of extrema of functions of one variable to the theory of problems of the type of the brachistochrone problem, i.e., to the calculus of variations, as this branch of mathematics was called in the 18th century. Historically, the year 1696 marks the birth of the calculus of variations.

This book consists of ten chapters, which include 181 theorems, 125 definitions, 261 solved examples, 575 answered problems, 41 figures and 87 remarks. Generally, these problems have the purpose of determining whether or not the concepts have been understood both theoretically and intuitively. The theorems, remarks, results, examples, and formulas are numbered separately in each chapter. For example, Theorem 8.3 indicates the third theorem within Chapter 8. The proofs begin with the empty square "□" and end with the Halmos box, "■". They are presented in such a way that they hold good for the corresponding multi-dimensional generalization almost without any changes.

The major features of our book which, from a methodological and conceptual point of view, make it unique are the following:

- The power conception of sets, countable and uncountable sets, the cardinality of the continuum; Dedekind completeness theorem for the set of real numbers;
- Products of series;
- Polynomials and interpolations, Lagrange and Newton interpolation formulas;
- Functions with bounded variation;
- Newton and Stieltjes integrals;
- Lebesgue measure theory in line and Lebesgue integrals;
- Additional applications: potential and kinetic energy, a body in the earth's gravitational field, escape velocity, nuclear reactions;
- The uncountable Cantor set with zero measure. Existence of non-measurable sets in line;
- The existence, at the end of each chapter, a plenty of problems with answers.

In the first chapter are given the main definitions and concepts of sets, some topological properties of such sets as metric spaces, the concepts of surjective, injective, and bijective functions, countability and uncountability of sets, cardinal number, and the cardinality of the continuum. These concepts are the very basis of mathematical analysis and in other fields of the mathematical sciences. For this reason, different theorems connected with the concepts of countability and uncountability of sets are proved, and it is shown that the cardinality of the set of functions defined on a closed interval is greater than the cardinality of the continuum. Furthermore, the Dedekind completeness theorem for the set of real numbers is formulated.

In the second chapter, the basic properties of monotone, strictly monotone, and Cauchy sequences are given in detail. Moreover, different types of convergence as well as different convergence tests (Cauchy, D'Alembert, Leibniz, Raabe) are given, and the harmonic series, the geometric series, and products of series are considered. By Riemann's Theorem, for any conditionally convergent series and an arbitrary number $S \in [-\infty, \infty]$ there is a rearrangement of the given series such that the resulting series converges to S. But the rearrangement of an absolutely convergent series yields another absolutely convergent series having the same sum as the original one. The convergence or divergence of the hyperharmonic series with general term $1/n^p$ is determined, depending on p. Lastly, it is also mentioned that the analogous properties of series with complex terms hold.

In the third chapter, the equivalency of the limit concepts of Cauchy and Heine is shown. Then continuous and uniformly continuous functions, the main properties of sequences of functions and series of functions, and the concepts of pointwise and uniform convergence are studied. It is shown that under some conditions a sequence of functions defined in a closed interval has a uniformly convergent subsequence of functions (Arzela's theorem). At the end of the chapter it is mentioned that some of the features of function sequences and series generalize to the case of complex-valued functions.

In the fourth chapter the important conceptions of mathematical analysis, such as the derivative and differential, with their mechanical and geometrical meanings, are given. Derivatives of product functions (Leibniz's formula), composite functions, and inverse functions are treated. Then the concept of the derivative is applied to study the free fall of a mass, linear motion at velocities near the speed of light, and applications in numerical analysis. It is also applied to curvature and evolvents.

In the fifth chapter, the theorems of Rolle, Lagrange, and Cauchy, the mean value theorem, and Fermat's theorem for local extreme values are presented. Taylor's formula, given with the remainder term in Lagrange's, Cauchy's, and Peano's forms, is given, and the reason is indicated why sometimes the Taylor series of a function does not converge to the function itself. Furthermore, the Maclaurin series expansions of some elementary functions are given and Euler's formulas are proved.

In the sixth chapter, the factorization problem for polynomials having repeated real and complex roots is studied, and Lagrange's and Newton's interpolation formulas are given. Then, a useful formula for differential calculus is derived by using Newton's interpolation formula. At the end of the chapter, the approximation by polynomials of a given function defined on a closed interval is studied.

In the seventh chapter, L'Hopital's rule and Taylor-Maclaurin's series are employed for the computation of limits of different indeterminate forms. By using the Taylor series at critical points, sufficient conditions for a local extremum are obtained. Moreover, the determination of whether the curve is bending upward or downward depends on the sign of higher order derivatives. Then, necessary and sufficient conditions for the existence of straight line and parabolic asymptotes to a curve are proved.

In the eighth chapter, we give the properties of indefinite integrals which are the basis of mathematical analysis. It is shown that sometimes the integrals of the elementary functions cannot be evaluated in terms of finite combinations of the familiar algebraic and elementary transcendental functions. But the methods of substitution and integration by parts can often be used to transform complicated integration problems into simpler ones. Moreover, the method of integrating rational functions that are expressible as a sum of partial fractions of the first and second types is presented. In particular, the Ostrogradsky method of integration is studied. The evaluation of integrals involving irrational algebraic functions, quadratic polynomials, trigonometric functions, etc., is also considered.

In the ninth chapter, firstly, Newton's and Riemann's integrals, the conditions for their equality, and mean value theorems are studied. In term of a partition of a given interval and its mesh, necessary and sufficient conditions for existence of the Riemann integral are formulated, and the Fundamental Theorem of Calculus is proved. Furthermore, with the use of the concept of measure zero, the Lebesgue criterion is formulated. It is also shown that the measure of the Cantor set (which has the cardinality of the continuum) is zero. Moreover, the Lebesgue integral, the integrability of Dirichlet's and Riemann's functions, functional series, the improper integral (in the sense of Cauchy principal values), the Stieltjes integral (with the use of a function of bounded variation), and numerical integration are studied.

Finally, the tenth chapter begins by using the length, area, and volume concepts to compute arc lengths, surface areas of surfaces of revolution, and volumes by the method of cylindrical shells in Cartesian and polar coordinates. Then, centroids of plane regions and curves, and the moments of a curve and of a solid are calculated. It is shown that the definite integral can be applied to the law of conservation of mechanical energy, to Einstein's theory of relativity, and to the determination of the escape velocity from the earth. Finally, problems connected with atomic energy, critical mass, and atomic reactors are considered. In fact, the basic cause of the destabilization and explosion of the reactor of Chernobyl's nuclear power station in the Ukraine in 1986 is explained.

This book could be all the more useful for the disciplines of mathematics, physics, and engineering, for the logical punctiliousness of these features. For this reason, it can be suggested as a text-book or a reference book for students and lecturers in faculties of mathematics, physics, and engineering, as well as engineers who want to extend their knowledge of mathematical analysis. References [1]–[36] may be consulted for discussed topics pertinent to calculus.

The book was written as a result of my experience lecturing at Azerbaijan State University, Istanbul University, and Istanbul Technical University. I express my gratitude to the Institute of Cybernetics of the National Academy of Sciences of Azerbaijan for providing a rich intellectual environment, and facilities indispensable for the writing of this textbook.

Professor Elimhan N. Mahmudov
Baku and Istanbul, September, 2012

Acknowledgements

During the work on this book, I have had the pleasure of discussing its various aspects with many talented people, colleagues, and friends, and my heartfelt thanks go out to them all. The material of this book has been used over the years for teaching advanced classes on differential and integral calculus attended mostly by my doctoral students and collaborators at Azerbaijan State University and Istanbul Technical University, Turkey. I'm indebted to the Institute of Cybernetics, headed by T.A. Aliev, Academician of the National Academy of Sciences of Azerbaijan. I am grateful to Prof. Cengiz Kahraman, (Department Chair) and other colleagues for their help and moral support during the preparation of this book. Especially valuable help was provided by Prof. M. Nahit Serarslan and Gülgün Kayakutlu. I thank my doctoral students Murat Engin Ünal and Kutay Tinç for their interest, tolerance, and endless patience in entering this book into the computer and answering my numerous questions on English. I'm very grateful to the publishing director Zeger Karssen, Project Manager Willie van Berkum, and supportive editor of Atlantis Press (Springer) Arjen Sevenster for their help during the preparation and publishing of this book. I don't have enough words to thank my grandchild Leyla who has inspired me at every stage of this project.

Baku and Istanbul, September, 2012   Elimhan N. Mahmudov

(Industrial Engineering Department,
Istanbul Technical University, Turkey;
Azerbaijan National Academy of Sciences,
Institute of Cybernetics, Azerbaijan).

Contents

About the author   iii

Preface   v

Acknowledgements   xi

1. Introduction to Numbers and Set Theory   1

    1.1   Sets and functions   1

    1.2   Cardinality   5

    1.3   The Main Properties of the Real Numbers   8

    1.4   Uncountable Sets 14

    1.5   Construction of the Complex Number Field 17

    1.6   Metric Spaces 21

    1.7   Problems 28

2. Sequences and Series   31

    2.1   The Basic Properties of Convergent Sequences 31

    2.2   Infinitely Small and Infinitely Large Sequences; Subsequences 40

    2.3   Cauchy Sequences 42

    2.4   Numerical Series; Absolute and Conditional Convergence 45

    2.5   Alternating series; Convergence tests 50

    2.6   Products of Series 59

    2.7   Problems 62

3. Limits and Continuity of Functions   **67**

3.1   Functions of a Real Variable 67

3.2   Limits of Functions 69

3.3   Local Properties of a Continuous Function 75

3.4   Basic Properties of Continuous Functions on Closed Intervals 78

3.5   Monotone and Inverse Functions 82

3.6   The Main Elementary Functions 84

3.7   Comparison of Infinitely Small and Infinitely Large Functions 91

3.8   Sequences and Series of Functions 92

3.9   Problems 101

4. Differential Calculus   **107**

4.1   Derivatives: Their Definition and Their Meaning for Geometry and Mechanics 107

4.2   Differentials 109

4.3   Derivatives of Composite and Inverse Functions 113

4.4   Derivatives of Sums, Products, and Quotients 116

4.5   Derivatives of Basic Elementary Functions 118

4.6   Derivatives of Trigonometric and Inverse Trigonometric functions 121

4.7   Higher Order Derivatives and Differentials 123

4.8   Derivatives of Functions Given by Parametric Equations 127

4.9   Curvature and Evolvents 129

4.10   Problems 139

5. Some Basic Properties of Differentiable Functions   **145**

5.1   Relative Extrema and Rolle's Theorem 145

5.2   The Mean Value Theorem and its Applications 147

5.3   Cauchy's Mean Value Theorem 150

5.4   Taylor's Formula 151

5.5   Lagrange's, Cauchy's, and Peano's Forms for the Remainder Term; Maclaurin's formula 153

5.6   Taylor Series; Maclaurin Series Expansion of Some Elementary Functions 155

5.7   Power Series 161

5.8   Problems 167

6. Polynomials and Interpolations 171

6.1 Factorization of Polynomials 171

6.2 Repeated Roots of Polynomials 174

6.3 Complex Roots of Polynomials 175

6.4 Polynomials and Interpolations 176

6.5 Problems 182

7. Applications of Differential Calculus to Limit Calculations and Extremum Problems 185

7.1 L'Hopital's Rule for the Indeterminate Form $\frac{0}{0}$ 185

7.2 L'Hopital's Rule for the Indeterminate Form $\frac{\infty}{\infty}$ 188

7.3 Taylor–Maclaurin Series and Limit Calculations 192

7.4 Sufficient Conditions for a Relative Extremum 194

7.5 Global Extrema of a Function on a Closed Interval 200

7.6 Higher Derivatives and Concavity 208

7.7 Asymptotes of Graphs; Sketching Graphs of Functions 211

7.8 Problems 216

8. The Indefinite Integral 223

8.1 The Antiderivative and the Indefinite Integral 223

8.2 Basic Properties of the Indefinite Integral 225

8.3 Integration by Substitution 227

8.4 Integration by Parts 230

8.5 Partial Fractions; Integrating Rational Functions 232

8.6 Integrals Involving Quadratic Polynomials 241

8.7 Integration by Transformation into Rational Functions 245

8.8 Problems 251

9. The Definite Integral 259

9.1 Newton's Integral 259

9.2 Mean Value Theorems for the Newton Integral 262

9.3 The Riemann Integral 264

9.4 Upper and Lower Riemann Sums and Their Properties 268

9.5   Necessary and Sufficient Conditions for the Existence of the Riemann
        Integral; Basic Properties of the Riemann Integral 272

9.6   The Class of Riemann Integrable Functions; Mean Value Theorems . . . 278

9.7   The Fundamental Theorem of Calculus; Term-wise Integration of Series . 282

9.8   Improper Integrals 286

9.9   Numerical Integration 296

9.10   Functions of Bounded Variation 301

9.11   Stieltjes Integrals 307

9.12   The Lebesgue Integral 311

9.13   Problems 329

10.   Applications of the Definite Integral   335

10.1   Arc length 335

10.2   Area and the Definite Integral 339

10.3   Volume and the Definite Integral 344

10.4   Surface Area 348

10.5   Centroids of Plane Regions and Curves 351

10.6   Additional Applications: Potential and Kinetic Energy, the Earth's Grav-
        itational Field and Escape Velocity, Nuclear Reaction 354

10.7   Problems 362

Bibliography   367

Index   369

Chapter 1

Introduction to Numbers and Set Theory

Here we will give the main definitions and concepts of sets, the concepts of surjective, injective, and bijective functions, countability and uncountability of sets, cardinal number, and the cardinality of the continuum. These are very important in mathematical analysis and in other fields of the mathematical sciences. For this reason, different theorems connected with the concepts of countability and uncountability of sets are proved and it is shown that the cardinality of the set of functions defined on a closed interval is greater than the cardinality of the continuum. Furthermore, the Dedekind completeness theorem for the set of real numbers is proved. Finally, some of the topological properties of metric spaces are treated.

1.1 Sets and functions

The concept of a set is fundamental in mathematics. A set can be thought of as a collection (assemblage, aggregate, class, family) of objects called members or elements of the set. Roughly speaking, a set is any identifiable collection of objects of any sort. We identify a set by stating what its members (elements, points) are. For instance, we can define the set of students in the classroom, the set of positive integers, the set of all triangles in a plane, etc. Obviously, these elements have the same common properties for the given set. In general, unless otherwise specified, we denote a set by a capital letter such as A, B, C, etc. If the object x is in a set A, then we say that x is an *element* (or a *member*) of A and write $x \in A$. If x is not an element of A, then we write $x \notin A$. A set having finitely many elements is called a *finite set*, otherwise it is called an *infinite set*. The *empty set* is the set having no elements and is denoted by $\varnothing$. It is a subset of any set $\varnothing \subset A$. If each element of a set A also belongs to a set B we call A a *subset* of B, written $A \subset B$, and read "A is contained in B." The fact that a set A consists of the elements x is denoted by $A = \{x\}$. If A is not equal to B, we write $A \neq B$. It is clear that $A \subset A$. If $A \subset B$, but $A \neq B$ and A is nonempty, we call A a *proper subset* of B.

E. Mahmudov, *Single Variable Differential and Integral Calculus*,
DOI: 10.2991/978-94-91216-86-2_1, © Atlantis Press 2013

Definition 1.1. If for given sets A and B we have $A \subset B$ and $B \subset A$, then we say that A and B are equal and write $A = B$.

Definition 1.2. If A_k $(k = 1, \ldots, n)$ are sets, their union S is the set of all elements x such that for at least one k, $x \in A_k$. Then S is denoted

$$S = \bigcup_{k=1}^{n} A_k.$$

If there are infinitely many sets, e.g., $k = 1, 2, 3, \ldots$, then we write ∞ instead of n.

Definition 1.3. The intersection Q of the sets A_k $(k = 1, \ldots, n)$ is the set of elements x such that for every k, $x \in A_k$. Then Q is denoted by

$$Q = \bigcap_{k=1}^{n} A_k.$$

For infinitely many sets we write $Q = \bigcap_{k=1}^{\infty} A_k$.

Example 1.1. Let $A = \{x : x \text{ is an odd integer}\}$ and $B = \{x : x^2 - 4x + 3 = 0\}$. We will show that $B \subset A$. Since B is the set of points satisfying $(x - 1)(x - 3) = 0$, we have $B = \{1, 3\}$. Thus every element of B belongs to A, and so $B \subset A$.

Example 1.2. If $A_1 = \{1\}, A_2 = \{1, 2, 3\}$, then $S = A_1 \cup A_2 = \{1, 2, 3\}$; $Q = A_1 \cap A_2 = \{1\}$.

Definition 1.4. The set consisting of all elements of A which don't belong to B is called the difference of A and B, and is denoted by $A \smallsetminus B$, that is,

$$A \smallsetminus B = \{x : x \in A \text{ and } x \notin B\}.$$

If M is a set which contains all the elements being considered in a certain context, we sometimes call M the universal set. If $B \subset M$, then $M \smallsetminus B$ is called the complement of B and is denoted by $\complement B$. A universe M can be represented geometrically by the set of points inside a rectangle, called the Venn diagram. If $B \subset A$, then $A \smallsetminus B$ is called the complement of B relative to A and is denoted $\complement_A B$. It follows that

$$B \cup \complement_A B = A; \quad B \cap \complement_A B = \varnothing; \quad \complement_A \varnothing = A; \quad \complement_A A = \varnothing.$$

Example 1.3. Let A be the set of positive integers and B be the set of positive and even integers. Then $A \smallsetminus B$ is the set of positive odd integers.

Theorem 1.1 (De Morgan's law[1]). *Let B_k be a family (finite or infinite) of subsets of A. Then the complement relative to A of the intersection (union) of the family B_k is the union (intersection) of the relative complements:*

$$\complement_A \bigcap_k B_k = \bigcup_k \complement_A B_k,$$

$$\complement_A \bigcup_k B_k = \bigcap_k \complement_A B_k.$$

☐ Let us prove the first of them (the second formula is proved analogously). The first requirement made by Definition 1.1 is that $\complement_A \bigcap_k B_k \subseteq \bigcup_k \complement_A B_k$. We prove this. Take an element $x \in \complement_A \bigcap_k B_k$, then by Definition 1.3 $x \notin \bigcap_k B_k$. Thus, at least for one k_0, $x \notin B_{k_0}$. Hence $x \in A \setminus B_{k_0} = \complement_A B_{k_0}$, and so $x \in \bigcup_k \complement_A B_k$. Consequently, $\complement_A \bigcap_k B_k \subseteq \bigcup_k \complement_A B_k$. Now, let $x \in \bigcup_k \complement_A B_k$. Again, at least for one k_0, we have $x \in \complement_A B_{k_0}$. This means that $x \in \complement_A \bigcap_k B_k$, and then $\bigcup_k \complement_A B_k \subseteq \complement_A \bigcap_k B_k$. Comparing the two relations obtained, we conclude the desired result by Definition 1.1. ∎

The following simple properties of sets are made plausible by considering a Venn diagram.

(1) $A \cup B = B \cup A$, $A \cap B = B \cap A$ (commutative law for unions and intersections).
(2) $(A \cup B) \cup C = A \cup (B \cup C)$, $(A \cap B) \cap C = A \cap (B \cap C)$ (associative law for unions and intersections).
(3) $A \cap (B \cup C) = (A \cap B) \cup (A \cap C)$ (distributive law).
(4) $(A \setminus B) \cap C = (A \cap C) \setminus (B \cap C)$.
(5) $A \subset A \cup B$; $B \subset A \cup B$; $A \cap B \subset A$; $A \cap B \subset B$.
(6) $A \subset B$ if and only if $A \cup B = B$ and $A \cap B = A$.
(7) $A \cup \varnothing = A$, $A \cap \varnothing = \varnothing$.

Let A and B be two sets. Then by a function (map) f from (or on) a set A to (or onto) a set B we mean a rule which assigns to each x in A a unique element $f(x)$ in B. We often express the fact that f is a function from A to B by writing

$$f : A \to B.$$

The set A is called the *domain* of f. The set of values $y = f(x)$ is called the *range* of f. That is, the range of f is the set defined by

$$f(A) = \{f(x) : x \in A\}.$$

---

[1]August De Morgan (1806-1871), French mathematician

Note that $f(A) \subset B$ but might not equal B.

We say f is *onto* or *surjective* if every $y \in B$ is of the form $f(x)$ for some $x \in A$. Thus the range of f is $f(A)$ and f is onto if and only if $f(A) = B$.

We say f is *one-to-one* or *injective* or *univalent* if, for every pair $x_1, x_2 \in A$, $x_1 \neq x_2$ implies $f(x_1) \neq f(x_2)$ (or if $f(x_1) = f(x_2)$ implies $x_1 = x_2$).

A function $f : A \to B$ is a *bijection* (or *one-to-one correspondence*) if it is injective and surjective.

Let $\mathbb{N} = \{1, 2, 3, \ldots\}$ and $\mathbb{Z} = \{0, \pm 1, \pm 2, \pm 3, \ldots\}$ be the set of natural numbers and the set of integers, respectively. Then it is not hard to see that the function $f : \mathbb{Z} \to N$, given by $y = f(x) = |x| + 1$, is onto, the function $f : \mathbb{Z} \to \mathbb{Z}$, given by $y = f(x) = 2x$ is one-to-one, and the function $f : \mathbb{Z} \to \mathbb{Z}$, defined by $y = f(x) = x + 1$, is a bijection.

If $f : A \to B$ is a bijection, that is, one-to-one and $f(A) = B$, then there is an inverse function $f^{-1} : B \to A$ defined by $y = f(x)$. In words, for every y in B there is a unique element x in A defined by the rule $x = f^{-1}(y)$. It is clear that $(f^{-1})^{-1} = f$, that is, f and f^{-1} are inverses each other. For example, if $f(x) = e^x$, $x \in \mathbb{R} = (-\infty, +\infty)$, then $f : \mathbb{R} \to (0, \infty)$ is a bijection, and so has an inverse.

Example 1.4. Let A be the set of positive integers, i.e., $A = \mathbb{N} = \{1, 2, 3, \ldots, n, \ldots\}$, and let $B = \{1, 2\}$. Then the function given by

$$y = f(x) = \frac{3 - (-1)^x}{2} = \begin{cases} 1, & \text{if } x \text{ is an even,} \\ 2, & \text{if } x \text{ is an odd} \end{cases}$$

is onto, but in general is not one-to-one. But if $A = \{1, 2\}$, then f is both one-to-one and onto.

Definition 1.5. Two sets, A and B, are *equivalent* if there exists a bijective function $f : A \to B$. In this case, we write $A \sim B$. Conversely, if the sets A and B are not equivalent, we write $A \nsim B$.

The idea is that the sets A and B have the same number of elements.

The equivalence relation $\sim$ has the following properties.

(1) $A \sim A$ (i.e. $\sim$ is *reflexive*).
(2) If $A \sim B$, then $B \sim A$ (i.e., $\sim$ is *symmetric*).
(3) If $A \sim B$ and $B \sim C$ then $A \sim C$ (i.e., $\sim$ is *transitive*).

□ The first claim is clear. For the second, let $f : A \to B$ be a bijection. Then the inverse function $f^{-1} : B \to A$ is also a bijection. For the third, let $f : A \to B$ be a bijection and let $g : B \to C$ be a bijection. Then the composite $g \circ f : A \to B$ is also a bijection. ■

Example 1.5. Let us denote $\mathbb{N} = \{1, 2, \ldots, n, \ldots\}$, $B = \{2, 4, \ldots, 2n, \ldots\}$. It is clear that the function $f : \mathbb{N} \to B$ given by $y = f(x) = 2x$ is the bijection required by Definition 1.4. Then it follows that $\mathbb{N} \sim B$.

Example 1.6. Let A be the set of integers, that is, $A = \mathbb{Z}$. Show that $A \sim \mathbb{N}$. Indeed it is clear that the needed bijection $f : \mathbb{N} \to A$ can be defined as

$$f(x) = \begin{cases} \dfrac{x}{2}, & \text{if } x \text{ is an even,} \\ -\dfrac{x-1}{2}, & \text{if } x \text{ is an odd.} \end{cases}$$

1.2 Cardinality

In this section we extend the idea of the number of elements in a set from finite sets to infinite sets.

Definition 1.6. A set A is *denumerable* if it is equivalent to the set of natural numbers $\mathbb{N} = \{1, 2, 3, \ldots\}$. A set A is *countable* if it is finite or denumerable. If a set is denumerable, we say that it has *cardinality d*, or, *cardinal number d*.

Thus, if for a positive integer n we denote $J_n = \{1, 2, \ldots, n\}$, then a set A is countable if either $A \sim J_n$ or $A \sim \mathbb{N}$.

Note that a sequence $x_1, x_2, x_3, \ldots$, or, $\{x_n\}$, can be considered as a function, given by the formula $f(n) = x_n$, $n \in \mathbb{N}$. Then a set A is denumerable if and only if its members can be enumerated in a sequence $\{x_n\}$. In general, this fails to hold for infinite sets.

Definition 1.7. With every set A we associate a notation $\overline{\overline{A}}$ called the *cardinal number* of A.

Suppose A and B are finite sets with numbers of elements n and m, respectively. Obviously, $n = m$ if and only if $A \sim B$. In this case we write $\overline{\overline{A}} = \overline{\overline{B}}$. On the other hand, $n > m$ ($n < m$) if and only if $A \not\sim B$ but $A_1 \sim B$ ($A \sim B_1$) for some subset $A_1 \subset A$ ($B_1 \subset B$). In the first and second cases we write $\overline{\overline{A}} > \overline{\overline{B}}$ and $\overline{\overline{A}} < \overline{\overline{B}}$, respectively. If $A = \varnothing$, we write $\overline{\overline{A}} = 0$.

The following definition extends the idea of the number of elements to infinite sets.

Definition 1.8. Suppose A and B are two infinite sets and $A_1 \subset A$ and $B_1 \subset B$ are subsets. Consider the following three cases.

(1) $B \sim A_1$ and A is nonequivalent to any subset of B.
(2) $A \sim B_1$ and B is nonequivalent to any subset of A.

(3) $A \sim B_1$ and $B \sim A_1$.

In these cases (1), (2), and (3), we write $\overline{\overline{A}} > \overline{\overline{B}}, \overline{\overline{A}} < \overline{\overline{B}}$, and $\overline{\overline{A}} = \overline{\overline{B}}$, respectively.

Remark 1.1. If A is a finite set, that is, $A \sim J_n$, then A is nonequivalent to all its proper subsets (if $A_1 \subset A$ then $A_1 \neq A$, $A_1 \neq \varnothing$). For infinite sets this is not the case. Indeed, in Example 1.5 we have seen that the set of even integers is denumerable (and similarly for the set of odd integers), that is $B \subset A$ and $A \sim B$.

Theorem 1.2. *Any subset of a countable set is countable.*

☐ Let A be countable and let $A = \{x_1, x_2, x_3, \ldots, x_n\}$ if A is finite, and $A = \{x_n\}$ if A is denumerable. Let E be any subset of A. Then we construct a subsequence $\{x_{n_K}\}$, $k = 1, 2, \ldots$ enumerating E by taking x_{n_K} to be the k-th member of the original sequence which is in E. Either this process never ends, in which case E is denumerable, or it does end in a finite number of steps, in which case E is finite. Note that in this case the function defined as $f(k) = x_{n_K}$ is both one-to-one and onto from $\mathbb{N}$ to E. ∎

Theorem 1.3. *Let $\{E_n\}$, $n \in \mathbb{N}$ be sequences of denumerable sets. Then*

$$A = \bigcup_{n \in \mathbb{N}} E_n$$

is denumerable.

☐ For every n, the set E_n is denumerable. Hence, its members can be enumerated in a sequence $E_k = \{x_1^k, x_2^k, x_3^k, \ldots\}$.
Now arrange the members of the each E_k in a row as shown:

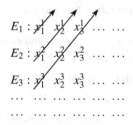

Then, obviously, we can write a sequence that contains all elements of the set A so that

$$x_1^1; \ x_1^2, \ x_2^1; \ x_1^3, \ x_2^2, \ x_3^1; \ x_1^4, \ x_2^3, \ x_3^2, \ x_4^1; \ \ldots \tag{1.1}$$

If the two of these sets have the same members, then they appear in sequence (1.1) more than once. This means that there is a subset J of the set of positive integers $\mathbb{N}$ such that $A \sim J$. It follows that the set A is at most denumerable (Theorem 1.2). On the other hand, since $E_1 \subset A$ and E_1 is denumerable, then A is denumerable. ∎

Corollary 1.1.

(1) *The union of finitely many denumerable sets is denumerable.*

(2) *The countable union of pairwise disjoint denumerable sets*

   $A_k \ (k \geqslant 2) \ (A_i \cap A_j = \varnothing, \ i \neq j)$ *is denumerable.*

A rational number is defined as $r = \dfrac{a}{b}$, where a and b are integers, and $b \neq 0$.

Theorem 1.4. *The set of all rational numbers is denumerable.*

□ It is clear that each positive rational number $r = m/n$ (m and n, positive integers) is contained in at least one of the sets E_k, $k \in \mathbb{N}$, where

$$E_1 = \{1, 2, 3, \ldots\},$$
$$E_2 = \left\{\frac{1}{2}, \frac{2}{2}, \frac{3}{2}, \ldots\right\},$$
$$E_3 = \left\{\frac{1}{3}, \frac{2}{3}, \frac{3}{3}, \ldots\right\}, \tag{1.2}$$
$$\cdots\cdots\cdots\cdots$$

In the k-th row of (1.2) are listed all positive rationals of the form m/k for some m. By Theorem 1.3, the set of positive rationals, $A = \bigcup_{n \in \mathbb{N}} E_n$, is denumerable. Because the set of all rational numbers is the union of the set of positive rational numbers, the set of negative rational numbers, and the set consisting of the number zero, it follows that it is denumerable. ∎

Theorem 1.5. *If a set A is infinite, then A contains a denumerable subset.*

□ Because $A \neq \varnothing$, there is at least one element x_1 in A. Because A is not finite, $A \neq \{x_1\}$, and so there exists x_2 in A, $x_2 \neq x_1$. Similarly, there exists x_3 in A, $x_3 \neq x_2$, $x_3 \neq x_1$. This process will never terminate, as otherwise $A \sim \{x_1, x_2, \ldots, x_n\}$ for some positive integer n. Therefore we have constructed a denumerable set $A_1 = \{x_1, x_2, x_3, \ldots\}$, where $A_1 \subset A$. ∎

The set of all subsets of a set A is usually denoted by $P(A)$. In particular, $\varnothing \in P(A)$ and $A \in P(A)$.

Theorem 1.6 (Cantor). *If A is any set, then $\overline{\overline{A}} < \overline{\overline{P(A)}}$.*

☐ Since $\overline{\overline{\varnothing}} = 0 < 1 = \overline{\overline{P(\varnothing)}}$, we may suppose, without loss of generality, that $A \neq \varnothing$. The function f defined on A by $f(x) = \{x\} \in P(A)$ is bijective (one-to-one and onto), and so $\overline{\overline{A}} \leqslant \overline{\overline{P(A)}}$. Suppose that $\overline{\overline{A}} = \overline{\overline{P(A)}}$. Then there is a bijective function $g : A \to P(A)$. Define a new set

$$B = \{x \in A \ : \ x \notin g(x)\}.$$

Because $B \subset A$, we have $B \in P(A)$. Hence, since g is onto $P(A)$, there exists an element $a \in A$ such that $g(a) = B$. Thus there are two alternatives: either $a \in B$ or $a \notin B$. In the first case ($a \in B$), we have $a \notin g(a) = B$ by the definition of B. Hence, $a \notin B$ and $a \notin g(a)$, which implies that $a \in B$. The obtained contradiction shows that $\overline{\overline{A}} \neq \overline{\overline{P(A)}}$. Hence $\overline{\overline{A}} < \overline{\overline{P(A)}}$. ∎

Thus, this theorem shows that there is no largest cardinal number. Indeed, applying Theorem 1.6 to $A = \mathbb{R}$, $P(\mathbb{R})$, $P(P(\mathbb{R})), \ldots$, we obtain an increasing sequence of cardinal numbers. Then the question arises, does there exist a set with "smallest" cardinality? We show that this set is just the set of positive integers, $\mathbb{N}$. Let M be an arbitrary infinite set. As in the proof of the Theorem 1.5, we can construct a subset $A_1 \subset A$ which is equivalent to the set of positive integers $\mathbb{N}$. Then it follows that $\overline{\overline{\mathbb{N}}} \leqslant \overline{\overline{M}}$. Thus, the set of natural numbers $\mathbb{N}$ is the smallest infinite set.

1.3   The Main Properties of the Real Numbers

We begin with a short construction of the real numbers, assuming the rational numbers as known. We will say that every nonterminating decimal fraction represents a real number. At first, it is more convenient to investigate nonnegative decimal fractions. Let $a \geqslant 0$ and $0 \leqslant \alpha_i \leqslant 9$ $(i = 1, 2, \ldots)$ be integers. If there exists a positive integer n such that $\alpha_i = 0$ for all $i > n$, then we say that the decimal fraction

$$x = a.\alpha_1 \alpha_2 \ldots \alpha_n \ldots \tag{1.3}$$

is represented by a terminating decimal expansion. Thus a terminating decimal fraction has the form

$$x = a.\alpha_1 \alpha_2 \ldots \alpha_n.$$

Note that any terminating decimal can be written in nonterminating form by adding zeros:

$$x = a.\alpha_1 \alpha_2 \ldots \alpha_n 000000 \ldots$$

Obviously, the terminating decimal expansion represents the rational number

$$x_n = a + \sum_{k=1}^{n} \alpha_k 10^{-k} \tag{1.4}$$

Definition 1.9. The rational number x_n given by formula (1.4) is called the lower *n-place accuracy approximation* of the real number (1.3). Similarly, the rational number

$$\bar{x}_n = x_n + 10^{-n} \tag{1.5}$$

is called the *upper n-place accuracy approximation* of the real number (1.3). It is easy to see that the rational numbers x_n are nondecreasing,

$$x_0 \leqslant x_1 \leqslant x_2 \leqslant \cdots,$$

and the upper approximations $\bar{x}_n$ are nonincreasing,

$$\bar{x}_0 \geqslant \bar{x}_1 \geqslant \bar{x}_2 \geqslant \cdots.$$

Definition 1.10. We say that a real number $x = a.\alpha_1\alpha_2 \ldots \alpha_n \ldots$ is *greater than* a real number $y = b.\beta_1\beta_2 \ldots \beta_n \ldots$ if there is an integer $n_0 \geqslant 0$ such that $x_{n_0} > \bar{y}_{n_0}$ and we denote $x > y$ or $y < x$.

If for two real numbers x and y neither $x > y$ nor $y > x$, we say that they are equal and write $x = y$.

Because the lower approximation of a real number x is nondecreasing and the upper approximation is nonincreasing it follows that $x_{n_0} > \bar{y}_{n_0}$ implies the inequality $x_n > \bar{y}_n$ for all $n > n_0$. Thus, simultaneous fulfillment of both $x > y$ and $x < y$ is impossible.

Note that taking $x_n = -a.\alpha_1 \ldots \alpha_n - 10^{-n}$, $\bar{x}_n = -a.\alpha_1\alpha_2 \ldots \alpha_n$ for negative numbers $x = -a.\alpha_0\alpha_1 \ldots \alpha_n \ldots$ $(x < 0)$ then, by analogy, "$<$" for the set of negative numbers can be defined.

Moreover, according to the definition it can be shown that two different decimal expansions $x = a.\alpha_1\alpha_2 \ldots \alpha_n \ldots$ and $y = b.\beta_1\beta_2 \ldots \beta_n \ldots$ define the same real number if and only if one of the expansions has the form

$$x = a.\alpha_1\alpha_2 \ldots \alpha_k 999 \ldots$$

and other has the form

$$x = a.\alpha_1\alpha_2 \ldots \alpha_k + 10^{-k}$$

Thus, some rational numbers may have two equivalent nonterminating expansions.

For example, the decimal expansions $5.7000\ldots$ and $5.6999\ldots$ represent equals.

Remark 1.2. It can be shown that every rational numbers can be represented by a decimal expansion which either terminates or repeats.

Example 1.7. These numbers are rational: $\frac{1}{5} = 0.2$; $\frac{23}{18} = 1.2777...$ (7 repeats itself to make a nonterminating expansion); $\frac{203}{165} = 1.2303030...$ (30 repeats); $\frac{23}{99} = 0.2323...$ (23 repeats); $\frac{1}{3} = 0.333...$ (3 repeats); $\frac{5}{7} = 0.71428571428571428...$ (571428 repeats).

Definition 1.11. A real number that is not rational is called *irrational*. More precisely, the decimal expansion of an irrational number is both nonterminating and nonrepeating.

Example 1.8. Since the decimal expansions of the real numbers $2.1010010001000010...$, $\sqrt{2}$, $\sqrt{3}$, $\pi = 3.14159265358...$, and $e = 2.7182818284...$ are nonterminating and nonrepeating, they represent irrational real numbers.

Note that any rational number is algebraic, i.e., is a solution of a polynomial equation of the form

$$a_0 x^n + a_1 x^{n-1} + \cdots + a_n = 0,$$

where $a_0, \ldots, a_n$ and $n > 0$ are integers. It should be pointed out that an irrational number may or may not be algebraic. For instance, $\sqrt{2}$ is algebraic, but the irrational π is not.

We now show the transitivity property is true for the relation $>$, i.e., the inequalities $x > y$ and $y > z$ imply $x > z$. Indeed, if $x > y$, there is a positive number n_0 such that $x_n \geqslant \bar{y}_n$ for all $n \geqslant n_0$. Similarly, if also $y > z$, there exists a natural number n_1 such that $y_n \geqslant \bar{z}_n$ for all $n \geqslant n_1$. Let N be the greater of n_0 and n_1. Then, obviously, $x_n \geqslant \bar{y}_n > y_n \geqslant \bar{z}_n$ for every $n \geqslant N$. It follows that $x_n > \bar{z}_n$ and so $x > z$. Finally, we have an ordering on the set of real numbers. For any pair of real numbers x and y, one of the relations $x = y$, $x > y$, $y > x$ is fulfilled.

According to our definitions for any real number x we have $\bar{x}_n \geqslant x \geqslant x_n$ for every $n = 0, 1, 2, \ldots$ Since $\bar{x}_n - x = 10^{-n}$ each of the numbers $\bar{x}_n, x_n$ is an approximation of number x accurate to 10^{-n}. In such a case we can then approximate, from above and below, x with the use of rational numbers $\bar{x}_n, x_n$ with any desired degree of accuracy simply by choosing n sufficiently large.

Theorem 1.7. *For any two real numbers x and y, if $x < y$, then there exists a rational number r such that $x < r < y$.*

□ Since $x < y$, there is an $n \geqslant 0$ such that $\bar{x}_n < y_n$, where $\bar{x}_n, y_n$ are the upper and lower n-place accuracy approximations for x and y, respectively. For the rational number $r = (\bar{x}_n + y_n)/2$, we have clearly the inequality $x \leqslant \bar{x}_n < r < y_n \leqslant y$. Consequently, $x < r < y$. ∎

It follows from this theorem that the rational numbers are dense in the set of real numbers, in the sense that between any two distinct real numbers there is a rational number.

Remark 1.3. Each real number is represented by precisely one point of the real line and each point of the real line represents precisely one real number. By convention, the positive numbers lie to the right of zero and the negative numbers to its left.

Below are given the main algebraic properties of the four operations: addition, multiplication, subtraction, and division. First we define the sum and product of two numbers x and y.

The sum $x + y$ of the real numbers x and y is that real z so that for all naturals n the inequality

$$x_n + y_n \leqslant z \leqslant \bar{x}_n + \bar{y}_n$$

holds, where x_n, y_n and $\bar{x}_n$, $\bar{y}_n$ are the lower and upper n-place accuracy approximations, respectively. It is not hard to show that there exists a unique number z satisfying all these inequalities simultaneously.

The product xy of the nonnegative numbers x and y is the number z such that

$$x_n y_n \leqslant z \leqslant \bar{x}_n \bar{y}_n$$

for all positive integers n. As with the sum, the product is uniquely defined. Now, it is easy to define subtraction and division of two real numbers.

For all real numbers x, y, z the following assertions are true.

1. If $x > y$ and $y > z$ then $x > z$ (transitivity).
2. $x + y = y + x$ (commutative law for addition).
3. $(x + y) + z = x + (y + z)$ (associative law for addition).
4. $x + 0 = x$ (zero, or, the additive identity).
5. $x + (-x) = 0$ (Negative, or, additive inverse).
6. $xy = yx$ (commutative law for multiplication).
7. $(xy)z = x(yz)$ (associative law for multiplication).
8. $x \cdot 1 = x$ (multiplicative identity).
9. $x \cdot \frac{1}{x} = 1$ $(x \neq 0)$ (reciprocal, or, multiplicative inverse).
10. $(x + y)z = xz + yz$ (distributive law).
11. $x > y$ implies $x + z > y + z$.
12. $x > y$ and $z > 0$ implies $xz > yz$.

In this book we will use the following standard notation for intervals:

$$[a,b] = \{x : a \leqslant x \leqslant b\}, \qquad (a,b) = \{x : a < x < b\},$$
$$(a,\infty) = \{x : x > a\}, \qquad [a,\infty) = \{x : x \geqslant a\}$$

with similar definitions for $[a,b), (a,b], (-\infty,a], (-\infty,a)$.

Moreover, the absolute value of a real number x is defined by

$$|x| = \begin{cases} x, & \text{if } x \geqslant 0, \\ -x, & \text{if } x < 0. \end{cases}$$

The definition of absolute value and the properties of inequalities listed above imply the following important facts for real numbers x and y.

1. $|xy| = |x| \, |y|$
2. $|x+y| \leqslant |x| + |y|$
3. $\big||x| - |y|\big| \leqslant |x-y|$

Definition 1.12. Let A be a subset of the set $\mathbb{R}$ of all real numbers. If there exists a number $M(m)$ such that $x \leqslant M \, (m \leqslant x)$ for all $x \in A$, then we say that the number $M(m)$ is an *upper (lower) bound* for A. A set A is *bounded* if there exist upper and lower bounds M and m for A.

The least upper bound and the greatest lower bound are denoted by $\sup A$ (supremum A) and $\inf A$ (infimum A), respectively.

It is easy to see that $M = \sup A \ (m = \inf A)$ if and only if:

1) $x \leqslant M \ (m \leqslant x)$, for all $x \in A$;
2) for every $\varepsilon > 0$ there is $x' \in A$ such that $x' > M - \varepsilon \ (x' < m + \varepsilon)$.

The set A may or may not contain the $\sup A$ and $\inf A$.

Example 1.9. If $A = [a,b)$, that is $a \leqslant x < b$, then $m \leqslant a, b \leqslant M$. Here, $\inf A = a \in A$, $\sup A = b \notin A$.

Example 1.10. Let $A = \{1, 2, 4, \ldots, 2n\}$. Then $m \leqslant 1, 2n \leqslant M$

$$\inf A = 1 \in A, \quad \sup A = 2n \in A$$

In general, for the existence of both $\sup A$ ($\inf A$) of a nonempty subset of real numbers $\mathbb{R}$, it is sufficient that the subset be bounded above (below). Of course, this is an important result, concerning the supremum (infimum) notion. Note, that in most advanced calculus texts (see, for example [7]) the completeness axiom of the real number system is stated as the least upper bound axiom.

Theorem 1.8 (Dedekind[2]). *Suppose A and B are two nonempty sets of real numbers satisfying the following conditions.*

1) *Every real number x belongs either to A or to B.*
2) *If $x \in A$ and $y \in B$ then $x < y$.*

Then there is a unique real number c such that $x \leqslant c$ for all $x \in A$ and $y \geqslant c$ for all $y \in B$. Besides, either $c = \sup A$ or $c = \inf B$.

☐   By the hypothesis of the theorem neither A nor B is nonempty, and so every $y \in B$ is an upper bound for A and any $x \in A$ is a lower bound for B. It follows that $\sup A$ and $\inf B$ exist and $\sup A \leqslant \inf B$. We show that $\sup A = \inf B$. Suppose $\sup A < \inf B$. Then there is a real number x' such that $\sup A < x' < \inf B$ that is $x' \notin A$, $x' \notin B$. This contradicts the condition 1). Therefore we obtain $\sup A = \inf B$. If we put $\sup A = \inf B = c$, then we deduce

$$x \leqslant \sup A = c = \inf B \leqslant y, \quad x \in A, \ y \in B.$$

Note that in the theorem either $c \in A$ or $c \in B$. Thus $c \notin A \cap B$ (by condition 2) $A \cap B = \varnothing$). If $c \in A$ then $c = \sup A$ and if $c \in B$, then $c = \inf B$. Next, we show that c is unique. If not, then let c and c' $(c < c')$ be two points satisfying the conditions of the theorem. Choose a point x such that $c < x < c'$. The inequalities $x < c'$ and $x > c$ imply $x \in A$, $x \in B$, respectively, that is $x \in A \cap B$. This contradiction proves that c is unique.   ■

The pair of sets, $\{A, B\}$, is called a *"Dedekind cut."* The intuitive idea of Dedekind's Theorem 1.8 is that the set of real numbers is complete and there are no "holes" in the real numbers. The analogous claim is not true of the field of rational numbers.

Example 1.11. First show that there is no a rational number a satisfying $a^2 = 2$. Suppose the contrary, i.e., that there is an $a = p^*/q^*$ so that $(p^*/q^*)^2 = 2$, where p^* and q^* are integers and $q^* \neq 0$. By dividing through by any common factors greater than 1, we obtain $(p/q)^2 = 2$, where p and q have no common factors. Then $p^2 = 2q^2$. Thus p^2 is even and so p must be even. Let $p = 2m$. Then $4m^2 = p^2 = 2q^2$ and so $2m^2 = q^2$. Therefore q^2 is even, and so q is even. Because both p and q are even, they must have the common factor 2, which is a contradiction. So $(p^*/q^*)^2 \neq 2$.

Let A be the set of positive rationals x satisfying $x^2 < 2$, and let B be the set of positive rationals x satisfying $x^2 > 2$. $(A \cap B = \varnothing)$.

---

[2]R. Dedekind (1831-1916), German mathematician.

We claim that $\sup A \notin A$ and $\inf B \notin B$.

More precisely, we prove that for any $a \in A$ there is a rational b in A, such that $a < b$ and for every $a \in B$ there exists a rational b in B such that $b < a$.

Let $a \in A$, that is, $a^2 < 2$. Choose a rational number h such that $h < \dfrac{2 - a^2}{2a + 1}$ and $0 < h < 1$. Let now $b = a + h$. Then $b > a$ and

$$b^2 = a^2 + (2a + h)h < a^2 + (2a + 1)h < a^2 + (2 - a^2) = 2.$$

It follows that $b \in A$.

Now suppose that $a \in B$, that is $a^2 > 2$. Take

$$b = a - \frac{a^2 - 2}{2a} = \frac{a}{2} + \frac{1}{a}.$$

Then

$$0 < b < a \quad \text{and} \quad b^2 = a^2 - (a^2 - 2) + \left(\frac{a^2 - 2}{2a}\right)^2 > a^2 - (a^2 - 2) = 2.$$

It follows that $b \in B$.

Consequently, in spite of the fact that between any two rationals there is a third rational, i.e., $a < \dfrac{a+b}{2} < b$, if $a < b$, there are "holes" in the set of rational numbers.

1.4   Uncountable Sets

In this section we will see that not all infinite sets are denumerable. First of all we prove Cantor's[3] Theorem, without proof.

Theorem 1.9 (Cantor). *Let $\{I_n\}$ be a sequence of closed intervals $I_n = [a_n, b_n]$, $n \in \mathbb{N}$, such that $I_n \supset I_{n+1}$ $(n = 1, 2, \ldots)$. Then if the lengths of the intervals tend to zero for infinitely large n, there is one and only one point ζ common to all the intervals:*

$$\zeta \in \bigcap_{n=1}^{\infty} I_n.$$

The intervals in such a case (that is, with $I_n \supset I_{n+1}$ $n = 1, 2, \ldots$) are called nested intervals.

The closure condition in the theorem is essential. Indeed, if $I_n = \left(0, \dfrac{1}{n}\right)$ $(n = 1, 2, 3, \ldots)$, then

$$\bigcap_{n=1}^{\infty} I_n = \varnothing$$

---

[3]G. Cantor (1845-1918), German mathematician.

and so there does not exist any point ζ common to all the intervals.

Definition 1.13. A set is *uncountable* if it is not countable.

Theorem 1.10. *The set $A = [0,1]$ is uncountable.*

□ On the contrary, suppose that the set $A = [0,1]$ is countable. Then the set of points of A can be represented as sequence

$$x_1, x_2, \ldots, x_n, \ldots$$

Divide the set A into three subintervals $\left[0, \dfrac{1}{3}\right]$, $\left[\dfrac{1}{3}\right]$, $\left[\dfrac{2}{3}, 1\right]$ with equal lengths. Let A_1 be one of these subintervals so that $x_1 \notin A_1$. We divide again the set A_1 into three equal subintervals and choose one of them, A_2 with $x_2 \notin A_2$. Inductively, we obtain a sequence of closed intervals $\{A_n\}$ such that for all n ($n = 1, 2, \ldots$)

$$A \supset A_1 \supset A_2 \supset \cdots \supset A_n \supset \cdots$$

and $x_n \notin A_n$. Obviously, the lengths of the subintervals A_n are $\dfrac{1}{3^n}$ and so tend to zero for infinitely large n. By Cantor's Theorem, there is one and only one point ζ common to all nested intervals A_n. But since ζ is in A_1, ζ must be in A. But the sequence $\{x_n\}$ does not contain the point ζ. Hence the set of points A cannot be represented by the sequence $\{x_n\}$. ∎

Corollary 1.2. *The set of all nonterminating decimal fractions is uncountable.*

In order to prove this, it is sufficient to represent the points of $A = [0,1]$ by nonterminating decimal fractions and then to apply Theorem 1.8.

Definition 1.14. If a set is equivalent to $A = [0,1]$, we say that it has the *cardinality of the continuum* (or, has cardinality c).

Theorem 1.11. *Every interval of the form $[a,b]$, (a,b), $(a,b]$, $[a,b)$ has the cardinality of the continuum.*

□ Firstly, let us prove that the intervals $[a,b]$ and $[0,1]$ are equivalent. Let $y \in [a,b]$ and $x \in [0,1]$. Indeed, the function defined by $y = f(x) = a + (b-a)x$ assigns to each point x in $[0,1]$ a unique element y in $[a,b]$, and conversely to each point $[a,b]$ assigns a unique point $x = \dfrac{y-a}{b-a}$ in $[0,1]$. Thus $[a,b] \sim [0,1]$. Next, observe that deleting either one or both endpoints of $[a,b]$ does not change its cardinality. Then all that remains is to apply Theorem 1.10. ∎

Theorem 1.12. *The set $\mathbb{R}$ of real numbers has the cardinality of the continuum.*

□ We show that the two sets, $\mathbb{R}$ and the open interval $(0,1)$, are equivalent, i.e., $\mathbb{R} \sim (0,1)$. It is easy to prove that the function $y = \tan(2x-1)\dfrac{\pi}{2}$ is one-to-one and onto; to each x in $(0,1)$ it assigns a unique real number y in R and for every real number y there is only one point x defined by $(2x-1)\dfrac{\pi}{2} = \arctan y$. Hence $\mathbb{R} \sim (0,1)$.   ∎

We have seen (Theorem 1.4) that the set of rational numbers has cardinality d. Then by Theorem 1.12, it follows that the set of irrational numbers has cardinality c. Therefore there are "more" irrationals than rationals.

Theorem 1.13. *The cardinality of the set of functions defined on $[0,1]$ is greater than cardinality c.*

□ Let Φ denote the set of functions defined on $[0,1]$. Take $A = [0,1]$. In order to prove the theorem it is sufficient to show that Φ is not equivalent to A and A is equivalent to some proper subset of Φ. First, we prove that $A \nsim \Phi$. On the contrary, suppose $A \sim \Phi$. Then for every $z \in [0,1]$ there is a function $f(x)$ defined on $[0,1]$, and this correspondence is one-to-one. Denote this function by $f_z(x) = f(x,z)$. Now, introduce an auxiliary function $\varphi(x) = f(x,x) + 1$. Observe that $\varphi(x) \in \Phi$ $(0 \leqslant x \leqslant 1)$ and $\varphi(x)$ assigns (bijectively) a unique number $\bar{z}$. In our notation, this means that $\varphi_{\bar{z}}(x) = F(x,\bar{z})$. It follows that $F(x,\bar{z}) \equiv F(x,x) + 1$ for all $x \in [0,1]$. Taking $x = \bar{z}$, we have $F(\bar{z},\bar{z}) = F(\bar{z},\bar{z}) + 1$, which is a contradiction. This implies $A \nsim \Phi$.

Now we show that A is equivalent to some proper subset of Φ. Consider the family of functions $\Phi_1 = \{y = ax + b : 0 \leqslant b \leqslant 1\}$, where a is some fixed number. It is clear that $\Phi_1 \subset \Phi$. On the other hand given any element $y = ax + b \in \Phi_1$ there is only one element $b \in A$. It follows that there is one-to-one and onto function between A and Φ_1. Hence $A \sim \Phi_1 \subset \Phi$, and so the cardinality of the set of functions defined on $[0,1]$ is greater than cardinality c.   ∎

In the proof of Theorem 1.13 it is taken into account that the family, Φ, of functions contains both continuous and discontinuous functions. The set of all continuous functions has cardinality c.[4]

The formulation of Theorem 1.13 implies the natural question: Is there a cardinal number between d and c? In other words, is there an infinite set $A \subset \mathbb{R}$ with no bijective function $f_1 : A \to \mathbb{N}$ and no bijective function $f_2 : A \to \mathbb{R}$? The assertion that the answer is *no* is called the Continuum Hypothesis.

---

[4]See, for example, N.N. Luzin, *The Theory of Functions of a Real Variable*, Uchpedgiz, 1948.

1.5 Construction of the Complex Number Field

Definition 1.15. An ordered pair (a,b) of a and b (in that order) is called a *complex number*. Two complex numbers (a,b) and (a_1,b_1) are equal to each other, i.e., $(a,b) = (a_1,b_1)$, if and only if $a = a_1$ and $b = b_1$. The sum and product of two complex numbers (a,b) and (a_1,b_1) is defined as follows:

$$(a,b) + (a_1,b_1) = (a+a_1, b+b_1), \quad (a,b)\cdot(a_1,b_1) = (aa_1 - bb_1, ab_1 + a_1b). \quad (1.6)$$

Let (a,b), (a_1,b_1), (a_2,b_2) be any three complex numbers. It can be shown that the algebraic properties of the operations of addition and multiplication given by Definition 1.15 satisfy the commutative, associative, and distributive laws. Indeed,

(1) $(a,b) + (a_1,b_1) = (a+a_1, b+b_1) = (a_1+a, b_1+b) = (a_1,b_1) + (a,b)$
    (commutative law for addition),

(2) $[(a,b) + (a_1,b_1)] + (a_2,b_2) = (a+a_1, b+b_1) + (a_2,b_2) = (a+a_1+a_2, b+b_1+b_2)$
    $$= (a,b) + (a_1+a_2, b_1+b_2) = (a,b) + [(a_1,b_1) + (a,b)]$$
    (associative law for addition),

(3) $(a,b)(a_1,b_1) = (aa_1 - bb_1, ab_1 + a_1b) = (a_1a - b_1b, b_1a + ba_1) = (a_1,b_1)(a,b)$
    (commutative law for multiplication),

(4) $[(a,b)(a_1,b_1)](a_2,b_2) = (aa_1 - bb_1, ab_1 + ba_1)(a_2,b_2)$
    $$= (aa_1a_2 - bb_1a_2 - ab_1b_2 - ba_1b_2, aa_1b_2 - bb_1b_2 + ab_1a_2 + ba_1a_2)$$
    $$= (a,b)(a_1a_2 - b_1b_2, a_1b_2 + b_1a_2) = (a,b)[(a_1,b_1)(a_2,b_2)]$$
    (associative law for multiplication),

and

(5) $[(a,b) + (a_1,b_1)](a_2,b_2) = (a+a_1, b+b_1)(a_2,b_2)$
    $$= (aa_2 + a_1a_2 - bb_2 - b_1b_2, ab_2 + a_1b_2 + ba_2 + b_1a_2)$$
    $$= (aa_2 - bb_2, ab_2 + ba_2) + (a_1a_2 - b_1b_2, a_1b_2 + b_1a_2)$$
    $$= (a,b)(a_2,b_2) + (a_1,b_1)(a_2,b_2)$$
    (distributive law).

Obviously, the subtraction operation is defined as

$$(a,b) - (a_1,b_1) = (a-a_1, b-b_1).$$

Then with the properties (1)–(5), the set of complex numbers forms a ring. If we define the division operation in this ring as follows,

(6) $$\frac{(a,b)}{(a_1,b_1)} = \left(\frac{aa_1 + bb_1}{a_1^2 + b_1^2}, \frac{a_1b - ab_2}{a_1^2 + b_1^2} \right), \quad (a_1,b_1) \neq (0,0),$$
    then we have that the set of complex numbers forms a field.

Moreover, for any real number we can write

$$(a_1,0)+(a_2,0)=(a_1+a_2,0), \quad (a_1,0)\cdot(a_2,0)=(a_1a_2,0).$$

It is understood that a complex number of the form $(a,0)$ and a real number a have the same properties and so we can take $(a,0)\equiv a$. Thus, setting $b=0$, we see that the set of real numbers can be regarded as a subset of the complex numbers.

We define the "imaginary" number by $i=(0,1)$. It follows from (1.6) that

$$i^2=(0,1)\cdot(0,1)=(-1,0)=-1, \quad \text{that is } i^2=-1, \ i=\sqrt{-1}.$$

Therefore,

$$z=(a,b)=(a,0)+(0,b)=a+bi,$$

and so $a+bi$ could be regarded as an alternative symbol for the ordered pair (a,b). Geometrically, a complex number can be viewed either as a point in the xy-plane or as a vector. The complex numbers corresponding to points on the imaginary axis $(a=0)$, are called pure imaginary numbers. Obviously,

$$z+(-z)=0, \quad z=a+ib, \quad -z=-a-ib,$$

$$z+0=z, \quad 0=0+i0,$$

$$z\cdot\frac{1}{z}=1, \quad \frac{1}{z}=\frac{a}{a^2+b^2}-i\frac{b}{a^2+b^2} \quad (z\neq 0).$$

The real number a is called the *real part* of z and the real number b, the *imaginary part* of z. Clearly, the formulae (1.6) for numbers $z_j=a_j+ib_j$ $(j=1, 2)$ take the form:

$$(a_1+ib_1)+(a_2+ib_2)=(a_1+a_2)+i(b_1+b_2),$$

$$(a_1+ib_1)\cdot(a_2+ib_2)=(a_1a_2-b_1b_2)+i(a_1b_2+a_2b_1).$$

If $z=a+ib$ is any complex number, then the *conjugate* of z, denoted by $\bar{z}$, is defined by $\bar{z}=a-ib$. The *modulus* (or *absolute value*) of a complex number z, denoted by $|z|$, is defined by $|z|=\sqrt{a^2+b^2}$. It is not hard to verify that

$$\overline{z_1+z_2}=\overline{z_1}+\overline{z_2}, \quad \overline{z_1\cdot z_2}=\overline{z_1}\cdot\overline{z_2}, \quad \left(\overline{\frac{z_1}{z_2}}\right)=\frac{\overline{z_1}}{\overline{z_2}},$$

$$|z|=|\bar{z}|, \quad z\cdot\bar{z}=a^2+b^2.$$

If $z=a+ib$ is a nonzero complex number, $r=|z|$, and φ measures the angle from the positive real axis to the vector z, then $a=r\cos\varphi$ and $b=r\sin\varphi$, so that $z=a+ib$ can be written as

$$z=r(\cos\varphi+i\sin\varphi). \tag{1.7}$$

This is called the *polar form* of z. The angle φ is called the *argument* of z and is denoted by $\varphi = \arg z$. Obviously, the argument φ is not determined uniquely, but there is only one value, $\varphi = \text{Arg} z$, of the argument in radians that satisfies $-\pi < \varphi \leqslant \pi$. This is called the *principal argument* of z (note the capitalization of "Arg"). If

$$z_1 = r_1(\cos\varphi_1 + i\sin\varphi_1), \quad z_2 = r_2(\cos\varphi_2 + i\sin\varphi_2),$$

then

$$\begin{aligned} z_1 z_2 &= r_1 r_2(\cos\varphi_1\cos\varphi_2 - \sin\varphi_1\sin\varphi_2) + i(\sin\varphi_1\cos\varphi_2 + \sin\varphi_2\cos\varphi_1) \\ &= r_1 r_2(\cos(\varphi_1 + \varphi_2) + i\sin(\varphi_1 + \varphi_2)). \end{aligned} \tag{1.8}$$

Thus we have shown that

$$r = |z_1 z_2| = |z_1||z_2| = r_1 r_2, \quad \arg(z_1 z_2) = \arg z_1 + \arg z_2.$$

In words, the product of two complex numbers is obtained by multiplying their moduli and adding their arguments. Similarly,

$$\frac{z_1}{z_2} = \frac{r_1}{r_2}[\cos(\varphi_1 - \varphi_2) + i\sin(\varphi_1 - \varphi_2)], \tag{1.9}$$

from which it follows that

$$\left|\frac{z_1}{z_2}\right| = \frac{|z_1|}{|z_2|} = \frac{r_1}{r_2}, \quad (z_2 \neq 0), \quad \arg\frac{z_1}{z_2} = \arg z_1 - \arg z_2.$$

In words, the quotient of two complex numbers is obtained by dividing their moduli and subtracting their arguments. In particular, since $1 = \cos 0 + i\sin 0$ it follows from (1.9) that for $z \neq 0$,

$$z^{-1} = r^{-1}[\cos(-\varphi) + i\sin(-\varphi)],$$

so that $|z^{-1}| = |z|^{-1}$, $\arg(z^{-1}) = -\arg z$.

If n is a positive integer and $z = r(\cos\varphi + i\sin\varphi)$, then formula (1.8) becomes

$$[r(\cos\varphi + i\sin\varphi)]^n = r^n(\cos n\varphi + i\sin n\varphi), \tag{1.10}$$

which is called *De Moivre's*[5] *formula*.

Note that, since $z^{-n} = (z^{-1})^n$, the formula (1.10) is true also for negative integers n. Indeed, the formula holds even when n is not an integer, but a fraction.

In order to calculate $z^n = (a + ib)^n$ for a positive integer n, we can use the binomial formula, where we take into account that

$$i^2 = -1, \quad i^3 = -i, \quad i^4 = 1, \ldots, \quad i^{4k} = 1, \quad i^{4k+1} = i, \quad i^{4k+2} = -1, \quad i^{4k+3} = -i.$$

---

[5]A. De Moivre (1667–1754), French mathematician.

Note that De Moivre's formula can be used to obtain roots of complex numbers. We denote an n-th root of z by $z^{1/n}$. If $z \neq 0$, then we can derive formulas for the n-th roots of z as follows. Let $\rho(\cos\theta + i\sin\theta)$ be an n-th root of $z = r(\cos\varphi + i\sin\varphi)$. Then, by formula (1.10),

$$[\rho(\cos\theta + i\sin\theta)]^n = r(\cos\varphi + i\sin\varphi), \quad \rho^n = r, \quad n\theta = \varphi + 2k\pi, \ k = 0, \pm 1, \pm 2, \ldots \quad .$$

Thus,

$$\sqrt[n]{r(\cos\varphi + i\sin\varphi)} = \sqrt[n]{r}\left(\cos\frac{\varphi + 2k\pi}{n} + i\sin\frac{\varphi + 2k\pi}{n}\right), \quad k = 0, \pm 1, \pm 2, \ldots \quad .$$

Although there are infinitely many values of k, it can be shown that $k = 0, 1, 2, \ldots, n-1$ produce distinct values of $\rho(\cos\theta + i\sin\theta)$ satisfying these equalities, but all other choices of k yield duplicates of these. Hence, there are exactly n different n-th roots of $z = r(\cos\varphi + i\sin\varphi)$, and these are given by

$$z^{\frac{1}{n}} = \sqrt[n]{r}\left(\cos\frac{\varphi + 2k\pi}{n} + i\sin\frac{\varphi + 2k\pi}{n}\right), \quad k = 0, 1, 2, \ldots, n-1 \qquad (1.11)$$

Because any real number A can be represented as

$$A = \begin{cases} |A|(\cos 0 + i\sin 0), & \text{if } A > 0, \\ |A|(\cos\pi + i\sin\pi), & \text{if } A < 0. \end{cases}$$

using (1.11), we can find the n-th roots of $x^n = A$ ($A \neq 0$) by

$$x = \sqrt[n]{A}\left(\cos\frac{2k\pi}{n} + i\sin\frac{2k\pi}{n}\right), \quad A > 0,$$

$$x = \sqrt[n]{|A|}\left(\cos\frac{\pi + 2k\pi}{n} + i\sin\frac{\pi + 2k\pi}{n}\right), \quad A < 0, \ k = 0, 1, 2, \ldots, n-1.$$

Example 1.12. Let $z_1 = 23 + i$, $z_2 = 3 + i$.

Then

$$z_1 \cdot z_2 = (23 + i)(3 + i) = 68 + 26i$$

and

$$\frac{z_1}{z_2} = \frac{(23 + i)(3 - i)}{(3 + i)(3 - i)} = \frac{70 - 20i}{10} = 7 - 2i.$$

Example 1.13. Let $z = 3\left(\cos\frac{\pi}{4} + i\sin\frac{\pi}{4}\right)$ be a complex number in polar form. Find z^4. By De Moivre's formula,

$$\left[3\left(\cos\frac{\pi}{4} + i\sin\frac{\pi}{4}\right)\right]^4 = 3^4(\cos\pi + i\sin\pi) = -81.$$

Example 1.14. Find $\sqrt[n]{1}$. Obviously, $1 = \cos 0 + i\sin 0$, and so

$$\sqrt[n]{1} = \cos\frac{2k\pi}{n} + i\sin\frac{2k\pi}{n}, \quad k = 0, 1, 2, \ldots, n-1.$$

1.6 Metric Spaces

Metric spaces play a fundamental role. If A and B are sets, then their Cartesian product $A \times B$ is the set of all ordered pairs (x, y) with $x \in A$ and $y \in B$. Thus

$$A \times B = \{(x, y) \; : \; x \in A, \; y \in B\}.$$

In particular, we write

$$\mathbb{R}^n = \underbrace{\mathbb{R} \times \cdots \times \mathbb{R}}_{n}$$

Definition 1.16. A *metric space* (X, ρ) is a set X together with a distance function (or *metric*) $\rho : X \times X \to \mathbb{R}$ such that for all $x_1, x_2, x_3 \in X$ the following hold:

1. $\rho(x_1, x_2) > 0$, $x_1 \neq x_2$ and $\rho(x_1, x_2) = 0$ if and only if $x_1 = x_2$,
2. $\rho(x_1, x_2) = \rho(x_2, x_1)$, $x_1, x_2 \in X$ (symmetry property),
3. $\rho(x_1, x_3) \leqslant \rho(x_1, x_2) + \rho(x_2, x_3)$, $x_i \in X$, $i = 1, 2, 3$ (triangle inequality).

Often the metric space is denoted by (X, ρ), to indicate that a metric space is determined by both the set X and the metric ρ.

Example 1.15. Let $X = \mathbb{R}$ be the set of real numbers. We define the metric $\rho(x_1, x_2) = |x_1 - x_2|$ for all $x_1, x_2 \in X$. Clearly, conditions 1–3 hold, and so (X, ρ) is a metric space.

Similarly, the set of complex numbers with the metric

$$\rho(z_1, z_2) = |z_1 - z_2|$$

is a metric space.

Example 1.16. Any set X with the metric given by

$$\rho(x_1, x_2) = \begin{cases} 1, & \text{if } x_1 \neq x_2, \\ 0, & \text{if } x_1 = x_2 \end{cases}$$

is a metric space.

Example 1.17. The set of bounded functions f (to say f is bounded means that there is a number M such that $|f(x)| \leqslant M$, $x \in [0, 1]$) defined on $[0, 1]$ with metric

$$\rho(f_1, f_2) = \sup_{x \in [0, 1]} |f_1(x) - f_2(x)|$$

is a metric space. Indeed, it is clear that $\rho(f_1, f_2) \geqslant 0$ and $\rho(f_1, f_2) = 0$, if $f_1(x) \equiv f_2(x)$. On the other hand, $\rho(f_1, f_2) = \rho(f_2, f_1)$. We prove that the triangle inequality is true. For all $x \in [a, b]$, we have

$$
\begin{aligned}
|f_1(x) - f(x)| &\leqslant |f_1(x) - f_2(x)| + |f_2(x) - f(x)| \\
&\leqslant \sup_{x \in [0,1]} |f_1(x) - f_2(x)| + \sup_{x \in [0,1]} |f_2(x) - f(x)| \\
&= \rho(f_1, f_2) + \rho(f_1, f).
\end{aligned}
$$

Then it follows that $\rho(f_1, f) \leqslant \rho(f_1, f_2) + \rho(f_2, f)$.

Example 1.18. Suppose that for a traveler, the metric from one point to another of some mountainous region of the earth's surface is defined as the time used on the road. Then we have a nonmetric space, because the symmetry property does not hold.

Remark 1.4. It is not hard to see that if (X, ρ) is a metric space, then for any subset $X_1 \subset X$, the space (X_1, ρ) is a metric space.

Remark 1.5. For a given set X, by defining different metrics ρ, there can be obtained different metric spaces (X, ρ).

Definition 1.17. If n is a positive integer, then an *ordered n-tuple* is a sequence of n real numbers $x = (x_1, \ldots, x_n)$. An ordered n-tuple $(x_1, \ldots, x_n)$ can be viewed either as a point or a vector, and $x_1, \ldots, x_n$ are the coordinates of x.

Two vectors, $x = (x_1, \ldots, x_n)$ and $y = (y_1, \ldots, y_n)$, are called equal if and only if $x_i = y_i$, $i = 1, \ldots, n$.

The sum $x + y$ and scalar multiple λx (λ is any scalar) are defined by

$$
\begin{aligned}
(x_1, \ldots, x_n) + (y_1, \ldots, y_n) &= (x_1 + y_1, \ldots, x_n + y_n), \\
\lambda (x_1, \ldots, x_n) &= (\lambda x_1, \ldots, \lambda x_n)
\end{aligned}
\tag{1.12}
$$

If x, y, z are vectors and α, β are real numbers, then the following are easy to verify.

1. $x + y = y + x$,
2. $(x + y) + z = x + (y + z)$,
3. $x + 0 = x$, $0 = (0, \ldots, 0)$,
4. $x + (-x) = 0$,
5. $1 \cdot x = x$,
6. $\alpha(\beta x) = (\alpha \beta) x$,
7. $(\alpha + \beta) x = \alpha x + \beta x$,
8. $\alpha(x + y) = \alpha x + \alpha y$.

Definition 1.18. The set of all ordered n-tuples supplied with the operations of sum $x + y$ and scalar multiple λx (1.12) is called the *vector space* $\mathbb{R}^n$.

If $x^1, \ldots, x^k$ is a given set of vectors, then the vector equation

$$\alpha_1 x^1 + \cdots + \alpha_k x^k = 0 \qquad (1.13)$$

has at least one solution, namely,

$$\alpha_1 = 0, \quad \alpha_2 = 0, \quad \ldots, \quad \alpha_k = 0$$

If the solution of (1.13) is unique, then the vectors are linearly independent. If there are other solutions, then the vectors $x^1, \ldots, x^k$ are linearly dependent. Any set of n linearly independent vectors of $\mathbb{R}^n$ is called a basis for $\mathbb{R}^n$. For example, the set of vectors $e_1 = (1, 0, \ldots, 0)$, $e_2 = (0, 1, 0, \ldots, 0), \ldots, e_n = (0, \ldots, 0, 1)$ is a basis (the standard basis) for $\mathbb{R}^n$. In fact, since $\alpha_1 e_1 + \cdots + \alpha_n e_n = (\alpha_1, \ldots, \alpha_n)$, the equation $\alpha_1 e_1 + \cdots + \alpha_n e_n = 0$ implies $\alpha_1 = 0$, $\alpha_2 = 0, \ldots, \alpha_n = 0$. Thus, the $e_1, \ldots, e_n$ are linearly independent, and every vector $x = (x_1, \ldots, x_n)$ can be expressed in the form $x = c_1 e_1 + \cdots + c_n e_n$ in exactly one way. It is easy to verify that all bases for the vector space $\mathbb{R}^n$ have the same number, n, of vectors. This number, n, is called the dimension of the vector space $\mathbb{R}^n$.

If $x = (x_1, \ldots, x_n)$ and $y = (y_1, \ldots, y_n)$ are any vectors in $\mathbb{R}^n$, then the Euclidean inner product $\langle x, y \rangle$ is defined by

$$\langle x, y \rangle = x_1 y_1 + \cdots + x_n y_n = \sum_{i=1}^{n} x_i y_i. \qquad (1.14)$$

The four main algebraic properties of the Euclidean inner product are listed below.

1. $\langle x, y \rangle = \langle y, x \rangle$,
2. $\langle x + y, z \rangle = \langle x, z \rangle + \langle y, z \rangle$,
3. $\langle \alpha x, y \rangle = \alpha \langle x, y \rangle$ (α is a real number),
4. $\langle x, x \rangle > 0$ ($x \neq 0$), $\langle x, x \rangle = 0$ if and only if $x = 0$

A space that is defined by (1.14) is called a Euclidean space.

One can define other inner products on $\mathbb{R}^n$. One simple class of examples is given by defining $\langle x, y \rangle = c_1 x_1 y_1 + \cdots + c_n x_n y_n = \sum_{i=1}^{n} c_i x_i y_i$ where $c_1, \ldots, c_n$ is any sequence of positive real numbers.

Below, we give one of the most important inequalities in Analysis, the Cauchy[6]–Schwarz[7]–Bunyakovsky[8] inequality

$$\langle x, y \rangle^2 \leqslant \langle x, x \rangle \langle y, y \rangle, \quad x, y \in \mathbb{R}^n \qquad (1.15)$$

[6] A.L. Cauchy (1789-1857), French mathematician.
[7] H.A. Schwarz (1843-1921), German mathematician.
[8] V. Bunyakovsky (1804-1889), Russian mathematician.

or, when expressed in terms of components,

$$\left(\sum_{i=1}^{n} x_i y_i\right)^2 \leqslant \left(\sum_{i=1}^{n} x_i^2\right)\left(\sum_{i=1}^{n} y_i^2\right).$$

Indeed, for all α, it follows from property 4 that

$$\langle \alpha x + y, \alpha x + y \rangle \geqslant 0$$

or

$$\alpha^2 \langle x, x \rangle + 2\alpha \langle x, y \rangle + \langle y, y \rangle \geqslant 0.$$

Thus, the discriminant must be nonpositive, i.e., $\langle x, y \rangle^2 - \langle x, x \rangle \langle y, y \rangle \leqslant 0$, which implies (1.15).

Definition 1.19. A normed vector space is a vector space together with a real-valued function called a norm. The *norm* of $x = (x_1, \ldots, x_n)$ is denoted by $\|x\|$. The following axioms are satisfied for all x in the vector space and all real numbers α:

(1) $\|x\| > 0$, $x \neq 0$ and $\|x\| = 0$ if and only if $x = 0$ (positivity),
(2) $\|\alpha x\| = |\alpha| \, \|x\|$ (homogeneity),
(3) $\|x + y\| \leqslant \|x\| + \|y\|$ for all x, y (triangle inequality).

The Euclidean space $\mathbb{R}^n$ with norm

$$\|x\| = \sqrt{\langle x, x \rangle} = \sqrt{\sum_{i=1}^{n} x_i^2}$$

is a normed space. Indeed, in term of this norm, the inequality (1.15) can be rewritten as follows

$$\|\langle x, y \rangle\| \leqslant \|x\| \, \|y\|. \tag{1.16}$$

It then suffices to verify the triangle inequality (3). With the use of (1.16), we have

$$\|x + y\|^2 = \langle x + y, x + y \rangle = \langle x, x \rangle + 2\langle x, y \rangle + \langle y, y \rangle$$
$$\leqslant \|x\|^2 + 2\|x\|\|y\| + \|y\|^2 = (\|x\| + \|y\|)^2,$$

or,

$$\|x + y\| \leqslant \|x\| + \|y\|.$$

Now, we rewrite (1.16) as follows:

$$-1 \leqslant \frac{\langle x, y \rangle}{\|x\| \|y\|} \leqslant 1 \quad (x \neq 0, \, y \neq 0)$$

Then we conclude that there is an angle ϕ ($0 \leqslant \phi \leqslant \pi$) such that

$$\cos\phi = \frac{\langle x, y \rangle}{\|x\|\|y\|}, \tag{1.17}$$

where Φ is the angle between x and y.

Remark 1.6. We easily see from (1.17) that Φ is, respectively, acute, obtuse, or $\frac{\pi}{2}$, if and only if $\langle x, y \rangle > 0$, $\langle x, y \rangle < 0$, or $\langle x, y \rangle = 0$.

The distance between two points $x = (x_1, \ldots, x_n) \in \mathbb{R}^n$ and $y = (y_1, \ldots, y_n) \in \mathbb{R}^n$ is defined as follows:

$$\rho(x, y) = \|x - y\| = \sqrt{\sum_{i=1}^{n} (x_i - y_i)^2},$$

and we now verify that the metric $\rho(x, y)$ satisfies Axioms 1–3 of Definition 1.16. Axioms 1 and 2 are verified at once. Moreover, by using the triangle inequality (3) of Definition 1.19, we obtain

$$\rho(x, z) = \|x - z\| = \|(x - y) + (y - z)\| \leqslant \|x - y\| + \|y - z\| = \rho(x, y) + \rho(y, z).$$

Consequently, $\mathbb{R}^n$ is a metric space. More generally, any normed space is also a metric space. There are other norms which we can define on $\mathbb{R}^n$. It is easy to check that

$$\|x\|_0 = \max_{1 \leqslant k \leqslant n} |x_k|, \quad \|x\|_1 = \sum_{k=1}^{n} |x_k|,$$

define norms on $\mathbb{R}^n$. For $1 \leqslant p < \infty$,

$$\|x\|_p = \left(\sum_{i=1}^{n} |x_i|^p \right)^{frac1p}$$

also defines a norm on $\mathbb{R}^n$, called the *p-norm*.

Similarly, it is easy to check that on the set of continuous real-valued functions defined on the interval $[0, 1]$,

$$\|f\| = \max_{0 \leqslant x \leqslant 1} |f(x)|$$

is a norm.

Definition 1.20. Let (X, ρ) be a metric space. The *open ball* of radius $\varepsilon > 0$ centered at $x_0 \in X$ is defined by

$$B_\varepsilon(x_0) = \{x \in X \,:\, \rho(x, x_0) < \varepsilon\}.$$

Note that the open balls in $\mathbb{R}$ are precisely the intervals of the form (a,b) with center $\dfrac{a+b}{2}$ and radius $\dfrac{b-a}{2}$.

Definition 1.21. Let $M \subset X$ be a subset of a metric space X. A point $x \in X$ is a *limit point* (or *accumulation point*) of M if for every open $\varepsilon > 0$ the open ball $B_\varepsilon(x)$, $(x \in X)$ centered at x contains at least one point $y \in M$ other than x.

A point $x \in M$ is an *isolated point* of M if some open ball centered at x contains no members of M other than x itself.

Example 1.19. Obviously, if $M = [a,b) \subset \mathbb{R}$, then there are no isolated points, and the set of limit points is $[a,b]$.

Example 1.20. If M is the set of rational numbers of the form $\left\{\frac{1}{n}\right\}$, $n = 1, 2, 3, \ldots$, then every point in M is an isolated point. The only limit point is 0 and $0 \notin M$.

Definition 1.22. A set $M \subset X$ is called *closed* if it contains all its limit points.

The closure of $M \subset X$ is the union of M and the set of limit points of M, and is denoted by $\overline{M}$.

Example 1.21. If $M = (a,b)$, then $\overline{M} = [a,b]$.

Clearly, the sets $M = \varnothing$ and $M = X$ are closed.

Example 1.22. Let M be a finite set. Obviously, the set of limit points is empty. Since $\varnothing \subset M$, it follows that M is closed.

Definition 1.23. Suppose that $M \subset X$. A point $x \in M$ is an *interior point* of M if there is some open ball centered at x with radius $\varepsilon > 0$, such that $B_\varepsilon(x) \subset M$. If every point of M is an interior point, then M is called an *open set*.

For example, if the metric space $X = \mathbb{R}$ is the set of real numbers (see Example 1.15), then an open ball $B_\varepsilon(x_0)$ is an open interval: $(x_0 - \varepsilon, x_0 + \varepsilon) \subset \mathbb{R}$ for every point $x_0 \in \mathbb{R}$ and so $\mathbb{R}$ is open. Here, every interval $[a,b]$ is closed and every interval (a,b) is open. But an interval of the form $(a,b]$ is neither closed nor open.

A family $\{G_\alpha\}$ of open sets is said to be an *open covering* of a set $M \subset X$ if every point of M belongs to some member of $\{G_\alpha\}$.

A set M is called *compact* if every open covering of M has a finite subcovering, that is, there are finite number of positive integers $\alpha_1, \alpha_2, \ldots, \alpha_n$ such that

$$M \subset G_{\alpha_1} \cup \cdots \cup G_{\alpha_n}.$$

For $M \subset \mathbb{R}^n$, compact is equivalent to closed and bounded.

Remark 1.7. More precisely, a set can be both open and closed. For instance, $\varnothing$ and $\mathbb{R}^n$ are both closed and open sets. On the other hand, the set $M = \left\{ \dfrac{1}{2}, \dfrac{2}{3}, \ldots, \dfrac{n}{n+1}, \ldots \right\}$ is neither closed nor open.

Theorem 1.14. *The union of finitely many closed sets is also closed.*

$\square$ Let $M = \bigcup\limits_{i=1}^{n} M_i$ where M_i $(i = 1, \ldots, n)$ are closed. We show that every limit point x of the set M is also a limit point of M_i for at least for one i $(i = 1, \ldots, n)$. Suppose the contrary, and let $x \notin M_i$, $i = 1, \ldots, n$. By Definition 1.21, for every i there is an open ball $B_{\varepsilon_i}(x)$ $(\varepsilon_i > 0)$ such that $B_{\varepsilon_i}(x) \not\subset M_i$. Put $\varepsilon = \min\{\varepsilon_1, \ldots, \varepsilon_n\}$. It follows that $B_\varepsilon(x) \not\subset M$, and so x is not a limit point of M. This is a contradiction, and hence at least for one I_0 the point x is contained in M_{i_0} (remember that M_i is closed for every i). Consequently, $x \in M$, and so M is closed. ∎

Remark 1.8. For infinitely many closed sets, the conclusion of the theorem may fail. Indeed, the example

$$(0,1) = \bigcup_{n=3}^{\infty} \left[\frac{1}{n}, \frac{n-1}{n} \right]$$

shows that a non-finite union of closed sets need not be closed, since the limit points 0 and 1 do not belong to the open interval $(0, 1)$.

Theorem 1.15. *If $x \in M \subset X$ is a limit point of M, then every open ball $B_\varepsilon(x)$ centered at x contains an infinite number of points from M.*

$\square$ On the contrary, suppose that there is an $\varepsilon > 0$ such that $B_\varepsilon(x)$ contains only a finite number of points $x^1, \ldots, x^m$ from M $(x \neq x^i, i = 1, \ldots, m)$. Let us put $\bar{\varepsilon} = \min\limits_{1 \leqslant i \leqslant m} \{\rho(x^i, x)\} > 0$. Since $\rho(x^i, x) \geqslant \bar{\varepsilon}$ $(i = 1, \ldots, m)$, the ball $B_{\bar{\varepsilon}}(x)$ does not contain any point of M. This contradicts the hypothesis that x is a limit point. Consequently, the ball $B_\varepsilon(x)$ contains infinitely many points of M. ∎

Theorem 1.16. *If a set $M \subset X$ is open then its complement $\complement M$ is closed.*

$\square$ Let $x \in M$. Since M is closed there is an $\varepsilon > 0$ such that $B_\varepsilon(x) \subset M$, while $\complement M \cap B_\varepsilon(x) = \varnothing$. It follows that the point x cannot be a limit point of $\complement M$. Consequently, none of the limit points of $\complement M$ are in M. ∎

Remark 1.9. Because the complement of a closed set is open, then by De Morgan's Law, Theorem 1.1, it follows that the intersection of finitely many open sets is also open. It should be pointed out that the intersection of infinitely many open sets need not be open.

For example, the intersection of the infinitely many open intervals $\left(-\dfrac{1}{n}, \dfrac{1}{n}\right)$ is closed interval, that is, $M = \bigcap\limits_{n=1}^{\infty} \left(-\dfrac{1}{n}, \dfrac{1}{n}\right) = \{0\}$.

1.7 Problems

Prove that:

(1) $1^3 + 2^3 + \cdots + n^3 = (1 + 2 + \cdots + n)^2$

(2) $1 + 2 + 2^2 + \cdots + 2^{n-1} = 2^n - 1$

(3) $(1+x)^n \geqslant 1 + nx \ (n > 1, x > -1)$

(4) $n! < \left(\dfrac{n+1}{2}\right)^n \ (n > 1)$

(5) $\dfrac{1}{2} \cdot \dfrac{3}{4} \cdots \dfrac{2n-1}{2n} < \dfrac{1}{\sqrt{2n+1}}.$

(6) $\inf\{x + y\} = \inf\{x\} + \inf\{y\}$

(7) $\sup\{x + y\} = \sup\{x\} + \sup\{y\}$

(8) $\inf\{xy\} = \inf\{x\} \cdot \inf\{y\}, \ x, y \geqslant 0$

(9) By using the definition of union and intersection of sets A, B, C, prove that:

$$A \cup B = B \cup A, \quad (A \cup B) \cup C = A \cup (B \cup C),$$

$$A \cap B = B \cap A, \quad (A \cap B) \cap C = A \cap (B \cap C).$$

(10) For sets A, B, and C, prove that

$$(A \cap B) \cup C = (A \cup C) \cap (B \cup C).$$

(11) Prove that $A \cap B = \varnothing$ if and only if either $A \smallsetminus B = A$ or $B \smallsetminus A = B$.

(12) Prove that the cardinal number of a finite set is not equal to the cardinal number of any of its proper subsets.

(13) Prove that the cardinal numbers of the set of points inside an ellipse is equal to that of a circle.

(14) Show that the set $\{1^2, 2^2, 3^2, \ldots, n^2, \ldots\}$ has cardinality d.

(15) Let Q be the set of rational numbers. In the form a function establish a correspondence between the set $[0, 1]$ and Q.

(16) Prove that the set of circles centered at the rational numbers and with rational radii has cardinality d.

(17) (a) Prove that the set of all irrational numbers in $[0, 1]$ is non-denumerable, and

(b) find its cardinal number.

(18) Let $a < b < c < d$ be real numbers. Determine a one-to-one and onto function from $[a, b]$ to $[c, d]$.

(19) Determine the cardinality of the algebraic numbers contained in $[0, 1]$.

(20) Prove that for sets A, B, and C, in each case, (a)–(c) $A \sim B$.

$$\text{(a)} \quad A \subseteq B, \ A = A \cap B, \ A \cup B = B$$
$$\text{(b)} \quad A \subseteq B, \ A \cap B = A \cup B$$
$$\text{(c)} \quad A \subseteq B \subseteq C, \ A \cup B = B \cup C.$$

(21) Prove that the set of all equations of the form

$$a_0 x^n + a_1 x^{n-1} + \cdots + a_{n-1} x + a_n = 0$$

as n ranges through all positive integers and $a_0, a_1, \ldots, a_n$ range through the field of rational numbers, has cardinality d.

(22) Prove that $(a, b) = \bigcup_{i=1}^{\infty} \left[a + \frac{1}{i}, b - \frac{1}{i} \right]$ and thus show that every open interval can be expressed as a countable union of closed intervals.

(23) Prove that the set of integers and the set of natural numbers have the same cardinality. Find a one-to-one and onto function between them.

(24) Let the function $f : \mathbb{R} \to \mathbb{R}$ be defined by

$$f(x) = \begin{cases} 1, & \text{if } x \in \mathbb{Q}, \\ 0, & \text{if } x \in \mathbb{R} \setminus \mathbb{Q} \end{cases}$$

where, as usual, $\mathbb{Q}$ is the set of rational numbers.

Find (a) $f(\pi)$, (b) $f(3,575757\ldots)$.   Answer: (a) 0,   (b) 1.

(25) Prove that a set M is closed if and only if $\overline{M} = M$.

(26) Show that the number $\sqrt{12}$ is not rational.

(27) If a and b are vectors in $\mathbb{R}^n$ with the Euclidean inner product, then show that

$$\langle a, b \rangle = \frac{1}{4} \|a + b\|^2 - \frac{1}{4} \|a - b\|^2.$$

(28) Let a, b, c, d be rational numbers. Show that the equality $a + b\sqrt{2} = c + d\sqrt{2}$ implies both $a = c$ and $b = d$.

(29) Let X be the metric space of rational numbers, $\rho(x, y) = |x - y|$, and M be the set of rational numbers x satisfying the inequality $2 < x^2 < 3$. Show that M is bounded and closed, but not compact.

(30) By using the De Moivre formula, express $\cos 3\phi$ and $\sin 3\phi$ in terms of $\cos\phi$ and $\sin\varphi$.

(31) Find the roots of the complex numbers

(a) $\sqrt[3]{2\left(\cos\dfrac{3}{4}\pi + i\sin\dfrac{3}{4}\pi\right)}$, (b) $\sqrt{i}$, (c) $\sqrt[3]{-8}$, (d) $\sqrt{1}$.

(32) For complex numbers z_1 and z_2, show that

$$\big||z_1| - |z_2|\big| \leqslant |z_1 - z_2|.$$

(33) Show that, in the set of real numbers, $\rho(x,y) = |\arctan x - \arctan y|$ is a metric.

(34) Give a distance function for a set consisting of two members.

(35) Let X be the set of bounded sequences of real numbers, that is, for every $x = \{\xi_i, \ldots, \xi_n, \ldots\}$ in X, there exists a K_x such that $|\xi_i| \leqslant K_x$ $(i = 1, 2, \ldots)$. Show that for all $x = \{\xi_i\} \in X$ and $y = \{\eta_i\} \in X$, the function defined by

$$\rho(x,y) = \sup_i |\xi_i - \eta_i|$$

is a metric.

Chapter 2

Sequences and Series

In this chapter we give the basic properties of monotone, strictly monotone, and Cauchy sequences in detail. In particular, we treat the convergence tests of Cauchy, D'Alembert, Leibniz, and Raabe. As examples, we consider the harmonic series, the geometric series, and products of series. The convergence or divergence of the hyperharmonic series with general term $1/n^p$ is determined, depending on p. We explain the concept of conditionally convergence and show that for any conditionally convergent series and any number $S \in [-\infty, \infty]$, there is a rearrangement of the given series such that the resulting series converges to S. But the rearrangement of an absolutely convergent series yields another absolutely convergent series having the same sum as the original one. Finally, it is pointed out that sequences and series with complex terms enjoy analogous properties.

2.1 The Basic Properties of Convergent Sequences

In the fifth century B.C., the Greek philosopher Zeno proposed the following paradox: in order for a runner to travel a given distance, the runner must first travel halfway, then half the remaining distance, then half the distance that still remains, and so on ad infinitum. But, Zeno argued, it is clearly impossible for a runner to accomplish these infinitely many tasks in a finite period of time, so motion from one point to another must be impossible. Zeno's paradox suggests the infinite subdivision of $[0, 1]$ into subintervals of length $1/2^n$ for each integer $n = 1, 2, 3, \ldots$. If the length of the interval is the sum of the lengths of the subintervals into which it is divided, then it would appear that

$$1 = \frac{1}{2} + \frac{1}{4} + \frac{1}{8} + \cdots + \frac{1}{2^n} + \cdots$$

Similarly, in order for a man to live his life he must first live half life, then half the remaining life, then half the life that still remains, and so on ad infinitum. But, if Zeno were correct, it would be impossible for a man to get from the time of birth to the time of death (T). This would imply that man is immortal. On the other hand, it is clear that:

$$T = \frac{T}{2} + \frac{T}{4} + \frac{T}{8} + \cdots + \frac{T}{2^n} + \cdots$$

E. Mahmudov, *Single Variable Differential and Integral Calculus*,
DOI: 10.2991/978-94-91216-86-2_2, © Atlantis Press 2013

Important examples of infinite sequences are the following.

(1) The equilateral polygonals inscribed in a circle.
(2) Arithmetic and geometric series.
(3) The sequence of rational numbers $x_1 = 0, 3$, $x_2 = 0, 33$, $x_3 = 0, 333, \ldots$.
(4) The Fibonacci sequence defined as follows:

$F_1 = 1$, $F_2 = 1$, $F_{n+1} = F_{n-1} + F_n$, $n \geqslant 2$. This is a recursively defined sequence—after the initial values are given, each following term is calculated from two of its predecessors.

Informally speaking, the term "sequence" in mathematics is used to describe an unending succession of numbers. Before going into details, let us give some necessary definitions.

Definition 2.1. If to every natural number $n \in \{1, 2, 3, \ldots\}$ there is associated real number x_n, then the set of real numbers

$$x_1, x_2, \ldots, x_n, \ldots \tag{2.1}$$

is called a *sequence* and x_n is the *n*-th *term of the sequence*. More precisely, a sequence of real numbers is a function defined on the set of all positive integers. Later on, we also write $\{x_n\}$ for the sequence.

Example 2.1. If $x_n = \dfrac{1}{n^3}$, then $\{x_n\} = \left\{1, \dfrac{1}{2^3}, \dfrac{1}{3^3}, \ldots, \dfrac{1}{n^3}, \ldots\right\}$ or, if $x_n = (-1)^n$, then $\{x_n\} = \{-1, 1, -1, \ldots, (-1)^n, \ldots\}$.

Remark 2.1. If the sequence

$$y_1, y_2, \ldots, y_n, \ldots \tag{2.2}$$

or, $\{y_n\}$ is another sequence, then from (2.1) and (2.2) can be formed the following sequences as a result of arithmetic operations: $\{x_n + y_n\}$, $\{x_n - y_n\}$, $\{x_n \cdot y_n\}$, $\left\{\dfrac{x_n}{y_n}\right\}$ $(y_n \neq 0)$.

Definition 2.2. The sequence $\{x_n\}$ is said to have the limit a if, given any $\varepsilon > 0$, there is a positive integer $N(\varepsilon)$ such that

$$|x_n - a| < \varepsilon, \quad \text{for all } n \geqslant N(\varepsilon). \tag{2.3}$$

(The conclusion $|x_n - a| < \varepsilon$ can be reformulated as $a - \varepsilon < x_n < a + \varepsilon$, of course.) We say that the sequence $\{x_n\}$ converges to the real number a, and we write

$$\lim_{n \to \infty} x_n = a \quad \text{or} \quad x_n \to a, \quad n \to \infty.$$

Thus, if we choose any positive number ε, the terms in the sequence will eventually be within ε units of a.

A sequence $\{x_n\}$ that does not have a finite limit is said to diverge.

Example 2.2. Let $x_n = 0,\overset{n}{\overbrace{3\ldots3}}$. We show that $\lim\limits_{n\to\infty} x_n = a = \dfrac{1}{3}$. For a given $\varepsilon > 0$, by the definition of limits, we must find an integer N, which could depend on ε, such that $\left| x_n - \dfrac{1}{3} \right| < \varepsilon$ when $n \geqslant N = N(\varepsilon)$. Obviously (see 1.4 of Chapter 1), for all integers n

$$0,3\ldots3 \leqslant \frac{1}{3} \leqslant 0,3\ldots3 + \frac{1}{10^n}.$$

Thus,

$$\left| x_n - \frac{1}{3} \right| \leqslant \frac{1}{10^n}.$$

Since $\dfrac{1}{10^n} \leqslant \dfrac{1}{10^N}$ for all $n \geqslant N$, we can choose N satisfying the inequality $\dfrac{1}{10^N} < \varepsilon$. Here, $N = \left[\!\left[\log \dfrac{1}{\varepsilon} \right]\!\right] + 1$, if $0 < \varepsilon \leqslant 1$ and $N = 1$, if $\varepsilon > 1$. Here, $[\![x]\!]$ is the integer part of the decimal fraction x.

Example 2.3. (a) If $x_n = \dfrac{1}{n}$, then we show that $\lim\limits_{n\to\infty} x_n = 0$. We need to show this: To each positive number ε, there corresponds an integer N such that for all $n \geqslant N$,

$$|x_n - 0| = |x_n| = \frac{1}{n} < \varepsilon.$$

It suffices to pick any fixed integer $N = N(\varepsilon) = \left[\!\left[\dfrac{1}{\varepsilon} \right]\!\right] + 1$. Then $n \geqslant N$ implies $1/n \leqslant 1/N < \varepsilon$ as desired.

(b) Let $x_n = \dfrac{1}{n^2}$. We show $\lim_{n\to\infty} x_n = 0$. Since $|x_n - 0| = \dfrac{1}{n^2}$ for every $\varepsilon > 0$, we can take $N(\varepsilon) = \left[\!\left[\dfrac{1}{\sqrt{\varepsilon}} \right]\!\right] + 1$. Then $|x_n - 0| < \varepsilon$ for all $n > N(\varepsilon)$.

Example 2.4. We show that the limit of the sequence $\{x_n\} = \{1 + (-1)^n\}$ does not exist, that is, we show that the sequence is divergent.

Suppose that the sequence is convergent. Then there exists a real number a and positive number $N = N(\varepsilon)$ such that for all $n > N(\varepsilon)$,

$$|x_n - a| < \varepsilon. \tag{2.4}$$

If we set $\varepsilon = 1$, then for all even integers $n > N(1)$,

$$|2 - a| < 1 \quad \text{or} \quad 1 < a < 3,$$

but for all odd numbers $n > N(1)$,

$$|a| < 1 \quad \text{or} \quad -1 < a < 1.$$

Obviously, there does not exist any such number a, simultaneously satisfying both these inequalities. This contradiction proves that the sequence diverges.

Theorem 2.1. *The limit of a convergent series $\{x_n\}$ is unique.*

☐   On the contrary, let us suppose that the sequence $\{x_n\}$ has two limits a and b ($a \neq b$). Then for any given $\varepsilon > 0$, there are positive numbers $N_1(\varepsilon)$ and $N_2(\varepsilon)$ such that $|x_n - a| < \varepsilon$ for all $n > N_1(\varepsilon)$ and $|x_n - b| < \varepsilon$ for all $n > N_2(\varepsilon)$. Let us choose $\varepsilon = \frac{|a-b|}{2}$ and $N(\varepsilon) = \max\{N_1(\varepsilon), N_2(\varepsilon)\}$. Then we have

$$|x_n - a| < \frac{|a-b|}{2}, \quad |x_n - b| < \frac{|a-b|}{2}, \quad n > N(\varepsilon).$$

These inequalities imply that

$$|a - b| = |a - x_n + x_n - b| \leqslant |x_n - a| + |x_n - b| < |a - b|,$$

and so $|a - b| < |a - b|$. By this contradiction, the proof of the theorem is completed. ■

Theorem 2.2. *A convergent sequence is bounded, that is, there is a constant $C > 0$ such that $|x_n| \leqslant C$, $n = 1, 2, 3, \ldots$*

☐   Assume that $x_n \to a$, $n \to \infty$, that is, given any number $\varepsilon > 0$, there exists an integer $N = N(\varepsilon)$ such that $|x_n - a| < \varepsilon$ for all $n > N$. Then we have $|x_n| = |x_n - a + a| \leqslant |x_n - a| + |a| < \varepsilon + |a|$. Clearly, $|x_n| \leqslant C = \max\{|a| + \varepsilon, |x_1|, \ldots, |x_n|\}$. ■

Remark 2.2. Obviously, the converse assertion to the theorem is not generally true. For example, the sequence $x_n = 1 + \{-1\}^n$ diverges (see Example 2.4), while $|x_n| \leqslant 2$. Consequently, bounded sequences may or may not converge.

Theorem 2.3. *Suppose that the sequences $\{x_n\}$ and $\{y_n\}$ converge to limits a and b, respectively, that is $x_n \to a$, $y_n \to b$, $n \to \infty$. Then the sequences $\{cx_n\}$, $\{x_n y_n\}$, $\{x_n \pm y_n\}$, $\left\{\dfrac{x_n}{y_n}\right\}$ ($y_n \neq 0$) converge, for c any constant, and moreover:*

(1) $\displaystyle \lim_{n\to\infty} cx_n = c \lim_{n\to\infty} x_n = c \cdot a$,

(2) $\displaystyle \lim_{n\to\infty}(x_n \pm y_n) = \lim_{n\to\infty} x_n \pm \lim_{n\to\infty} y_n = a \pm b$,

(3) $\displaystyle \lim_{n\to\infty}(x_n \cdot y_n) = \lim_{n\to\infty} x_n \cdot \lim_{n\to\infty} y_n = a \cdot b$,

(4) $\displaystyle \lim_{n\to\infty} \frac{x_n}{y_n} = \frac{\displaystyle\lim_{n\to\infty} x_n}{\displaystyle\lim_{n\to\infty} y_n} = \frac{a}{b}$ ($b \neq 0$).

☐ (1) Let $\varepsilon > 0$ be any positive number and $\varepsilon_0 = \dfrac{\varepsilon}{|c|}$. Then there is a positive integer $N = N(\varepsilon_0)$ such that $|x_n - a| < \varepsilon_0$ for all $n > N$. Consequently, $|cx_n - ca| = |c||x_n - a| < \varepsilon$, that is, $\lim_{n\to\infty} cx_n = ca$.

(2) For any $\varepsilon > 0$, there exist integers $N_1(\varepsilon/2)$ and $N_2(\varepsilon/2)$ such that $|x_n - a| < \varepsilon/2$ for all $n > N_1(\varepsilon/2)$ and $|y_n - b| < \varepsilon/2$ for all $n > N_2(\varepsilon/2)$.

Then,

$$|(x_n \pm y_n) - (a \pm b)| \leqslant |x_n - a| + |y_n - b| < \frac{\varepsilon}{2} + \frac{\varepsilon}{2} = \varepsilon, \ n > N(\varepsilon) = \max\left\{N_1\left(\frac{\varepsilon}{2}\right), N_2\left(\frac{\varepsilon}{2}\right)\right\}$$

and the proof of part (2) is complete.

(3) By Theorem 2.2, there exists a constant $C > 0$ such that $|x_n| \leqslant C \ (n = 1, 2, \ldots)$ and $|b| \leqslant C$. For any $\varepsilon > 0$ there exist positive integers $N_1\left(\dfrac{\varepsilon}{2C}\right)$ and $N_2\left(\dfrac{\varepsilon}{2C}\right)$ such that

$$|x_n - a| < \frac{\varepsilon}{2C} \text{ when } n > N_1\left(\frac{\varepsilon}{2C}\right) \text{ and } |y_n - b| < \frac{\varepsilon}{2C} \text{ when } n > N_2\left(\frac{\varepsilon}{2C}\right).$$

Let $N = N(\varepsilon)$ be the greater of N_1 and N_2.

Then

$$|x_n y_n - ab| = |x_n y_n - x_n b + x_n b - ab| \leqslant |x_n||y_n - b| + |b||x_n - a| < \frac{C \cdot \varepsilon}{2C} + \frac{C \cdot \varepsilon}{2C} = \varepsilon$$

(4) To begin with, let us show that $\lim_{n\to\infty} \dfrac{1}{y_n} = \dfrac{1}{b}$. Again by the definition of the limit, given any number $\varepsilon = |b|/2$ there exists an integer $N_1(\varepsilon)$ such that $|y_n - b| < |b|/2$ for all $n > N_1(\varepsilon)$. Then, under the condition $n > N_1(\varepsilon)$,

$$|y_n| = |b + (y_n - b)| \geqslant |b| - |y_n - b| > |b| - \frac{|b|}{2} = \frac{|b|}{2}.$$

Moreover, let $N(\varepsilon)$ be the greater of $N_1(\varepsilon)$ and $N_2(\varepsilon)$. Then

$$\left|\frac{1}{y_n} - \frac{1}{b}\right| = \frac{|y_n - b|}{|y_n||b|} < \frac{2}{|b|^2}\varepsilon\frac{|b|^2}{2} = \varepsilon.$$

Now, by part (3) of the theorem, we have $\lim_{n\to\infty} \dfrac{x_n}{y_n} = \lim_{n\to\infty} x_n \cdot \lim_{n\to\infty} \dfrac{1}{y_n} = \dfrac{a}{b}$. ∎

Theorem 2.4 (Comparison Test). *Let $\{x_n\}$, $\{y_n\}$ and $\{z_n\}$ be sequences. Then*

(1) *(Limit inequality) If $x_n \leqslant y_n$ for all $n > N$ and $\{x_n\}$, $\{y_n\}$ are convergent, then*

$$\lim_{n\to\infty} x_n \leqslant \lim_{n\to\infty} y_n.$$

(2) *(Sandwich or squeeze property) If $x_n \leqslant y_n \leqslant z_n$ for all $n > N$ and the sequences $\{x_n\}$, $\{z_n\}$ have the same limits, then the sequence $\{y_n\}$ is convergent and the equalities*

$$\lim_{n\to\infty} x_n = \lim_{n\to\infty} y_n = \lim_{n\to\infty} z_n \text{ hold.}$$

☐   (1) Let $\lim\limits_{n\to\infty} x_n = a$ and $\lim\limits_{n\to\infty} y_n = b$ where $a > b$. The sequences are convergent, and
so for any $\varepsilon = \dfrac{a-b}{2}$ there exist integers $N_1(\varepsilon)$ and $N_2(\varepsilon)$ such that $|x_n - a| < \dfrac{a-b}{2}$ for
all $n > N_1(\varepsilon)$ and $|y_n - b| < \dfrac{a-b}{2}$ for all $n > N_2(\varepsilon)$. If $N_0(\varepsilon)$ is the greater of $N_1(\varepsilon)$ and
$N_2(\varepsilon)$, then for all $n > N_0(\varepsilon)$

$$x_n - y_n = (x_n - a) - (y_n - b) + (a - b)$$
$$\geqslant -|x_n - a| - |y_n - b| + (a - b)$$
$$> -\frac{a-b}{2} - \frac{a-b}{2} + (a - b) = 0.$$

This contradiction proves that part (1) is true.

(2) If $\lim\limits_{n\to\infty} x_n = \lim\limits_{n\to\infty} z_n = a$, then for any $\varepsilon > 0$ there exist positive numbers $N_1 = N_1(\varepsilon)$,
$N_2 = N_2(\varepsilon)$ such that $|x_n - a| < \varepsilon$ for all $n > N_1$ and $|z_n - a| < \varepsilon$ for all $n > N_2$. In particular,
since $a - \varepsilon < x_n$ and

$$z_n < a + \varepsilon \text{ for all } n > N_0(\varepsilon) \quad (N_0(\varepsilon) = \max\{N_1(\varepsilon), N_2(\varepsilon)\}),$$

for all integers $n > \overline{N}(\varepsilon) = \max\{N, N_1, N_2\}$ the inequality $a - \varepsilon < x_n \leqslant y_n \leqslant z_n < a + \varepsilon$ is
satisfied. In other words, $\lim\limits_{n\to\infty} y_n = a$.   ∎

Remark 2.3. If in part (1) of the theorem the strict inequality $x_n < y_n$ holds, this does not
necessarily imply the strict inequality of their limits, that is, in general $\lim\limits_{n\to\infty} x_n \leqslant \lim\limits_{n\to\infty} y_n$.
For example, if $x_n = \dfrac{1}{n^2 + 1}$, $y_n = \dfrac{1}{n}$, then $\dfrac{1}{n^2 + 1} < \dfrac{1}{n}$: but $\lim\limits_{n\to\infty} x_n = \lim\limits_{n\to\infty} y_n = 0$.
      The same is true for sequences $x_n = a$ or $y_n = b$, $n = 1, 2, 3, \ldots$ (a, b are constants).
Indeed, if $x_n = 0$ and $y_n = \dfrac{1}{n}$ then $\dfrac{1}{n} > 0$. By passing to the limit, we have

$$\lim\limits_{n\to\infty} y_n = \lim\limits_{n\to\infty} \frac{1}{n} = 0 = \lim\limits_{n\to\infty} x_n.$$

Example 2.5. Find $\lim\limits_{n\to\infty} \dfrac{1}{n^2} \sin^2 n$. At once it follows from the inequality $0 \leqslant \sin^2 n \leqslant 1$ that

$$0 \leqslant \frac{1}{n^2} \sin^2 n \leqslant \frac{1}{n^2}.$$

If we set $x_n = 0$, $z_n = \dfrac{1}{n^2}$, then by the Comparison Theorem we have $\lim\limits_{n\to\infty} x_n = \lim\limits_{n\to\infty} z_n = 0$,
and as a result,

$$\lim\limits_{n\to\infty} \frac{1}{n^2} \sin^2 n = 0.$$

Example 2.6. Suppose that y_n satisfies the inequality $1 - \dfrac{1}{n^2} \leqslant y_n \leqslant \cos \dfrac{1}{n}$. Find the limit $\lim\limits_{n \to \infty} y_n$. Clearly,

$$\lim_{n \to \infty} x_n = \lim_{n \to \infty} \left(1 - \frac{1}{n^2} \right) = 1 \text{ and } \lim_{n \to \infty} z_n = \lim_{n \to \infty} \cos \frac{1}{n} = 1.$$

By the Comparison Test, we then have $\lim\limits_{n \to \infty} y_n = 1$.

Example 2.7. We compute the limit of the well-known sequence $y_n = \sqrt[n]{n} - 1$ $(y_n \geqslant 0)$. Using the Binomial Formula, it is easy to see that

$$n = (1 + y_n)^n \geqslant \frac{n(n-1)}{n} y_n^2.$$

Thus,

$$y_n \leqslant \sqrt{\frac{2}{n-1}} \quad (n \geqslant 2).$$

Consequently, we have

$$x_n \leqslant y_n \leqslant z_n, \quad x_n = 0, \quad z_n = \sqrt{\frac{2}{n-1}}.$$

By part (2) of Theorem 2.4, since $\lim\limits_{n \to \infty} x_n = \lim\limits_{n \to \infty} z_n = 0$, we have $\lim\limits_{n \to \infty} y_n = 0$, or $\lim\limits_{n \to \infty} \sqrt[n]{n} = 1$.

Example 2.8. For the sequence $y_n = \dfrac{a^n}{n!}$ $(a > 0)$, find $\lim\limits_{n \to \infty} y_n$.

Let N be a positive integer satisfying the inequality $N \geqslant 2a$. Clearly, for $n > N$, i.e., for $n > 2a$, we have $\dfrac{a}{n} < \dfrac{1}{2}$. Thus,

$$0 < \frac{a^n}{n!} = \frac{a^N a^{n-N}}{N!(N+1)\cdots n} = \frac{a^n}{N!} \cdot \frac{a}{N+1} \cdot \frac{a}{N+2} \cdots \frac{a}{n} < \frac{a^n}{N!} \left(\frac{1}{2} \right)^{n-N}.$$

As a result,

$$0 = x_n < \frac{a^n}{n!} < z_n = \frac{a^n}{N!} \left(\frac{1}{2} \right)^{n-N}.$$

Here $\lim\limits_{n \to \infty} x_n = \lim\limits_{n \to \infty} z_n = 0$, and by the Comparison Test, the limit sought is zero:

$$\lim_{n \to \infty} \frac{a^n}{n!} = 0.$$

Definition 2.3. A sequence $\{x_n\}$ is called, respectively, *increasing, nondecreasing, decreasing*, or *nonincreasing* if

$$x_{n+1} > x_n, \quad x_{n+1} \geqslant x_n, \quad x_{n+1} < x_n, \quad \text{or } x_{n+1} \leqslant x_n, \quad n = 1, 2, 3, \ldots.$$

A sequence that is either nondecreasing or nonincreasing is called *monotone*, and a sequence that is increasing or decreasing is called *strictly monotone*. It is clear that a strictly monotone sequence is monotone, but not conversely.

Theorem 2.5. *Every bounded monotone sequence converges.*

☐ For definiteness, we prove the theorem for a nondecreasing sequence $\{x_n\}$ (the proofs for the other cases are analogous). By hypothesis, $A = \{x_n\}$ is bounded, and so the supremum $\sup A = a$ exists. Thus, there is a number $x_n \in A$ such that $x_n \leqslant a$, $a - \varepsilon < x_N \leqslant a$. Observe that $x_n \geqslant x_N$ for all integers $n > N$ (the sequence is nondecreasing), and consequently, $a - \varepsilon < x_n \leqslant a$ or $a - \varepsilon < x_n < a + \varepsilon$ for all $n > N$. Finally, $|x_n - a| < \varepsilon$, $n > N$.
∎

By using this theorem, it can be shown that the limit of the sequence $x_n = \dfrac{a^n}{n!}$ $(a > 0)$ is equal to zero. Obviously, for integers n satisfying the inequality $n > a - 1$, we have the relation $x_{n+1} = x_n \cdot \dfrac{a}{x_n + 1}$, and then $x_{n+1} > x_n$. Moreover, the sequence $\{x_n\}$ is bounded from below by zero. Finally, by Theorem 2.5, the sequence x_n is convergent, that is, $x_n \to x$, where x is a real number.

Thus, we have

$$x = \lim_{n \to \infty} x_{n+1} = \lim_{n \to \infty} x_n \cdot \lim_{n \to \infty} \frac{a}{n+1} = x \cdot 0 = 0.$$

Example 2.9. In the theory of numerical analysis, in order to calculate the square root of the positive number a, the following recurrence sequence is considered:

$$x_{n+1} = \frac{1}{2}\left(x_n + \frac{a}{x_n}\right), \quad n = 1, 2, \ldots \tag{2.5}$$

Here, x_1 is an initial approach and can be any positive number. It is easy to see from (2.5) that $x_n > 0$ $(n = 1, 2, \ldots)$, that is, the sequence $\{x_n\}$ is bounded below by zero. On the other hand, for a positive number $t > 0$ the inequality $(t - 1)^2 \geqslant 0$ gives us that $t + \dfrac{1}{t} \geqslant 2$. Now, let us put $t = \dfrac{x_n}{\sqrt{a}} > 0$. Then it follows from (2.5) that

$$x_{n+1} = \frac{\sqrt{a}}{2}\left(\frac{x_n}{\sqrt{a}} + \frac{\sqrt{a}}{x_n}\right) = \frac{\sqrt{a}}{2}\left(t + \frac{1}{t}\right) \geqslant \sqrt{a} \quad (n = 1, 2, \ldots).$$

Thus, for integers $n \geqslant 2$, the inequality $x_n \geqslant \sqrt{a}$ holds. Next, let us show that the sequence $\{x_n\}$ $(n \geqslant 2)$ is nonincreasing. We can rewrite (2.5) as follows:

$$\frac{x_n + 1}{x_n} = \frac{1}{2}\left(1 + \frac{a}{x_n^2}\right). \tag{2.6}$$

Since $x_n \geqslant \sqrt{a}$ $(n \geqslant 2)$, from equality (2.6) we can derive that $\dfrac{x_{n+1}}{x_n} \leqslant 1$ or $x_{n+1} \leqslant x_n$. Then from Theorem 2.5, the sequence $\{x_n\}$ is convergent. By Theorem 2.4, $\lim\limits_{n \to \infty} x_n = x \geqslant \sqrt{a}$. Finally, by passing to the limit in (2.5), we deduce that

$$x = \frac{1}{2}\left(x + \frac{a}{x}\right),$$

which implies that $x = \sqrt{a}$.

Example 2.10. (a) Find the limit of the sequence $x_n = \sqrt{a + \sqrt{a + \sqrt{a + \cdots + \sqrt{a}}}}$, where $a > 0$ and the number of radical signs is n. Here, as a recursively defined sequence, we can set

$$x_{n+1} = \sqrt{a + x_n}, \text{ for } (n = 1, 2, \ldots), \quad x_1 = \sqrt{a}. \tag{2.7}$$

Firstly, let us show that the sequence (2.7) is increasing and bounded. We do this by mathematical induction. Observe that $x_1 = \sqrt{a} < \sqrt{a + \sqrt{a}} = x_2$. Suppose that $x_n < x_{n+1}$. The inductive step is then to prove that $x_{n+1} < x_{n+2}$. By (2.7), the assertion $x_{n+1} < x_{n+2}$ is equivalent to

$$x_{n+1} = \sqrt{a + x_n} < \sqrt{a + x_{n+1}} = x_{n+2}.$$

But $x_n < x_{n+1}$ by hypothesis, and so $\sqrt{a + x_n} < \sqrt{a + x_{n+1}}$ as required for the inductive step. We next prove that it is bounded (for example, by $\sqrt{a} + 1$). Clearly, $x_1 = \sqrt{a} < \sqrt{a} + 1$. We proceed by mathematical induction and so assume that for some integer n the inequality

$$x_n < \sqrt{a} + 1 \tag{2.8}$$

holds, whereas $x_{n+1} = \sqrt{a + x_n}$, it follows from (2.8) that

$$x_{n+1} < \sqrt{a + \sqrt{a} + 1} < \sqrt{a + 2\sqrt{a} + 1} = \sqrt{(\sqrt{a} + 1)^2} = \sqrt{a} + 1.$$

Then, by Theorem 2.5, the sequence $\{x_n\}$ approaches some limit point, x. In order to find x, we square both sides of (2.7). By passing to the limit as n approaches infinity, we have

$$x_{n+1}^2 = a + x_n,$$

$$\lim_{n \to \infty} x_{n+1}^2 = x^2 = a + x = a + \lim_{n \to \infty} x_n.$$

The quadratic equation here, $x^2 - x - a = 0$, has a unique positive root $(x > 0)$ $x = \frac{1}{2}(1 + \sqrt{1 + 4a})$, which is the required expression for the limit.

(b) Let $\{F_n\}$ be the Fibonacci sequence. Assume that $a = \lim_{n \to \infty} \left(\frac{F_{n+1}}{F_n} \right)$ exists. We show that $a = \frac{1}{2}(1 + \sqrt{5})$. Taking into account that $F_{n+1} = F_n + F_{n-1}$, the equation

$$a = \lim_{n \to \infty} \left(\frac{F_{n+1}}{F_n} \right)$$

becomes

$$a = \lim_{n \to \infty} \left(1 + \frac{F_{n-1}}{F_n} \right) = 1 + \frac{1}{a}.$$

Thus $a^2 = a + 1$ and so $a = \frac{1}{2}(1 + \sqrt{5})$ (note $a > 0$).

2.2   Infinitely Small and Infinitely Large Sequences; Subsequences

Definition 2.4. If, for a given sequence $\{x_n\}$, $\lim\limits_{n\to\infty} x_n = 0$, then the sequence is said to be an *infinitely small* (or *infinitesimal*) *sequence.*

Observe that if $x_n \to a$, then $y_n = x_n - a$ is an infinitely small sequence. Conversely, if $x_n = a + y_n$ and y_n is an infinitely small sequence, then $x_n \to a$.

Moreover, the limit of a sequence $\{x_n\}$ is said to be $+\infty$ $(-\infty)$ if for any positive real number M there is an integer $N = N(M)$ such that $|x_n| > M$ $(x_n < -M)$ for all $n > N$, and is then written

$$\lim_{n\to\infty} y_n = +\infty \quad (\lim_{n\to\infty} x_n = -\infty) \quad \text{or} \quad x_n \to +\infty \quad (x_n \to -\infty).$$

Sometimes the sequences such that $\lim\limits_{n\to\infty} |x_n| = +\infty$ are called *infinitely large sequences.*

Theorem 2.6. *If $\{x_n\}$ and $\{y_n\}$ are two infinitely small sequences, then their sum and product are infinitely small sequences.*

☐   By the definition of infinitely small sequences, $\lim\limits_{n\to\infty} x_n = 0$ and $\lim\limits_{n\to\infty} y_n = 0$. Then, by Theorem 2.3, $\lim\limits_{n\to\infty} (x_n + y_n) = \lim\limits_{n\to\infty} x_n + \lim\limits_{n\to\infty} y_n = 0$ and $\lim\limits_{n\to\infty} x_n \cdot y_n = \lim\limits_{n\to\infty} x_n \cdot \lim\limits_{n\to\infty} y_n = 0$. ■

Theorem 2.7. *The product of a bounded and an infinitely small sequence is an infinitely small sequence.*

☐   Let $\{x_n\}$ be bounded and let $\{y_n\}$ be infinitely small. By hypothesis, for some real number $C > 0$, the inequality $|x_n| \leqslant C$ $(n = 1, 2, \ldots)$ holds and $\lim\limits_{n\to\infty} y_n = 0$. Because $\{y_n\}$ is infinitely small, given any number $\varepsilon > 0$ there is an integer $N = N(\varepsilon)$ such that $|y_n| < \frac{\varepsilon}{C}$ for all $n > N$. For such integers,

$$|x_n y_n| = |x_n||y_n| < C \cdot \frac{\varepsilon}{C}.$$

In other words, $\lim\limits_{n\to\infty} x_n y_n = 0$. ■

Theorem 2.8. *If $\lim\limits_{n\to\infty} |x_n| = +\infty$ and $x_n \neq 0$ $(n = 1, 2, \ldots)$, then $\{y_n\} = \left\{\dfrac{1}{x_n}\right\}$ is an infinitely small sequence.*

☐   By hypothesis, for any real number $\varepsilon > 0$ there is an integer N such that $|x_n| > 1/\varepsilon$ for all $n > N$. This implies $|y_n| < \varepsilon$ for all $n > N$, that is, $\lim\limits_{n\to\infty} y_n = 0$. ■

Example 2.11. Suppose $x_n = n^2$, $y_n = 1/n$. Obviously, $\{x_n\}$ and $\{y_n\}$ are infinitely large and infinitesimal sequences, respectively. Observe that $\left\{\dfrac{1}{x_n}\right\} = \left\{\dfrac{1}{n^2}\right\}$ and $\left\{\dfrac{1}{y_n}\right\} = n$. The first is an infinitesimal sequence, but the second is an infinitely large sequence.

Definition 2.5. Let $\{x_n\}$ be a sequence and let $k_1 < k_2 < \cdots < k_n < \cdots$ be an increasing sequence of natural numbers. Then $\{x_{k_n}\} = \{x_{k_1}, x_{k_2}, \ldots, x_{k_n}, \ldots\}$ is said to be a *subsequence* of the sequence.

Theorem 2.9. *If* $\lim_{n \to \infty} x_n = a$, *then the limit of an arbitrary subsequence* $\{x_{k_n}\}$ *is the same a.*

☐ By hypothesis, for any given $\varepsilon > 0$ there is an integer $N = N(\varepsilon)$ such that $|x_n - a| < \varepsilon$ for any $n > N$. Then from the inequality $k_n \geqslant n$, it follows that $|x_{k_n} - a| < \varepsilon, n > N$, that is, $x_{k_n} \to a$. ■

Theorem 2.10 (Bolzano[1]–Weierstrass[2]). *Every infinite and bounded set in Euclidean space has at least one limit point.*

☐ Let for simplicity M be an infinite and bounded set of the real axis. Let I_1 be a closed interval including M, i.e., $I_1 = [a, b] \supset M$. Divide I_1 into two equal subintervals. Suppose I_2 is one of them, containing infinitely many points of M. Then again we can divide the interval I_2 into two equal subintervals such that one of them, say I_3, contains infinitely many points belonging to M. For $k = 2, 3, 4, \ldots$, repeat this bisection on I_k and choose I_{k+1} the same way. We have $I_j = [a_j, b_j]$ and $I_1 \supset I_2 \supset \cdots \supset I_n \supset \cdots$. By Cantor's Theorem (Theorem 1.7), since $\lim_{j \to \infty} (b_j - a_j) = \lim_{j \to \infty} \dfrac{b_j - a_j}{2^j} = 0$, the intersection $\bigcap_{n=1}^{\infty} I_n$ contains exactly one point, say x. At the same time, x is a limit point of the set M: we see this as follows: since $\lim_{n \to \infty} I_n = 0$ for any small number $\varepsilon > 0$, there is an integer n such that $I_n \subset (x - \varepsilon, x + \varepsilon)$. In other words, any ε-neighborhood of the point x contains infinitely many points of M. ■

Corollary 2.1. *If* $\{x_n\}$ *is bounded, then* $\{x_n\}$ *has a convergent subsequence.*

☐ At first, assume that $\{x_n\}$ has only a finite number of different points. Then in our constructed subsequence, at least one of those points can be repeated:

$$x = x_k = x_{k_1} = \cdots = x_{k_n} = \cdots$$

Therefore, the point x is a limit of such a subsequence $\{x_{k_n}\}$.

Now, suppose that the sequence $\{x_n\}$ has infinitely many different points. By hypothesis, the sequence is bounded, hence there is at least one limit point, say x. For any positive real number ε_1 ($\varepsilon_1 > 0$), the ε_1-neighborhood of x contains infinitely many different points of the sequence. Let us denote by x_{k_1} one such point, and form the ε_2 ($\varepsilon_2 < \varepsilon_1$)-neighborhood of x that does not contain x_{k_1}. Similarly, in this ε_2-neighborhood of x we can

---

[1]B. Bolzano (1781-1848), Czech mathematician.
[2]K. Weierstrass (1815-1897), German mathematician.

take another point $x_{k_2} \neq x$, $k_2 > k_1$. Continuing this process, we form ε_i-neighborhoods of the point x, where $\varepsilon_i \to 0$. Finally, we take the subsequence $x_{k_1}, x_{k_2}, \ldots, x_{k_n}, \ldots$ of $\{x_n\}$, the limit point of which is the point x. ∎

Remark 2.4. As is seen from the proof, this assertion is true for an infinite bounded sequence in the Euclidean space $\mathbb{R}^n$.

Let us denote by E the set of limit points of subsequences of the bounded sequence $\{x_n\}$. Obviously the set E is nonempty and bounded. In fact, if $x_{k_n} \to x$ ($x \in E$), then from the inequality $|x_{k_n}| \leqslant C$ we have $|x| \leqslant C$, where C is a positive constant. Since E is bounded, $\sup E$ and $\inf E$ exist.

Definition 2.6. The numbers $x^* = \sup E$ and $x_* = \inf E$ are called the *superior* and *inferior* *limits* of the sequence $\{x_n\}$, respectively. Further, we use the symbols

$$x^* = \varlimsup_{n \to \infty} x_n \text{ and } x_* = \varliminf_{n \to \infty} x_n.$$

If $\{x_n\}$ and $\{y_n\}$ are two sequences, we have

$$\varliminf_{n \to \infty} x_n + \varliminf_{n \to \infty} y_n \leqslant \varliminf_{n \to \infty} (x_n + y_n) \leqslant \varlimsup_{n \to \infty} x_n + \varliminf_{n \to \infty} y_n \leqslant \varlimsup_{n \to \infty} (x_n + y_n) \leqslant \varlimsup_{n \to \infty} x_n + \varlimsup_{n \to \infty} y_n$$

Example 2.12. Find the inferior and superior limits of the sequence $\{x_n\} = \{1 + (-1)^n\}$.

Obviously, for even ($n = 2k$) and odd ($n = 2k+1$) numbers $\{x_{2k}\} = \{2\}$ and $\{x_{2k+1}\} = \{0\}$, respectively. Clearly, $E = \{2, 0\}$. Here, $\varlimsup_{n \to \infty} x_n = 2$, $\varliminf_{n \to \infty} x_n = 0$ because $\sup E = 2$ and $\inf E = 0$.

2.3 Cauchy Sequences

Definition 2.7. A sequence $\{x_n\}$ is a *Cauchy*[3] *sequence* if for every $\varepsilon > 0$ there exists an integer $N(\varepsilon)$ such that $|x_m - x_n| < \varepsilon$ for all $m, n > N(\varepsilon)$.

Let $x_n = \dfrac{1}{n^2}$. For $m, n > N = N(\varepsilon)$, say $m > n$, then we have

$$|x_m - x_n| = \frac{1}{n^2} - \frac{1}{m^2} < \frac{1}{n^2} \leqslant \frac{1}{N^2}.$$

For every $\varepsilon > 0$, we can take an integer $N > \dfrac{1}{\sqrt{\varepsilon}}$. Then $m, n > N$ implies $|x_m - x_n| \leqslant \dfrac{1}{N^2} < \varepsilon$, so $\{x_n\}$ is a Cauchy sequence.

[3]A. Cauchy (1789–1857), French mathematician.

Warning. This is stronger than claiming that, beyond a certain point in the sequence, consecutive terms are within ε of one another.

For example, consider the sequence $x_n = \sqrt{n}$. It is not hard to see that

$$|x_{n+1} - x_n| = \left(\sqrt{n+1} - \sqrt{n}\right) \cdot \frac{\sqrt{n+1} + \sqrt{n}}{\sqrt{n+1} + \sqrt{n}} = \frac{1}{\sqrt{n+1} + \sqrt{n}}.$$

Therefore $|x_{n+1} - x_n| \to 0$ as $n \to \infty$. But $|x_m - x_n| = \sqrt{m} - \sqrt{n}$ if $m > n$, and so for any integer $N = N(\varepsilon)$ we can choose $n = N$ and $m > N$ such that $|x_m - x_n|$ is as large as desired. Thus the sequence $\{x_n\}$ is not a Cauchy sequence.

Theorem 2.11. *If a sequence $\{x_n\}$ is a Cauchy sequence, then $\{x_n\}$ is bounded.*

☐ Let $\varepsilon > 0$ be any positive number and suppose $\{x_n\}$ is a Cauchy sequence, that is, for any there is an integer such that $|x_n - x_m| < \varepsilon$ for all. Let us take. Then $|x_n| = |x_{N+1} + (x_n - x_{N+1})| \leqslant |x_{N+1}| + |x_n - x_{N+1}| < |x_{N+1}| + \varepsilon$ for all $n > N$ Let $C = \max\{|x_{N+1}| + \varepsilon, |x_1|, \ldots, |x_n|\}$. Then for all $n > N$, the inequality $|x_n| < |x_{N+1}| + \varepsilon$ implies the desired inequality $|x_n| \leqslant C$ $(n = 1, 2, \ldots)$. ∎

Theorem 2.12. *A sequence $\{x_n\}$ converges if and only if $\{x_n\}$ is a Cauchy sequence.*

☐ At first, let $\{x_n\}$ be convergent and $x_n \to x$. We show that $\{x_n\}$ is a Cauchy sequence. For a given $\varepsilon > 0$, we can choose an integer $N = N(\varepsilon)$, such that $|x_n - x| < \varepsilon/2$ whenever $n > N$. Then for all $n, m > N$, we have

$$|x_n - x_m| \leqslant |x_n - x| + |x_m - x| < \frac{\varepsilon}{2} + \frac{\varepsilon}{2} = \varepsilon,$$

i.e., $\{x_n\}$ is a Cauchy sequence.

Conversely, let $\{x_n\}$ be a Cauchy sequence. We prove that $\{x_n\}$ is convergent.

Observe that, by Theorem 2.11, the sequence $\{x_n\}$ is bounded. Moreover, by Corollary 2.1, it has a subsequence $\{x_{k_n}\}$ that converges, say to x. Next, we show that $\lim\limits_{n \to \infty} x_n = x$.

Let $\varepsilon > 0$ be arbitrary. Since $\{x_n\}$ is a Cauchy sequence, there is an integer n_0 such that $m, n > N_0$ yields $|x_n - x_m| < \varepsilon/2$. Moreover, $k_n \geqslant n$ and so from the last inequality, setting $m = k_n$ for all $n > N_0$, we have $|x_n - x_{k_n}| < \varepsilon/2$. Since the sequence $\{x_{k_n}\}$ is convergent, i.e., $x_{k_n} \to x$, there is an integer N_1 such that $|x_{k_n} - x| < \varepsilon/2$ for all $n > N_1$. Let $N = \max\{N_0, N_1\}$. Then for all $n > N$ we have

$$|x_n - x| \leqslant |x_n - x_{k_n}| + |x_{k_n} - x| < \frac{\varepsilon}{2} + \frac{\varepsilon}{2} = \varepsilon.$$

That is, $x_n \to x$. ∎

Let us apply this theorem to show the divergence of the sequence $\{x_n\}$, where $x_n = 1 + \frac{1}{2} + \cdots + \frac{1}{n}$. Let us take, in Definition 2.7, $m = 2n$. Then

$$|x_m - x_n| = |x_{2n} - x_n| = \frac{1}{n+1} + \cdots + \frac{1}{2n} \geqslant n \cdot \frac{1}{2n} = \frac{1}{2}$$

and so if we take $\varepsilon = \frac{1}{2}$, then there is no integer N such that for all $m, n > N$ $(m = 2n)$ $|x_m - x_n| < \frac{1}{2}$. This means that $\{x_n\}$ is divergent.

Definition 2.8. The notion of a Cauchy sequence makes sense in the more general context of a metric space (X, ρ). At first note that if $\{x_n\} \subset X$ and $x \in X$, then $x_n \to x$ if and only if $\rho(x_n, x) \to 0$. Then, by analogy, $\{x_n\}$ is a Cauchy sequence if $\rho(x_n, x_m) \to 0$ as $m, n \to \infty$.

A metric space (X, ρ) is complete if every Cauchy sequence in X has a limit in X. At once we note that the Euclidean space $\mathbb{R}^n$ is complete.

Below we will see that the set of rational numbers with the corresponding metric is not a complete space.

Remark 2.5. We remark that Theorem 2.12 is sometimes valid in the more general case of a metric space (X, ρ). In this case, in Definition 2.7, the distance function $\rho(x_n, x_m) < \varepsilon$ is used. For example, Theorem 2.12 is valid for the Euclidean space $\mathbb{R}^n$. But it is not true for arbitrary metric spaces, i.e., it happens that a Cauchy sequence might not converge to an element of the space in question. For example, let X be the set of all rational numbers. The distance between two points r_1, $r_2 \in X$ is given by $\rho(r_1, r_2) = |r_1 - r_2|$. Clearly X is a metric space, and $r_n = \left(1 + \frac{1}{n}\right)^n$ $(n = 1, 2, \ldots)$ are rational numbers. Observe that the sequence $\{r_n\}$ is a Cauchy sequence, but its limit point e $(e = 2,71828...)$ is an irrational number. Finally, this sequence $\{r_n\}$ can not be considered as a convergent sequence within the space X. In order to prove the convergence $r_n \to e$, we need Theorem 2.5. Let $x_n = \left(1 + \frac{1}{n}\right)^{n+1}$. We show that $\{x_n\}$ is a monotone decreasing sequence. We have

$$\frac{x_n}{x_{n-1}} = \left(\frac{1 + \frac{1}{n}}{1 + \frac{1}{n-1}}\right)^n \left(1 + \frac{1}{n}\right) = \left(1 - \frac{1}{n^2}\right)^n \left(1 + \frac{1}{n}\right)$$

$$< \left(1 - \frac{1}{n^2}\right)^n \left(1 + \frac{1}{n^2}\right)^n = \left(1 - \frac{1}{n^4}\right)^n < 1$$

that is, $x_n < x_{n-1}$.

Moreover, $x_n > 1$ $(n = 1, 2, \ldots)$ and so, by Theorem 2.5, $x_n \to e$ converges:

$$\lim_{n \to \infty} r_n = \lim_{n \to \infty} \left(1 + \frac{1}{n}\right)^n = \lim_{n \to \infty} \frac{\left(1 + \frac{1}{n}\right)^{n+1}}{1 + \frac{1}{n}} = \frac{\lim_{n \to \infty} \left(1 + \frac{1}{n}\right)^{n+1}}{\lim_{n \to \infty} \left(1 + \frac{1}{n}\right)} = e.$$

Note that in terms of sequences there is another definition of compactness than the one given in Section 1.6; a subset M of a metric space (X, ρ) is compact if every sequence in M has a subsequence which converges to an element of M. Compactness turns out to be a stronger condition than completeness, though in some arguments one notion can be used in place of the other.

2.4   Numerical Series; Absolute and Conditional Convergence

Suppose $\{x_n\}$ is an infinite sequence of real numbers. Then an infinite numerical series is an expression of the form,

$$\sum_{k=1}^{\infty} x_k = x_1 + x_2 + \cdots + x_n + \cdots \tag{2.9}$$

The number x_n is called the n-th term of the series. An example of an infinite series is the series

$$\sum_{n=1}^{\infty} \frac{1}{2^n} = \frac{1}{2} + \frac{1}{4} + \frac{1}{8} + \cdots + \frac{1}{2^n} + \cdots$$

This was mentioned in Section 2.1. The n-th term of this infinite series is $x_n = \frac{1}{2^n}$.

To say what is meant by such an infinite sum, let us introduce the partial sums of the series (2.9). The n-th partial sum S_n of the series (2.9) is the sum of its first n terms

$$S_n = x_1 + \cdots + x_n.$$

Thus, each infinite series has associated with it an infinite sequence of partial sums

$$S_1, S_2, S_3, \ldots, S_n, \ldots.$$

Definition 2.9 (Sum of an infinite series). We say that the infinite series (2.9) *converges* (or *is convergent*), *with sum S*, provided that the limit of its sequence of partial sums $\{S_n\}$

$$\lim_{n \to \infty} S_n = S$$

exists (and of course, is finite). Otherwise, we say that the series (2.9) *diverges* (or *is divergent*). If a series diverges then it has no sum.

Thus an infinite sum is a limit of finite sums

$$\lim_{n\to\infty} S_n = S = \sum_{k=1}^{\infty} x_k$$

provided that this limit exists.

Theorem 2.13 (The n-th term test for divergence). *If the infinite series $\sum_{k=1}^{\infty} x_k$ is convergent, then the limit of the n-th term is equal to zero, i.e., $\lim_{n\to\infty} x_n = 0$.*

□ Observe that the n-th term x_n is the difference

$$x_n = S_n - S_{n-1}. \tag{2.10}$$

Under the assumption that the series is convergent,

$$\lim_{n\to\infty} S_n = \lim_{n\to\infty} S_{n-1} = S. \tag{2.11}$$

Taking into account (2.11), and then passing to the limit in (2.10), we have:

$$\lim_{n\to\infty} x_n = \lim_{n\to\infty} (S_n - S_{n-1}) = S - S = 0.$$

■

Consequently, if $\lim_{n\to\infty} x_n \neq 0$, then the series $\sum_{k=1}^{\infty} x_k$ diverges. In other words, the condition $x_n \to 0$ is necessary but not sufficient for convergence. That is, a series may satisfy the condition $\lim_{n\to\infty} x_n = 0$ and yet diverge. An important example of a divergent series with terms that approach zero is the harmonic series below.

Example 2.13. The following series is called the harmonic series

$$\sum_{k=1}^{\infty} \frac{1}{k} = 1 + \frac{1}{2} + \frac{1}{3} + \cdots + \frac{1}{n} + \cdots. \tag{2.12}$$

Since each term of the harmonic series is positive, its sequence of partial sums $\{S_n\}$ is monotone increasing. We shall prove that the harmonic series diverges, by showing that there are arbitrarily large partial sums. We have

$$S_2 = 1 + \frac{1}{2} > \frac{1}{2} + \frac{1}{2} = \frac{2}{2},$$

$$S_4 = S_2 + \frac{1}{3} + \frac{1}{4} > S_2 + \left(\frac{1}{4} + \frac{1}{4}\right) = S_2 + \frac{1}{2} > \frac{3}{2},$$

$$S_8 = S_4 + \frac{1}{5} + \frac{1}{6} + \frac{1}{7} + \frac{1}{8} > S_4 + \left(\frac{1}{8} + \frac{1}{8} + \frac{1}{8} + \frac{1}{8}\right) = S_4 + \frac{1}{2} > \frac{4}{2},$$

$$\cdots$$

$$\cdots$$

$$S_{2^n} > \frac{n+1}{2}.$$

Now, if C is an arbitrarily large number and n is such that $\dfrac{n+1}{2} > C$, then $S_{2^n} > C$. This means that the sequence of partial sums is unbounded, i.e., $S_n \to \infty$.

Note that if the sequence of partial sums of the series (2.9) diverges to infinity, then we say that the series itself diverges to infinity, and we write $\sum\limits_{n=1}^{\infty} x_n = \infty$.

Theorem 2.14 (Sum of the geometric series). *If $|x| \geqslant 1$ then the geometric series $\sum\limits_{k=1}^{\infty} ax^{k-1}$ ($a \neq 0$) diverges and if $|x| < 1$ then the geometric series is convergent and its sum is $\dfrac{a}{1-x}$.*

☐ The elementary identity
$$1 + x + x^2 + \cdots + x^n = \frac{1 - x^{n+1}}{1 - x}$$
follows after multiplying left and right hand sides by $1 - x$.

Therefore, for $x \neq 1$, we have $S_n = \dfrac{a(1 - x^n)}{1 - x} = \dfrac{a}{1-x} - \dfrac{a}{1-x} x^n$, and if $|x| < 1$, then
$$S = \lim_{n \to \infty} S_n = \frac{a}{1-x}.$$

If $|x| = 1$, that is, $x = 1$ or $x = -1$, then the n-th partial sums of the geometric series are $S_n = n \cdot a$ and $S_n = a - a + \cdots + (-1)^{n-1}a$, respectively. In the first case, $\lim\limits_{n \to \infty} S_n = \lim\limits_{n \to \infty} na = \pm\infty$ (the sign depending on whether a is positive or negative). This proves divergence. In the second case, the sequence of partial sums is
$$a, 0, a, 0, a, 0, \ldots$$
which diverges. Finally, if $|x| > 1$ then $\lim\limits_{n \to \infty} x_n = \lim\limits_{n \to \infty} ax^{k-1} = \infty$, and so the series diverges.

■

Theorem 2.15. *If $\sum\limits_{k=1}^{\infty} x_k$ and $\sum\limits_{k=1}^{\infty} y_k$ are convergent series, that is, $\sum\limits_{k=1}^{\infty} x_k = A$ and $\sum\limits_{k=1}^{\infty} y_k = B$ with A, B real numbers, then the sum of these series, $\sum\limits_{k=1}^{\infty} (x_k + y_k)$, is convergent and equal to $C = A + B$. Furthermore, if α is a nonzero constant, then the series $\sum\limits_{k=1}^{\infty} x_k$ and $\sum\limits_{k=1}^{\infty} \alpha x_k$ either both converge or both diverge. In the case of convergence, the sums are related by*
$$\sum_{k=1}^{\infty} \alpha x_k = \alpha \sum_{k=1}^{\infty} x_k = \alpha \cdot A.$$

☐ Let A_n, B_n, and C_n be the n-th partial sums of the series $\sum\limits_{k=1}^{\infty} x_k$, $\sum\limits_{k=1}^{\infty} y_k$, and $\sum\limits_{k=1}^{\infty} (x_k + y_k)$, respectively. Then by part (2) of Theorem 2.3,
$$C = \lim_{n \to \infty} C_n = \lim_{n \to \infty} (A_n + B_n) = \lim_{n \to \infty} A_n + \lim_{n \to \infty} B_n = A + B$$

On the other hand, by part (1) of Theorem 2.3,

$$\lim_{n\to\infty} \alpha A_n = \alpha \lim_{n\to\infty} A_n = \alpha A.$$

∎

Further, we note that convergence or divergence is unaffected by deleting a finite number of terms from the beginning of a series, that is, for any positive integer N, the series

$$\sum_{k=1}^{\infty} x_k = x_1 + x_2 + \cdots + x_n + \cdots$$

and

$$\sum_{k=N}^{\infty} x_k = x_N + x_{N+1} + x_{N+2} + \cdots$$

both converge or both diverge.

Theorem 2.16 (Cauchy's Test). *For the convergence of the series* (2.9) *it is necessary and sufficient that for any* $\varepsilon > 0$ *there is an integer* N *such that*

$$\left| \sum_{k=n+1}^{n+p} x_k \right| < \varepsilon \qquad (2.13)$$

for all $n > N$ *and all natural numbers* p.

☐   Let S_n be n-th partial sum of the series (2.9). By Theorem 2.12, the sequence $\{S_n\}$ converges if and only if $\{S_n\}$ is a Cauchy sequence, that is, for any number $\varepsilon > 0$, there is an integer N such that $|S_m - S_n| < \varepsilon$ for all $m, n > N$. Then by setting $m = n + p$ $(p \geqslant 1)$ in the last inequality, we have $|S_{n+p} - S_n| = \left| \sum_{k=n+1}^{n+p} x_k \right| < \varepsilon$.   ∎

Definition 2.10. A series (2.9) is said to *converge absolutely* if the series of absolute values $\sum_{k=1}^{\infty} |x_k|$ converges.

Definition 2.11. A series (2.9) is said to be *conditionally convergent* if it is convergent, but not absolutely convergent, that is, the series $\sum_{k=1}^{\infty} |x_k|$ is divergent.

Theorem 2.17. *If the series* $\sum_{k=1}^{\infty} |x_k|$ *converges, so does the series* $\sum_{k=1}^{\infty} x_k$.

☐   The series $\sum_{k=1}^{\infty} |x_k|$ is convergent and so, by Cauchy's test (Theorem 2.16), for any $\varepsilon > 0$ there is an integer N such that $\left| \sum_{k=n+1}^{n+p} x_k \right| < \varepsilon$ for all $n > N$ and all positive integers p.

Then, obviously,

$$\left| \sum_{k=n+1}^{n+p} x_k \right| \leqslant \sum_{k=n+1}^{n+p} |x_k| < \varepsilon.$$

Hence, the series (2.9) converges. ∎

Theorem 2.18. (1) *If (2.9) is a series with positive terms, and if there is a constant K such that $S_n \leqslant K$ for every n, then the series converges and the sum of the series S satisfies $S \leqslant K$. If no such K exists then the series diverges.*

(2) *If (2.9) is a convergent series with nonnegative terms, then the series $\sum_{k=1}^{\infty} y_k$ obtained from the series (2.9) by rearranging and renumbering its terms is also convergent having the same sum.*

□ (1) If an infinite series has positive terms, $x_k \geqslant 0$ $(k = 1, 2, \ldots)$, then the partial sums S_n form an increasing sequence, that is $S_1 < S_2 < S_3 \cdots < S_n < \cdots$. Then, since $S_n \leqslant K$, according to Theorem 2.4 the sequence of partial sums will converge to a limit S, satisfying $S \leqslant K$. If no such constant exists, then $\lim_{n \to \infty} = \infty$.

(2) Let S be the sum of the series (2.9), that is $\lim_{n \to \infty} S_n = S$ and y_n be the partial sum of the series $\sum_{k=1}^{\infty} y_k$. Then for a fixed k we have

$$Y_k = y_1 + y_2 + \cdots + y_k = x_{n_1} + x_{n_2} + \cdots + x_{n_k}$$

Thus denoting $N = \max\{n_1, \ldots, n_k\}$ we can write $Y_k \leqslant S_k \leqslant S$, which implies that the series $\sum_{k=1}^{\infty} y_k$ converges. So if Y is its sum we derive that $Y \leqslant S$. Reasoning vice versa, we conclude that $Y = S$ ∎

Theorem 2.19 (Comparison Test). *Suppose $|x_n| \leqslant y_n$ for all $n \geqslant n_0$, where n_0 is some natural number. Then the following assertions hold.*

(1) *If the "bigger series" $\sum_{k=1}^{\infty} y_n$ converges, then the "smaller series" $\sum_{k=1}^{\infty} x_k$ converges absolutely.*

(2) *If the "smaller series" $\sum_{k=1}^{\infty} |x_k|$ diverges, then the "bigger series" $\sum_{k=1}^{\infty} y_n$ also diverges.*

□ (1) If the series $\sum_{k=1}^{\infty} y_k$ is convergent, then the Cauchy test is satisfied and hence, for any $\varepsilon > 0$ there is a natural number $N(\varepsilon)$ such that $0 \leqslant \left| \sum_{k=n+1}^{n+p} y_k \right| < \varepsilon$ for all $n > N(\varepsilon)$

and arbitrary natural number p. Then for all integers $n > \max\{N(\varepsilon), N_0\}$, the inequality $|x_n| \leqslant y_n$ holds, and so

$$\sum_{k=n+1}^{n+p} |x_k| \leqslant \sum_{k=n+1}^{n+p} y_k < \varepsilon.$$

Thus, the Cauchy test is satisfied by the series $\sum_{k=1}^{\infty} |x_k|$. Consequently, the series $\sum_{k=1}^{\infty} |x_k|$ converges and so $\sum_{k=1}^{\infty} x_k$ converges absolutely.

(2) This part is really just an alternative phrasing of part (1). In fact, if the series $\sum_{k=1}^{\infty} |x_k|$ diverges, then $\sum_{k=1}^{\infty} y_k$ must diverge since the convergence of $\sum_{k=1}^{\infty} y_k$ would imply the convergence of $\sum_{k=1}^{\infty} |x_k|$, contrary to the hypothesis. ∎

2.5 Alternating series; Convergence tests

Definition 2.12. A series whose terms are alternately positive and negative is called an *alternating series*. Such a series has the form:

$$\sum_{k=1}^{\infty} (-1)^{k+1} a_k = a_1 - a_2 + a_3 - \cdots + (-1)^{n+1} a_n + \cdots \tag{2.14}$$

where $a_k > 0, k = 1, 2, \ldots$

Theorem 2.20 (Leibniz[4]). *An alternating series* (2.14) *converges if the following two conditions are satisfied:*

(1) *The sequence* $\{a_n\}, n = 1, 2, \ldots$ *is nonincreasing,*

(2) $\lim_{n \to \infty} a_n = 0.$

☐ Let $\{S_n\}$ be the sequence of n-th partial sums of (2.14). Consider the subsequences $\{S_{2m}\}$ and $\{S_{2m+1}\}$. Since $a_{2m+1} \geqslant a_{2m+2}$, the inequality $S_{2m+2} = S_{2m} + a_{2m+1} - a_{2m+2} \geqslant S_{2m}$ holds, which implies that the "even" subsequence $\{S_{2m}\}$ is nondecreasing. On the other hand, $S_{2m+1} = S_{2m-1} - a_{2m} + a_{2m+1} \leqslant S_{2m-1}$, and so the "odd" subsequence $\{S_{2m+1}\}$ is nonincreasing. Besides, $S_{2m+1} = S_{2m} + a_{2m+1} \geqslant S_{2m}$ and then, since $S_{2m} \leqslant S_{2m+1} \leqslant S_{2m-1} \leqslant \cdots \leqslant S_1 = a_1$, the "even" subsequence $\{S_{2m}\}$ is bounded above. According to Theorem 2.5, the subsequence $\{S_{2m}\}$ converges, say $S_{2m} \to S$. We next show that the sequence $\{S_{2m+1}\}$ also has the limit S. In fact, taking into account that $a_n \to 0$, we have $\lim_{n \to \infty} S_{2m+1} = \lim_{n \to \infty} S_{2m} + \lim_{n \to \infty} a_{2m+1} = S$. Then the convergence of both $S_{2m} \to S$ and $S_{2m+1} \to S$ implies that $S_n \to S$, so the series converges. ∎

---

[4]G. Leibniz (1646-1716), German mathematician.

Remark 2.6. (a) It is not hard to see that Theorem 2.20 is valid in the nonincreasing case $a_n \geqslant a_{n+1} \geqslant 0$ $(n = 1, 2, \ldots)$, $n > N_0$, where n_0 is a positive integer.

(b) If an alternating series violates condition (2) of the alternating series test, then the series must diverge by the divergence test (Theorem 2.13). However, if condition (2) is satisfied, but (1) not, the series may either converge or diverge.

Example 2.14. (a) The series $\sum_{k=1}^{\infty} \dfrac{(-1)^{k+1}}{k}$ is called the alternating harmonic series. In the notation of Definition 2.12, $a_k = \dfrac{1}{k} > 0$. Clearly, $\dfrac{1}{k} = a_k > a_{k+1} = \dfrac{1}{k+1}$ $(k = 1, 2, \ldots)$, and $\lim_{k \to \infty} a_k = \lim_{k \to \infty} \dfrac{1}{k} = 0$. Thus the series converges by the alternating series (Leibniz) test. We emphasize that the series of absolute values is the harmonic series, which is diverges (Example 2.13). Hence, the alternating harmonic series is conditionally convergent (see Definition 2.11).

(b) The series $\sum_{k=1}^{\infty} \dfrac{(-1)^{k+1}}{2k-1} = 1 - \dfrac{1}{3} + \dfrac{1}{5} - \dfrac{1}{7} + \cdots$ satisfies the hypothesis of Theorem 2.20, and so converges. In Section 5.6 we will see that the sum of this series is $\dfrac{\pi}{4}$.

Theorem 2.21 (The Ratio Test of D'Alembert[5]). *If, for all $n \geqslant n_0$ (n_0 an integer),*

$$\left| \frac{x_{n+1}}{x_n} \right| \leqslant \rho < 1 \quad (x_n \neq 0), \tag{2.15}$$

then the series (2.9) converges absolutely, and if

$$\left| \frac{x_{n+1}}{x_n} \right| \geqslant \rho > 1, \tag{2.16}$$

then the series (2.9) diverges.

☐ It follows from condition (2.15) that for all $n > N_0$,

$$|x_n| \leqslant \rho |x_{n-1}| \leqslant \rho^2 |x_{n-2}| \leqslant \cdots \leqslant \rho^{n-N_0} |x_{N_0}|.$$

Consider the series $\sum_{k=1}^{\infty} \rho^{k-N_0} |x_{N_0}|$ $(0 < \rho < 1)$, which according to Theorem 2.14, is convergent. Then, taking into account the inequality $|x_n| \leqslant \rho^{n-N_0} |x_{N_0}|$, by Theorem 2.19 we conclude that the series $\sum_{k=1}^{\infty} |x_k|$, and so the series $\sum_{k=1}^{\infty} x_k$, are convergent. Thus, the first part of the theorem is proved. On the other hand, it follows from condition (2.16) that for all $n > N_0$,

$$|x_n| \geqslant \rho |x_{n-1}| \geqslant \cdots \geqslant \rho^{n-N_0} |x_{N_0}| > |x_{N_0}| > 0,$$

and so $|x_n| > |x_{N_0}| > 0$. This implies that $\lim_{n \to \infty} x_n \neq 0$, and so, by Theorem 2.13, the series $\{x_n\}$ diverges. ∎

---

[5]J. D'Alembert (1717-1783), French mathematician.

Corollary 2.2. *The limiting case* $\lim\limits_{n\to\infty}\left|\dfrac{x_{n+1}}{x_n}\right| = \rho$ *of the D'Alembert ratio test consists of the following:*

(1) If $\rho < 1$, *the series* (2.9) *converges absolutely.*

(2) If $\rho > 1$ *or if* $\rho = \infty$, *then the series* (2.9) *diverges.*

(3) If $\rho = 1$, *no conclusion about convergence can be drawn from this test.*

Example 2.15. (a) For the series

$$\frac{3}{10} + \frac{3}{10^2} + \frac{3}{10^3} + \cdots + \frac{3}{10^n} + \cdots,$$

the n-th partial sum is

$$S_n = \frac{3}{10} + \frac{3}{10^2} + \frac{3}{10^3} + \cdots + \frac{3}{10^n}.$$

Multiplying both sides of this equality by $1/10$, we have

$$\frac{1}{10}S_n = \frac{3}{10^2} + \frac{3}{10^3} + \cdots + \frac{3}{10^n} + \frac{3}{10^{n+1}}.$$

Then, by subtracting the second equality from the first, we can define the n-th partial sum $S_n = \dfrac{1}{3}\left(1 - \dfrac{1}{10^n}\right)$. Thus, the sum of the series is $S = \lim\limits_{n\to\infty} S_n = \dfrac{1}{3}$.

(b) Let us see whether the series $\sum\limits_{k=1}^{\infty}\dfrac{1}{k(k+1)}$ converges or diverges. Since the n-th partial sum

$$S_n = \sum_{k=1}^{n}\frac{1}{k(k+1)} = \sum_{k=1}^{n}\left(\frac{1}{k} - \frac{1}{k+1}\right) = 1 - \frac{1}{n+1},$$

the limit $\lim\limits_{n\to\infty} S_n = 1$. Notice that, since $x_n = \frac{1}{n(n+1)}$, the limit that occurs in D'Alembert's test is $\lim\limits_{n\to\infty}\left|\dfrac{x_{n+1}}{x_n}\right| = \lim\limits_{n\to\infty}\dfrac{n}{n+2} = 1$.

On the other hand we showed that (see Example 2.13) the Harmonic series is divergent and since $x_k = \dfrac{1}{k}$, the limit occuring in the D'Alembert ratio test is $\lim\limits_{n\to\infty}\left|\dfrac{x_{n+1}}{x_n}\right| = \lim\limits_{n\to\infty}\dfrac{n}{n+1} = 1$, also. Consequently, in both cases $\rho = 1$, yet one of them converges, while the other diverges. Hence, no conclusion can be drawn from D'Alembert's test when $\rho = 1$.

Example 2.16. The series $\sum\limits_{k=1}^{\infty}\dfrac{k^k}{k!}$ diverges by the D'Alembert ratio test since $\rho =$

$$\lim_{k\to\infty}\frac{x_{k+1}}{x_k} = \lim_{k\to\infty}\frac{(k+1)^{k+1}}{(k+1)!}\cdot\frac{k!}{k^k} = \lim_{k\to\infty}\frac{(k+1)^k}{k^k} = \lim_{k\to\infty}\left(1 + \frac{1}{k}\right)^k = e.$$ Thus $\rho = e > 1$

and the series diverges.

Theorem 2.22 (The Root Test of Cauchy). *If, for all $n \geqslant n_0$ (n_0 is some integer),*

$$\sqrt[n]{|x_n|} \leqslant \rho < 1, \tag{2.17}$$

then the series (2.9) converges absolutely.

On the other hand, if for all $n \geqslant n_0$,

$$\sqrt[n]{|x_n|} \geqslant \rho > 1, \tag{2.18}$$

then the series diverges.

□ Firstly, it follows from condition (2.17) that $|x_n| \leqslant \rho^n$. Thus the series $\sum\limits_{k=1}^{\infty} |x_k|$ is eventually dominated by the convergent geometric series $\sum\limits_{k=1}^{\infty} \rho^k$ ($0 < \rho < 1$). Hence, $\sum\limits_{k=1}^{\infty} |x_k|$ converges, and so the series $\sum\limits_{k=1}^{\infty} x_k$ converges absolutely. On the other hand, if (2.18) holds, then $|x_n| > \rho^n > 1$ ($n \geqslant n_0$) and hence $\lim\limits_{n \to \infty} x_n \neq 0$. Therefore the n-th term divergence test (Theorem 2.13) implies that the series $\sum\limits_{k=1}^{\infty} x_k$ diverges. ■

Corollary 2.3. *Suppose that $\lim\limits_{n \to \infty} \sqrt[n]{|x_n|} = \rho$. Then the infinite series $\sum\limits_{k=1}^{\infty} x_k$*
(1) *Converges absolutely if $\rho < 1$,*
(2) *Diverges if $\rho > 1$.*

If $\rho = 1$, the root test is inconclusive.

Even though the ratio test is generally simpler to apply than the root test, there are certain series for which the root test succeeds while the ratio test fails.

Example 2.17. (a) In the Example 2.15 it was shown that the series $\sum\limits_{k=1}^{\infty} \dfrac{1}{k(k+1)}$ converges. Now, applying the root test, we can see that this test is inconclusive. In fact,

$$\lim_{n \to \infty} \sqrt[n]{\frac{1}{n(n+1)}} = \lim_{n \to \infty} \sqrt[n]{\frac{1}{n}} \cdot \sqrt[n]{\frac{1}{n+1}} = 1 \text{ (see Example 2.7), that is, } \rho = 1.$$

(b) Consider the series $\sum\limits_{k=0}^{\infty} \dfrac{1}{2^{k+(-1)^k}} = \dfrac{1}{2} + \dfrac{1}{1} + \dfrac{1}{8} + \dfrac{1}{4} + \dfrac{1}{32} + \cdots$. Then

$$\frac{x_{n+1}}{x_n} = \begin{cases} \dfrac{1}{2} & \text{if } n \text{ is even,} \\ \dfrac{1}{8} & \text{if } n \text{ is odd.} \end{cases}$$

Therefore, the limit required for the ratio test does not exist. On the other hand,

$$\lim_{n \to \infty} \sqrt[n]{|x_n|} = \lim_{n \to \infty} \left(\frac{1}{2^{n+(-1)^n}} \right)^{1/n} = \lim_{n \to \infty} \frac{1}{2} \left(\frac{1}{2^{(-1)^n/n}} \right) = \frac{1}{2},$$

so the series converges by the root test.

Example 2.18. (a) The series $\sum\limits_{k=1}^{\infty} \left(\dfrac{3k-2}{k+1} \right)^k$ diverges by the root test since

$$\rho = \lim_{n\to\infty} \sqrt[n]{\left(\frac{3n-2}{n+1} \right)^n} = \lim_{n\to\infty} \frac{3n-2}{n+1} = 3 > 1.$$

(b) Obviously the series $\sum\limits_{k=1}^{\infty} k$ is divergent. Nevertheless, $\rho = \lim\limits_{n\to\infty} \sqrt[n]{|x_n|} = \lim\limits_{n\to\infty} \sqrt[n]{n} = 1.$

By setting $x_n = n$, the proof of the next theorem can be obtained from Kummer's[6] theorem[7]

Theorem 2.23 (Raabe[8] test). *Let* $\sum\limits_{k=1}^{\infty} x_k$ *be a series with positive terms and suppose that*

$$\lim_{n\to\infty} \left[\frac{x_n}{x_{n+1}} - 1 \right] \cdot n = q.$$

Then:

(1) *if* $q > 1$, *the series converges;*

(2) *if* $q < 1$, *the series diverges.*

Example 2.19. Let us verify by the Raabe test that the series $\sum\limits_{k=1}^{\infty} \dfrac{1 \cdot 3 \cdot 5 \cdots (2k-1)}{2 \cdot 4 \cdot 6 \cdots 2k}$ diverges. We can easily establish that the quotient occuring in the test is

$$\frac{x_k}{x_{k+1}} = \frac{2k+2}{2k+1}.$$

Then,

$$q = \lim_{n\to\infty} \left[\frac{2n+2}{2n+1} - 1 \right] n = \frac{1}{2}.$$

Since $q < 1$, the series diverges.

Theorem 2.24 (The Limit Comparison Test). *Let* $\sum\limits_{k=1}^{\infty} x_k$ *and* $\sum\limits_{k=1}^{\infty} y_k$ *be series with positive terms and assume that* $\rho = \lim\limits_{k\to\infty} \dfrac{x_k}{y_k}$. *If* ρ *is finite and* $\rho \neq 0$, *then either the series both converge or they both diverge.*

☐   By the definition of limit, for any $\varepsilon > 0$ there is a positive integer N such that

$$\left| \frac{x_k}{y_k} - \rho \right| < \varepsilon \text{ for all } k > N \text{ or } \rho - \varepsilon < \frac{x_k}{y_k} < \rho + \varepsilon \text{ for all } k > N.$$

Setting $\varepsilon = \rho/2$ (ρ a positive number) in the last inequality, we have

$$\frac{1}{2}\rho y_k < x_k < \frac{3}{2}\rho y_k, \quad k > N.$$

---

[6]E. Kummer (1810–1893), German mathematician

[7]See, Marian Nureshan, *A Concrete Approach to Classical Analysis*, Springer New, York, 2009.

[8]J. Raabe (1801–1859), German mathematician

Now, if $\sum\limits_{k=1}^{\infty} y_k$ converges, then $\sum\limits_{k=1}^{\infty} x_k$ is eventually dominated by the convergent series $\sum\limits_{k=1}^{\infty} \frac{3}{2}\rho y_k$, so by Theorem 2.15, the series $\sum\limits_{k=1}^{\infty} x_k$ converges. If $\sum\limits_{k=1}^{\infty} y_k$ diverges, then $\sum\limits_{k=1}^{\infty} x_k$ eventually dominates the divergent series $\sum\limits_{k=1}^{\infty} \frac{1}{2}\rho y_k$, so Theorem 2.15 implies that $\sum\limits_{k=1}^{\infty} x_k$ also diverges. Thus the convergence of either series implies the convergence of the other.

∎

Example 2.20. Show that the series $\sum\limits_{k=1}^{\infty} \frac{1}{k-\frac{1}{4}}$ diverges. Let us take $x_k = \frac{1}{k-\frac{1}{4}}$ and $y_k = \frac{1}{k}$. Then, by Theorem 2.24, since

$$\rho = \lim_{k\to\infty} \frac{k}{k-\frac{1}{4}} = \lim_{k\to\infty} \frac{1}{1-\frac{1}{4k}} = 1,$$

the series is divergent.   ∎

Let $\sum\limits_{k=1}^{\infty} x_k$ be a series and $\{m_k\}$ be an arbitrary sequence of natural integers such that every natural number take place only once. In other words $\{m_k\}$ is some rearrangement of the natural numbers. Setting $y_k = x_{m_k}$, consider the new numerical series $\sum\limits_{k=1}^{\infty} y_k$. The series $\sum\limits_{k=1}^{\infty} y_k$ is obtained as a result of rearrangement of the terms of the series $\sum\limits_{k=1}^{\infty} x_k$, and contains each of the terms of the first series once only.

A very important property of a sum of a finite number of summands is the commutative property, that is, a rearrangement of the terms does not affect their sum. The following example shows that this is not the case for arbitrary series.

Example 2.21. Consider the alternating harmonic series (Example 2.14 (a)): denote its sum by S. Rearrange this series according to the rule: two negative terms after a positive one. Then we can write

$$\left(1 - \frac{1}{2} - \frac{1}{4}\right) + \left(\frac{1}{3} - \frac{1}{6} - \frac{1}{8}\right) + \left(\frac{1}{5} - \frac{1}{10} - \frac{1}{12}\right) + \cdots. \qquad (2.19)$$

Denote by S_n the n-th partial sum of series (2.19). Then we have

$$S_{3n} = \sum_{k=1}^{n} \left(\frac{1}{2k-1} - \frac{1}{4k-2} - \frac{1}{4k}\right) = \sum_{k=1}^{n} \left(\frac{1}{4k-2} - \frac{1}{4k}\right)$$

$$= \frac{1}{2} \sum_{k=1}^{n} \left(\frac{1}{2k-1} - \frac{1}{2k}\right) = \frac{1}{2} S_{2n}$$

Finally, we can write

$$S_{3n} = \frac{S_{2n}}{2}, \quad S_{3n-1} = \frac{S_{2n}}{2} + \frac{1}{4n}, \quad S_{3n-2} = \frac{S_{2n}}{2} + \frac{1}{(4n-2)}. \qquad (2.20)$$

and then from (2.20) we deduce that

$$\lim_{n\to\infty} S_{3n} = \lim_{n\to\infty} S_{3n-1} = \lim_{n\to\infty} S_{3n-2} = \frac{S}{2}.$$

Thus we have found that the series (2.19) converges and its sum is $\frac{S}{2}$. ∎

But the next theorem of Cauchy shows that for absolutely convergent series, any series obtained from it by rearrangement has the same sum as the original series.

Theorem 2.25 (Cauchy). *The rearrangement of an absolutely convergent series supplies another absolutely convergent series having the same sum as the original one.*

☐ Let $\sum_{k=1}^{\infty} x_k$ be an absolutely convergent series. Consider the positive and negative terms of this series:

$$x_k^+ = \begin{cases} x_k, & \text{if } x_k \geq 0, \\ 0, & \text{if } x_k < 0, \end{cases} \qquad x_k^- = \begin{cases} -x_k, & \text{if } x_k \leq 0, \\ 0, & \text{if } x_k > 0. \end{cases} \tag{2.21}$$

Obviously, $x_k = x_k^+ - x_k^-$. Take the following two series with nonnegative terms $\sum_{k=1}^{\infty} x_k^+$ and $\sum_{k=1}^{\infty} x_k^-$. Since $x_k^+ \leq |x_k|$ and $x_k^- \leq |x_k|$, these series are convergent. Let $\sum_{k=1}^{\infty} y_k$ be the rearranged series of $\sum_{k=1}^{\infty} x_k$. In view of (2.21) by analogy, for it construct the series positive and negative parts $\sum_{k=1}^{\infty} y_k^+$, $\sum_{k=1}^{\infty} y_k^-$, respectively. Then by using the Theorem 2.18 it is easy to see that

$$\sum_{k=1}^{\infty} x_k = \sum_{k=1}^{\infty} (x_k^+ - x_k^-) = \sum_{k=1}^{\infty} x_k^+ - \sum_{k=1}^{\infty} x_k^- = \sum_{k=1}^{\infty} y_k^+ - \sum_{k=1}^{\infty} y_k^- = \sum_{k=1}^{\infty} (y_k^+ - y_k^-) = \sum_{k=1}^{\infty} y_k.$$

∎

Theorem 2.26 (Riemann[9]). *Let $\sum_{k=1}^{\infty} x_k$ be a conditionally convergent series. Choose an $S \in [-\infty, \infty]$. Then there is a rearrangement of the series such that the resulting series converges to S.*

☐ We split the series $\sum_{k=1}^{\infty} x_k$ into two series $\sum_{k=1}^{\infty} x_k^+$ and $\sum_{k=1}^{\infty} x_k^-$ as introduced in the proof of previous theorem (see (2.21)). Show that if $\sum_{k=1}^{\infty} x_k$ be a conditionally convergent series, both series $\sum_{k=1}^{\infty} x_k^+$ and $\sum_{k=1}^{\infty} x_k^-$ generated by it are divergent, where $x_n^+ \to 0$, $x_n^- \to 0$ as $n \to \infty$. By Theorem 2.25 it is not hard to see, that for series $\sum_{k=1}^{\infty} x_k$ to be absolutely convergent it

---

[9]G. Riemann (1826–1866), German mathematician

is necessary and sufficient that series $\sum\limits_{k=1}^{\infty} x_k^+$ and $\sum\limits_{k=1}^{\infty} x_k^-$ generated by it be convergent. By condition of theorem our series is a conditionally convergent series. Then it follows, that at least one of the series $\sum\limits_{k=1}^{\infty} x_k^+$ and $\sum\limits_{k=1}^{\infty} x_k^-$ generated by it is divergent, that is, $\sum\limits_{k=1}^{\infty} x_k^+ = \infty$.

Let us study the behavior of the right-hand side of sum $\sum\limits_{k=1}^{n} x_k^- = \sum\limits_{k=1}^{n} x_k^+ - \sum\limits_{k=1}^{n} x_k$ as $n \to \infty$.

Since $\sum\limits_{k=1}^{\infty} x_k$ is convergent, the second sum tends to infinite number. The first sum increases to ∞. Consequently, the sum $\sum\limits_{k=1}^{\infty} x_k^-$ increases to ∞ as $n \to \infty$. As a result if one of the series $\sum\limits_{k=1}^{\infty} x_k^+$, $\sum\limits_{k=1}^{\infty} x_k^-$ is divergent, the other is divergent, too. We remark, that since the series $\sum\limits_{k=1}^{\infty} x_k$ is convergent, the sequences $\{x_n^+\}$ and $\{x_n^-\}$ tend to zero. It remains to show, that for any $S \in [-\infty, \infty]$ we can construct a series

$$x_1^+ + x_2^+ + \cdots + x_{m_1}^+ - x_1^- - x_2^- - \cdots - x_{n_1}^-$$
$$+ x_{m_1+1}^+ + x_{m_2+2}^+ + \cdots + x_{m_2}^+ - x_{n_1+1}^- - \cdots - x_{n_2}^- + \cdots \qquad (2.22)$$

whose sum is equal to S. Series (2.22) contains all the terms of $\sum\limits_{k=1}^{\infty} x_k^+$ and $\sum\limits_{k=1}^{\infty} x_k^-$ only once. At first we assume, that S is a real finite number. In this case the indices $m_1, m_2, \ldots$ and $n_1, n_2, \ldots$ can be chosen as the smallest natural numbers for which the corresponding inequalities

$$\beta_1 = \sum_{k=1}^{m_1} x_k^+ > S, \qquad \beta_2 = \beta_1 - \sum_{k=1}^{n_1} x_k^- < S,$$
$$\beta_3 = \beta_2 + \sum_{k=m_1+1}^{m_2} x_k^+ > S, \qquad \beta_4 = \beta_3 - \sum_{k=n_1+1}^{n_2} x_k^- < S, \qquad (2.23)$$

are fulfilled. Since the series $\sum\limits_{k=1}^{\infty} x_k^+$ and $\sum\limits_{k=1}^{\infty} x_k^-$ have positive terms and diverge, at the p-th step of this construction we indeed can choose the natural numbers m_p and n_p satisfying the p-th inequality. Then, taking into account the inequalities (2.23) and the fact that $x_n^+ \to 0$, $x_n^- \to 0$ as $n \to \infty$ it follows that the series thus constructed converges to S. Note that in the case $S = +\infty$ we can replace S in the right-hand side of the inequalities (2.23) by a divergent sequence of the form $2, 1, 4, 3, 5, \ldots$. ∎

Theorem 2.27 (Dirichlet[10]). (1) *Let a series $\sum\limits_{k=0}^{\infty} x_k$ be given, and suppose that the sequence of n-th partial sums $\{A_n\}$ of this series is bounded.*

(2) *Suppose that the sequence $\{y_n\}$ $(n = 0, 1, 2, \ldots)$, is nondecreasing and $\lim\limits_{n \to \infty} y_n = 0$.*

Then the series $\sum\limits_{k=0}^{\infty} x_k y_k$ converges.

---

[10]P.G. Le Jeune-Dirichlet (1805–1859), German mathematician

☐ Let $A_n = \sum\limits_{k=0}^{n} x_k$, $n \geqslant 0$ and $A_{-1} = 0$. Then for $0 \leqslant p \leqslant q$, we have

$$\sum_{n=p}^{q} x_n y_n = \sum_{n=p}^{q} (A_n - A_{n-1}) y_n = \sum_{n=p}^{q} A_n y_n - \sum_{n=p-1}^{q-1} A_n y_{n+1}$$

$$= \sum_{n=p}^{q-1} A_n (y_n - y_{n+1}) + A_q y_q - A_{p-1} y_p. \tag{2.24}$$

On the other hand, by condition (1) of the theorem, there is a positive constant C such that $|A_n| \leqslant C$ ($n = 0, 1, 2, \ldots$). For any number $\varepsilon > 0$ there is a positive integer $N = N(\varepsilon)$ such that $y_n \leqslant \dfrac{\varepsilon}{2C}$. Then, since $y_n - y_{n+1} \geqslant 0$ for $N \leqslant p \leqslant q$, it follows from (2.24) that

$$\left| \sum_{n=p}^{q} x_n y_n \right| = \left| \sum_{n=p}^{q-1} A_n (y_n - y_{n+1}) + A_q y_q - A_{p-1} y_p \right| \leqslant C \left| \sum_{n=p}^{q-1} (y_n - y_{n+1}) + y_q + y_p \right|.$$

But $\left| \sum\limits_{n=p}^{q-1} (y_n - y_{n+1}) + y_q + y_p \right| = 2y_p$ and so $\left| \sum\limits_{n=p}^{q} x_n y_n \right| \leqslant 2C y_p \leqslant 2C y_N \leqslant \varepsilon$.

Then by the Cauchy criterion (Theorem 2.16), the series $\sum\limits_{k=0}^{\infty} x_k y_k$ converges. ■

Theorem 2.28 (Cauchy's Condensation Test). *Let $\{x_n\}$ be a nonincreasing series which is bounded below by 0. Then the series $\sum\limits_{n=1}^{\infty} x_n$ converges if and only if the series*

$$\sum_{k=0}^{\infty} 2^k x_{2^k} = x_1 + 2x_2 + 4x_4 + \cdots \tag{2.25}$$

converges.

☐ By Theorem 2.5, it suffices to consider the boundedness of the sequences $\{A_n\}$. ($A_n = x_1 + \cdots + x_n$). Let S_k be the n-th partial sum of the series (2.25):

$$S_k = x_1 + 2x_2 + \cdots + 2^k x_{2^k}$$

For $n < 2^k$

$$A_n \leqslant x_1 + (x_2 + x_3) + \cdots + (x_{2^K} + \cdots + x_{2^{K+1}} - 1) \leqslant x_1 + 2x_2 + \cdots + 2^k x_{2^k} = S_k.$$

On the other hand, for $n > 2^k$,

$$A_n \geqslant x_1 + x_2 + (x_3 + x_4) + \cdots + \left(x_{2^{k-1}+1} + \cdots + x_{2^k} \right)$$

$$\geqslant \frac{1}{2} x_1 + x_2 + 2x_4 + \cdots + 2^{k-1} x_{2^k} = \frac{1}{2} s_k.$$

From the obtained inequalities, $A_n \leqslant S_k$ and $2A_n \geqslant S_k$ we have that the sequences $\{A_n\}$ and $\{S_k\}$ are simultaneously either bounded or unbounded. ■

The harmonic series is a special case, with $p = 1$, of a class of series called the *p-series*, or, *hyperharmonic* series, $\sum_{n=1}^{\infty} \frac{1}{n^p}$. We will determine the convergence or divergence of the hyperharmonic series using the previous theorem. (Later, in Section 9.8, applying the integral test, we will again investigate the convergence of this series.)

Theorem 2.29. *The hyperharmonic series $\sum_{n=1}^{\infty} \frac{1}{n^p}$ converges if and only if $p > 1$.*

□ If $p \leqslant 0$, its divergence follows by the n-th term test for convergence (Theorem 2.13). If $p > 0$, then by Theorem 2.28, setting $x_{2^k} = \frac{1}{2^{kp}}$, we are led to the series

$$\sum_{k=0}^{\infty} 2^k \frac{1}{2^{kp}} = \sum_{k=0}^{\infty} 2^{k(1-p)}.$$

Obviously, $2^{(1-p)} < 1$ if and only if $1 - p < 0$, and the result follows by comparison with the geometric series (Theorem 2.14), taking $a = 1$ and $x = 2^{1-p}$. ∎

2.6 Products of Series

Definition 2.13. Let $\sum_{k=1}^{\infty} x_k$ and $\sum_{k=1}^{\infty} y_k$ be two series. Then the series $\sum_{k=1}^{\infty} z_k$, where

$$z_k = \sum_{i=0}^{k} x_i y_{k-i} = x_0 y_k + x_1 y_{k-1} + \cdots + x_k y_0, \tag{2.26}$$

is called the product of the two given series.

Theorem 2.30. *Suppose the series $\sum_{k=0}^{\infty} x_k$ and $\sum_{k=0}^{\infty} y_k$ are convergent and their sums are A and B, respectively, and one of them, say $\sum_{k=0}^{\infty} x_k$, converges absolutely. Then the series $\sum_{k=0}^{\infty} z_k$ converges and has the sum $C = AB$.*

□ Let us denote $A_n = \sum_{k=0}^{n} x_k$, $B_n = \sum_{K=0}^{n} y_K$, and $C_n = \sum_{k=0}^{n} z_k$. By Definition 2.13,

$$C_n = x_0 y_0 + (x_0 y_1 + x_1 y_0) + \cdots + (x_0 y_n + x_1 y_{n-1} + \cdots + x_n y_0)$$

$$= x_0 B_n + x_1 B_{n-1} + \cdots + x_n B_0 = x_0(B + R_n) + x_1(B + R_{n-1}) + \cdots + x_n(B + R_0)$$

$$= A_n B + x_0 R_n + x_1 R_{n-1} + \cdots + x_n R_0,$$

where $R_n = B_n - B \to 0$.

Therefore,

$$C_n - A_n B = x_0 R_n + \cdots + x_n R_0 = r_n. \tag{2.27}$$

Now, we should show that $\lim_{n \to \infty} r_n = 0$.

Let us denote $Q = \sum\limits_{k=0}^{\infty} |x_k|$. Since $R_n \to 0$, for a given $\varepsilon > 0$ we can choose a positive integer N such that $|R_n| < \varepsilon/(2Q)$ for all $n > N$.

For $n > N$, we have

$$|r_n| \leqslant |R_0 x_n + \cdots + R_N x_{n-N}| + |R_{N+1} x_{n-N-1} + \cdots + R_n x_0|$$

$$\leqslant |R_0 x_n + \cdots + R_N x_{n-N}| + \frac{\varepsilon}{2Q}(|x_{n-N-1}| + \cdots + |x_0|)$$

$$\leqslant |R_0 x_n + \cdots + R_N x_{n-N}| + \frac{\varepsilon}{2}. \tag{2.28}$$

Since $x_n \to 0$, for fixed N we can find an integer n_0 such that for all $n > N_0$,

$$|R_0 x_n + \cdots + R_N x_{n-N}| \leqslant \frac{\varepsilon}{2}. \tag{2.29}$$

Then the inequalities (2.28) and (2.29) imply that $|r_n| < \varepsilon$ for all $n > \max\{N, N_0\}$. Thus $r_n \to 0$. Finally, by passing to the limit in (2.27) we have $C = A \cdot B$. ∎

Remark 2.7. In the above proved Theorem 2.30, it is essential that one of the series converges absolutely. For example, the series $\sum\limits_{k=0}^{\infty} \frac{(-1)^k}{\sqrt{k+1}}$ converges by Theorem 2.20, since it is an alternating series, $\lim\limits_{k\to\infty} \frac{1}{\sqrt{k}} = 0$, and $a_{k+1} = \frac{1}{\sqrt{k+1}} > \frac{1}{\sqrt{k}} = a_k$ $(k = 1, 2, \ldots)$. But the convergence is not absolute. In fact, $\frac{1}{k} < \frac{1}{\sqrt{k}}$ and by the Comparison Test (Theorem 2.19), the series $\sum\limits_{k=0}^{\infty} \left| \frac{(-1)^k}{\sqrt{k+1}} \right|$ diverges. Setting $x_k = y_k = \frac{(-1)^k}{\sqrt{k+1}}$, we form the product series of the given series with itself,

$$\sum\limits_{n=0}^{\infty} z_n = 1 - \left(\frac{1}{\sqrt{2}} + \frac{1}{\sqrt{2}} \right) + \left(\frac{1}{\sqrt{3}} + \frac{1}{\sqrt{2}\sqrt{2}} + \frac{1}{\sqrt{3}} \right)$$

$$- \left(\frac{1}{\sqrt{4}} + \frac{1}{\sqrt{3}\sqrt{2}} + \frac{1}{\sqrt{2}\sqrt{3}} + \frac{1}{\sqrt{4}} \right) + \cdots$$

so that

$$z_n = (-1)^n \sum\limits_{k=0}^{n} \frac{1}{\sqrt{(n-k+1)(k+1)}}.$$

Since

$$(n-k+1)(k+1) = \left(\frac{n}{2} + 1 \right)^2 - \left(\frac{n}{2} - k \right)^2 \leqslant \left(\frac{n}{2} + 1 \right)^2,$$

we have

$$|z_n| \geqslant \sum\limits_{k=0}^{n} \frac{2}{n+2} = \frac{2(n+2)}{n+2},$$

so that $\lim\limits_{n\to\infty} z_n \neq 0$, and by the n-th term divergence test (Theorem 2.13), the series $\sum\limits_{n=0}^{\infty} z_n$ diverges.

Definition 2.14. Let $\{x_n\}$ be a numerical series. Then an expression of the form

$$\prod_{k=1}^{\infty} x_k = x_1 x_2 \cdots x_k \cdots \qquad (2.30)$$

is called the *infinite product* of the numbers x_n, $n = 1, 2, 3, \ldots$. The number $P_n = \prod_{k=1}^{\infty} x_k = x_1 x_2 \cdots x_n$ is called the *n-th partial product* of the infinite product (2.30).

Definition 2.15. The infinite product (2.30) is said to be convergent if the sequence of n-th partial sums $\{P_n\}$ converges, i.e., $P_n \to P$ $(P \neq 0)$. In this case we write

$$\prod_{k=1}^{\infty} x_k = P.$$

Theorem 2.31. *If the infinite product (2.30) converges, then $\lim\limits_{n \to \infty} x_n = 1$.*

$\square$ According to Definition 2.15, the infinite product converges if $P_n \to P$ $(P \neq 0)$. Clearly, the limit of the subsequence is the same, P. But $\dfrac{P_n}{P_{n-1}} = x_n$, and so

$$\lim_{n \to \infty} x_n = \frac{\lim\limits_{n \to \infty} P_n}{\lim\limits_{n \to \infty} P_{n-1}} = 1.$$

$\blacksquare$

Remark 2.8. It follows from the theorem that there is an integer N_0 such that $x_n > 0$ for all $n > N_0$. Hence, without loss of generality, we may assume $x_k > 0$ for all k.

Moreover, the convergence of the infinite product with $x_k > 0$ can be reduced to the convergence of the series

$$\sum_{k=1}^{\infty} y_k, \quad \text{where} \ \ y_k = \ln x_k. \qquad (2.31)$$

In reality, we observe that if P_n and y_n are the n-th partial product and n-th partial sum of the infinite product (2.30) and (2.31), respectively, then

$$Y_n = \ln x_1 + \ln x_2 + \cdots + \ln x_n = \ln x_1 x_2 \cdots x_n = \ln P_n$$

and so $P_n = e^{Y_n}$. If now there is the limit $\lim\limits_{n \to \infty} y_n = Y$, then $P = e^Y$. $\blacksquare$

The following example shows that, generally, the converse statement of Theorem 2.31 does not hold. It occurs because the condition $\lim\limits_{n \to \infty} x_n = 1$ is only a necessary condition for convergence of infinite products, and is not a sufficient condition.

Example 2.22. (1) Consider the infinite product $\prod\limits_{k=1}^{\infty} \left(1 + \dfrac{1}{k}\right)$.

The n-th partial product is

$$P_n = \prod_{k=1}^{\infty} \left(1 + \frac{1}{k}\right) = \prod_{k=1}^{\infty} \frac{k+1}{k} = \frac{2}{1} \cdot \frac{3}{2} \cdots \frac{n+1}{n} = 1.$$

Evidently, $P_n \to +\infty$ when $n \to \infty$, and so the infinite product diverges. Nevertheless, $\lim_{n \to \infty} x_n = \lim_{n \to \infty} \left(1 + \frac{1}{n}\right) = 1.$

(2) Test for convergence the infinite product $\prod_{k=1}^{\infty} \left(1 - \frac{1}{(k+1)^2}\right)$. The n-th partial product is

$$P_n = \prod_{k=1}^{n} \left(1 - \frac{1}{(k+1)^2}\right) = \prod_{k=1}^{n} \frac{k(k+2)}{(k+1)^2} = \frac{1 \cdot 3}{2^2} \cdot \frac{2 \cdot 4}{3^2} \cdot \frac{3 \cdot 5}{4^2} \cdots \frac{n(n+2)}{(n+1)^2} = \frac{n+2}{2(n+1)}.$$

Obviously, $\lim_{n \to \infty} P_n = \frac{1}{2}$. Thus, the infinite product converges. Naturally, the limit of the n-th partial product is equal to one: $\lim_{n \to \infty} x_n = \lim_{n \to \infty} \left(1 - \frac{1}{(n+1)^2}\right) = 1.$

Remark 2.9. Generally, the definitions, concepts, and properties of the sequences and series of real numbers are also valid for sequences and series of complex numbers (with the possible exception of some properties connected with comparison operations and the order relations on the real line). For example, it is known that necessary and sufficient conditions for the convergence of sequences of complex numbers $\{z_n\}$ ($z_n = a_n + ib_n$) consist of the same conditions as for the real series $\{a_n\}$ and $\{b_n\}$ ($n = 1, 2, \ldots$). Since convergence of the series of real numbers is given by means of sequences, the convergence of series $\sum_{k=0}^{\infty} z_n$ is defined by analogy.

2.7 Problems

Using the definition of limits prove the assertions in (1)–(5).

(1) $\lim_{n \to \infty} \dfrac{n}{n+1} = 1.$

(2) $\lim_{n \to \infty} \sqrt[n]{a} = 1 \ (a > 0).$

(3) $\lim_{n \to \infty} \dfrac{1}{\sqrt[n]{n!}} = 0.$

(4) $\lim_{n \to \infty} \dfrac{n^k}{a^n} = 0 \ (a > 1).$

(5) $\lim_{n \to \infty} \dfrac{2^n}{n!} = 0.$

(6) (a) For all numerical sequences $\{x_n\}$ and $\{y_n\}$, prove that

$$\overline{\lim_{n \to \infty}} \, (x_n + y_n) \leqslant \overline{\lim_{n \to \infty}} \, x_n + \overline{\lim_{n \to \infty}} \, y_n).$$

(b) Prove that if for a sequence $\{x_n\}$, any sequence whatever, $\{y_n\}$, would provide that

$$\overline{\lim_{n\to\infty}}\,(x_n + y_n) = \overline{\lim_{n\to\infty}}\,x_n + \overline{\lim_{n\to\infty}}\,y_n,$$

then the sequence $\{x_n\}$ must be convergent.

(7) Prove that for the existence of the limit (finite or infinite) of a sequence $\{x_n\}$, the condition $\underline{\lim_{n\to\infty}}\,x_n = \overline{\lim_{n\to\infty}}\,x_n$ is necessary and sufficient.

(8) Prove that $\lim_{n\to\infty} \left(\dfrac{1}{2} \cdot \dfrac{3}{4} \cdots \dfrac{2n-1}{2n}\right) = 0.$
Suggestion: See Example 5 of Section 1.7.

(9) (a) Prove that $\lim_{n\to\infty} \left(1 + 1 + \dfrac{1}{2!} + \dfrac{1}{3!} + \cdots + \dfrac{1}{n!}\right) = e.$

Suggestion: Use the fact that $\lim_{n\to\infty} \left(1 + \dfrac{1}{n}\right)^n = e.$

(b) Prove that for a positive integer p, $\lim_{n\to\infty} \dfrac{1^p + 2^p + \cdots + n^p}{n^{p+1}} = \dfrac{1}{p+1}.$

(10) Prove that if some subsequence of the monotone sequence $\{x_n\}$ converges, then the sequence $\{x_n\}$ is convergent.

(11) If $x_n \to a$, then investigate the limit $\lim_{n\to\infty} \dfrac{x_{n+1}}{x_n}.$

(12) Find the limits $\underline{\lim_{n\to\infty}}\,x_n$ and $\overline{\lim_{n\to\infty}}\,x_n$ for the sequence $x_n = \dfrac{n}{n+1}\sin^2\dfrac{n\pi}{4}.$

Answer: 0; 1.

(13) Prove the Stolz theorem, if (a) $y_{n+1} > y_n$ $(n = 1, 2, \ldots)$, (b) $\lim_{n\to\infty} y_n = +\infty$, and

(c) $\lim_{n\to\infty} \dfrac{x_n}{y_n} = \lim_{n\to\infty} \dfrac{x_{n+1} - x_n}{y_{n+1} - y_n}$ exists, then $\lim_{n\to\infty} \dfrac{x_n}{y_n} = \lim_{n\to\infty} \dfrac{x_{n+1} - x_n}{y_{n+1} - y_n}.$

(14) Prove that if $\{x_n\}$ converges, then the sequence $t_n = \dfrac{1}{n}(x_1 + \cdots + x_n)$ $(n = 1, 2, \ldots)$ converges and $\lim_{n\to\infty} t_n = \lim_{n\to\infty} x_n.$
Suggestion: Use Problem (13).

(15) Let the sequence $\{x_n\}$ satisfies the condition $0 \leqslant x_{m+1} \leqslant x_m + x_n$ $(m, n = 1, 2, \ldots)$. Prove that $\lim_{n\to\infty} \dfrac{x_n}{n}$ exists.

(16) Prove that if $\lim_{n\to\infty} x_n = +\infty$, then $\lim_{n\to\infty} \dfrac{x_1 + \cdots + x_n}{n} = +\infty.$

(17) Let the sequence $\{x_n\}$ be defined as follows:

$$x_1 = a, \quad x_2 = b, \quad x_n = \dfrac{x_{n-1} + x_{n-2}}{2} \quad (n = 3, 4, \ldots).$$

Find $\lim_{n\to\infty} x_n.$

Answer: $\dfrac{1}{3}(a + 2b).$

(18) If the sequence $\{x_n\}$ converges and $x_n > 0$ $(n = 1, 2, \ldots)$, prove that

$$\lim_{n\to\infty} \sqrt[n]{x_1 \cdots x_n} = \lim_{n\to\infty} x_n.$$

(19) Find the limit $\lim\limits_{n\to\infty} \left(\dfrac{1}{n+1} + \cdots + \dfrac{1}{2n} \right)$.

Answer: $\ln 2$.

(20) Let the sequence $\{x_n\}$ be defined as $x_0 > 0$, $x_{n+1} = \dfrac{1}{2}\left(x_n + \dfrac{9}{x_n} \right)$ $(n = 0, 1, 2, \ldots)$.

Prove that $\lim\limits_{n\to\infty} x_n = 3$.

Find the sums of the series in problems (21)–(24).

(21) $1 - \dfrac{1}{2} + \dfrac{1}{4} - \dfrac{1}{8} + \cdots + \dfrac{(-1)^{n-1}}{2^{n-1}} + \cdots$.

Answer: $\dfrac{2}{3}$.

(22) $\left(\dfrac{1}{2} + \dfrac{1}{3} \right) + \left(\dfrac{1}{2^2} + \dfrac{1}{3^2} \right) + \cdots + \left(\dfrac{1}{2^n} + \dfrac{1}{3^n} \right) + \cdots$.

Answer: $\dfrac{3}{2}$.

(23) $\dfrac{1}{1 \cdot 2} + \dfrac{1}{2 \cdot 3} + \cdots + \dfrac{1}{n(n+1)} + \cdots$.

Answer: 1.

(24) $\dfrac{1}{2} + \dfrac{3}{2^2} + \dfrac{5}{2^3} + \cdots + \dfrac{2n-1}{2^n} + \cdots$.

Answer: 3.

(25) Determine whether the series $\sum\limits_{n=1}^{\infty} \sin nx$ converges or diverges.

Suggestion: For the values $x \neq k\pi$ (k integer) $\lim\limits_{n\to\infty} \sin nx \neq 0$.

Determine whether the series $\sum\limits_{n=1}^{\infty} x_n$ converges or diverges.

(26) $x_n = \dfrac{1}{\sqrt{(2n-1)(2n+1)}}$.

Answer: Diverges.

(27) $x_n = \dfrac{n}{2n-1}$.

Answer: Diverges.

(28) $x_n = \ln\left(\dfrac{n+1}{n} \right)$.

Answer: Diverges.

(29) $x_n = \dfrac{1}{(2n-1)^2}$.

Answer: Converges.

Suppose that the series $\sum\limits_{n=1}^{\infty} x_n^2$ and $\sum\limits_{n=1}^{\infty} y_n^2$ are convergent. Determine whether the following series converge or diverge.

(30) $\sum\limits_{n=1}^{\infty} \dfrac{|x_n|}{n}$.

Answer: Converges.

(31) $\displaystyle\sum_{n=1}^{\infty} |x_n y_n|$.

   Answer: Converges.

(32) $\displaystyle\sum_{n=1}^{\infty} (x_n + y_n)^2$.

   Answer: Converges.

Use Cauchy's criterion to determine the convergence of each series $\displaystyle\sum_{n=1}^{\infty} x_n$ in (33)–(35).

(33) $x_n = \dfrac{\sin nx}{2^n}$.

(34) $x_n = \dfrac{\cos x^n}{n^2}$.

   Suggestion: $\dfrac{1}{n^2} < \dfrac{1}{n(n-1)} = \dfrac{1}{n-1} - \dfrac{1}{n}$ $(n = 2, 3, \ldots)$.

(35) $x_n = \dfrac{a^n}{10^n} \cdot |a| < 10$.

Use the convergence tests to determine whether the series $\displaystyle\sum_{n=1}^{\infty} x_n$ converges or diverges, where:

(36) $x_n = \dfrac{n!}{n^n}$.

   Answer: Converges.

(37) $x_n = \dfrac{(n!)^2}{2^{n^2}}$.

   Answer: Converges.

(38) $x_n = \underbrace{\sqrt{2 + \sqrt{2 + \sqrt{2 + \cdots + \sqrt{2}}}}}_{n}$.

   Answer: Converges.

   Suggestion: $\sqrt{2} = 2\cos\dfrac{\pi}{4}$.

(39) $x_n = \dfrac{1}{\sqrt[n]{\ln n}}$.

   Answer: Diverges.

(40) $\displaystyle\sum_{n=1}^{\infty} \dfrac{(-1)^{\lfloor \ln n \rfloor}}{n}$.

   Answer: Diverges.

(41) $\displaystyle\sum_{n=1}^{\infty} \sin n^2$.

   Answer: Diverges.

   Suggestion: Show that $\lim_{n\to\infty} \sin n^2 \neq 0$.

Determine whether the series in problems (42)–(44) converge absolutely or converge conditionally.

(42) $\displaystyle\sum_{n=2}^{\infty} \frac{\sin \frac{n\pi}{12}}{\ln n}$.

Answer: Converges conditionally.

(43) $\displaystyle\sum_{n=1}^{\infty} (-1)^{n-1} \frac{2^n \sin^{2n} x}{n}$.

Answer: (a) Converges absolutely, if $|x - \pi k| < \dfrac{\pi}{4}$ (k integer).

(b) Converges conditionally, if $x = \pi k \pm \dfrac{\pi}{4}$.

(44) $\displaystyle\sum_{n=1}^{\infty} \frac{(-1)^{n-1}}{n^p}$.

Answer: (a) Converges absolutely, if $p > 1$.

(b) Converges conditionally, if $0 < p \leqslant 1$.

(45) Show that the product of the divergent series

$$1 - \sum_{n=1}^{\infty} \left(\frac{3}{2}\right)^n \text{ and } 1 + \sum_{n=1}^{\infty} \left(\frac{3}{2}\right)^{n-1} \left(2^n + \frac{1}{2^{n+1}}\right)$$

converges absolutely.

Chapter 3

Limits and Continuity of Functions

In this chapter, we show the equivalency of two different definitions of limit, Cauchy's and Heine's. Then we classify the different kinds of discontinuity points which functions may possess. Moreover, we clarify the distinction between the concepts of continuous and uniformly continuous. Then we look at the main properties of sequences of functions and infinite series of functions: in particular, that the limit of a uniformly covergent sequence of continuous functions is still continuous, which might not happen for a sequence of continuous functions which is merely pointwise convergent. We lay out the hypotheses necessary in order to prove that a sequence of functions defined on a closed interval has a uniformly convergent subsequence (Arzela's theorem). Results such as Cauchy's criterion, the usual comparison tests, and Arzela's theorem, are valid for complex-valued functions as well.

3.1 Functions of a Real Variable

In Section 1.1, we considered functions f from a set A to a set B which includes the values taken by f. In this chapter, as usual, unless otherwise clear from the context, we consider functions f for which A and B are subsets of the real numbers. That is, we consider real-valued functions of one (real) variable. As before, when we describe the function f by writing a formula $y = f(x)$, we call x the independent variable and y the dependent variable. The set of values $y = f(x)$ is called the range of f. The set of all those numbers x for which $y = f(x)$ is defined is called the domain of definition of f (and is often denoted by $\mathrm{dom} f$). In this section we will be more concerned with the domain of a function than with its range. In general, in what follows the domain of definition of f will often be denoted by Ω which is one of the intervals of the form (a,b), $(a,b]$, $[a,b)$, $[a,b]$. Here, instead of a and b, the symbols $-\infty$ and $+\infty$ can be taken, respectively. The graph of the function f is the set of all points in the plane of the form $(x, f(x))$, where x is in the domain of f.

E. Mahmudov, *Single Variable Differential and Integral Calculus*,
DOI: 10.2991/978-94-91216-86-2_3, © Atlantis Press 2013

Example 3.1. For the function f with the formula $y = f(x) = \sqrt{9 - x^2}$, the domain is the set of all points satisfying the inequality $|x| \leqslant 3$. This means that $\text{dom} f = \{x : -3 \leqslant x \leqslant 3\}$ and the range of f is the interval $0 \leqslant y \leqslant 3$.

Example 3.2. For the so-called Dirichlet function

$$y = f(x) \equiv D(x) = \begin{cases} 0, & \text{if } x \text{ is irrational,} \\ 1, & \text{if } x \text{ is rational.} \end{cases}$$

$\text{dom} f = \{x : -\infty < x < +\infty\}$ and the range of f consists of the points 0 and 1.

Example 3.3. Let

$$y = f(x) \equiv \text{sgn} x = \begin{cases} +1, & \text{if } x > 0, \\ 0, & \text{if } x = 0, \\ -1, & \text{if } x < 0. \end{cases}$$

Then $\text{dom} f = \{x : -\infty < x < +\infty\}$ and the range of f is the set which consists of the points $-1, 0$, and $+1$.

Example 3.4. For the greatest integer function $y = f(x) = [\![x]\!]$, defined as the upper bound of the set of all integers which are less than or equal to x, $\text{dom} f = \{x : -\infty < x < +\infty\}$ and the range of f is the set of all integers.

Example 3.5. The function $y = n! = 1 \cdot 2 \cdot \ldots \cdot n$ is defined on the set of positive integers.

Generally, functions can be defined by equations of the form $y = f(x)$, or graphically, or by a table, where to each of a list of values, x_i $(i = 1, 2, \ldots, n)$, of the independent variable, is assigned a corresponding value, y_i. An equation of the form $y = f(x)$ is said to define y explicitly as a function of x. An equation that is not of the form $y = f(x)$, but can be written in this form, is said to define y implicitly as a function of x. A graphical representation of a function is that curve in the plane which is the graph of $y = f(x)$, and a curve in the plane is the graph of f for some function f if and only if no vertical line intersects the curve more than once. In the tabular representation of a function, an extension, if needed, can be defined in the intervals between the given points x_i $(i = 1, 2, \ldots, n)$ by interpolation methods.

A function of the form $y = a_0 x^n + a_1 x^{n-1} + \cdots + a_n$, where $a_0, a_1, \ldots, a_n$ $(a_0 \neq 0)$ are constants and n is a positive integer, is called a polynomial function of x. A function that is expressible as a ratio of two polynomials is called a rational function. The domain of such a function consists of all x where the denominator differs from zero. The rational functions belong to the class of functions called explicit algebraic functions. These are functions that can be evaluted using finitely many additions, subtractions, multiplications, divisions, and root extractions. It is not hard to see that such functions are *algebraic* also in the wider sense that they satisfy an equation of the form

$$P_0(x)y^n + P_1(x)y^{n-1} + \cdots + P_n(x) = 0,$$

where $P_i(x)$ $(i = 0, 1, \ldots, n)$ are polynomials in x. A function that is not algebraic is said to be transcendental. The class of transcendental functions include trigonometric expressions, logarithms, exponential function, and hyperbolic functions. For example, the functions $y = 10^x$, $y = \cos x$, $y = \log x$ are trancendental functions.

3.2   Limits of Functions

Let f be a function defined on a set A of real numbers, and let a be a limit point of A, (see Definition 1.21).

Definition 3.1 (Cauchy). If for any $\varepsilon > 0$ there is a $\delta = \delta(\varepsilon)$ such that for every $x \in A$ satisfying the inequality

$$0 < |x - a| < \delta \tag{3.1}$$

we have

$$|f(x) - L| < \varepsilon, \tag{3.2}$$

then we say $f(x)$ converges to L (or has the limit L) as x tends to a, i.e., $x \to a$. This is denoted by

$$\lim_{x \to a} f(x) = L \quad \text{(or in the concise form } f(x) \to L, x \to a).$$

What this means, very roughly, is that $f(x)$ tends to get closer and closer to L as x gets closer and closer to a. Conversely, if $\lim_{x \to a} f(x) \neq L$, there is an $\varepsilon > 0$ such that for every $\delta > 0$ there is an $x \in A$ with $0 < |x - a| < \delta$ and $|f(x) - L| \geq \varepsilon$.

Definition 3.2 (Heine[1]). We say that the function f has the limit L as $x \to a$ if for every sequence $x_n \to a$, $n \to \infty$ $(x_n \neq a)$, $\{x_n\} \subset A$, we have $f(x_n) \to L$.

---

[1] H. Heine (1821-1881), German mathematician.

Definition 3.3 (One-sided limits). If in the previous definitions $A = \Omega$ is an open interval with a as a left or right endpoint, we write

$$L = \lim_{x \to a+0} f(x) = f(a+0) \quad \left(L = \lim_{x \to a-0} f(x) = f(a-0) \right)$$

and refer to the limit as the right-hand limit as x approaches a from the right or the left-hand limit as x approaches a from the left.

For example, if $f(x) = \dfrac{x}{|x|}$ $(x \neq 0)$ and $a = 0$, then

$$\lim_{x \to a+0} f(x) = \lim_{x \to 0+0} \frac{x}{x} = 1, \quad \lim_{x \to a-0} f(x) = \lim_{x \to 0-0} \frac{x}{-x} = -1.$$

Definition 3.4. Let a be a limit point of a set A. Then a function f defined on the set A has the limit $+\infty$ $(-\infty)$, or diverges to $+\infty$ $(-\infty)$, as x tends to a, if for every real number $E > 0$ there is a $\delta = \delta(E) > 0$ such that for every $x \in A$, $0 < |x - a| < \delta(\varepsilon)$ implies $f(x) > E$ $(f(x) < -E)$. This is denoted by

$$\lim_{x \to a} f(x) = +\infty \quad \left(\lim_{x \to a} f(x) = -\infty \right).$$

As in Definition 3.3, we similarly define the cases $\lim\limits_{x \to a \pm 0} f(x) = \pm\infty$.

Definition 3.5. Let f be a function defined on an open interval $(a, +\infty)$ $((-\infty, a))$ We say that $f(x)$ converges to L (or has limit L) as $x \to +\infty$ $(x \to -\infty)$ if for every $\varepsilon > 0$, there is a $\Delta = \Delta(\varepsilon) > 0$ such that for every $x > \Delta$, $x \in (a, +\infty)$ $(x < -\Delta, x \in (-\infty, a))$, it follows that $|f(x) - L| < \varepsilon$. This is denoted $\lim\limits_{x \to +\infty} f(x) = L$ ($\lim\limits_{x \to -\infty} f(x) = L$) or $f(x) \to L$, $x \to +\infty$ $(f(x) \to L, x \to -\infty)$.

Remark 3.1. The limit (finite or infinite) of a function f as $x \to a$ or $x \to \infty$ need not exist. For example, $y = \cos x$, $-\infty < x < +\infty$ has no limit as $x \to \infty$. Also, the function $y = \sin \dfrac{1}{x}$ $(x \neq 0)$ has no limit if $x \to 0$.

Example 3.6. Prove that $\lim\limits_{x \to 2} (3x + 2) = 8$. We use the definition of Cauchy. For any $\varepsilon > 0$, the inequality $|(3x + 2) - 8| < \varepsilon$ implies $|x - 2| < \dfrac{\varepsilon}{3}$. Then, evidently, $\delta = \dfrac{\varepsilon}{3}$.

Example 3.7. Let us show that $\lim\limits_{x \to 2} \dfrac{1}{(x - 2)^2} = +\infty$. For sufficiently large $E > 0$, if $\dfrac{1}{(x - 2)^2} > E$, then $(x - 2)^2 < \dfrac{1}{E}$, $|x - 2| < \dfrac{1}{\sqrt{E}} = \delta$.

Example 3.8. Prove that $\lim\limits_{x \to \infty} \left(2 + \dfrac{1}{x} \right) = 2$. Again, for arbitrary real $\varepsilon > 0$, from the inequality $\left| \left(2 + \dfrac{1}{x} \right) - 2 \right| < \varepsilon$ we have $|x| > \dfrac{1}{\varepsilon}$. Thus the required $\Delta = \Delta(\varepsilon)$ is $\Delta = \dfrac{1}{\varepsilon}$.

Theorem 3.1. *Definitions 3.1 and 3.2 are equivalent.*

☐ (1) First we show that Definition 3.1 implies Definition 3.2. Suppose, then, that (3.1) and (3.2) hold. If $x_n \to a$ ($x_n \neq a$, $\{x_n\} \subset A$), we have to show $f(x_n) \to L$. By (3.1), there is a $\delta > 0$ (depending on ε) such that $0 < |x_n - a| < \delta$ for all $n > N$ and so, by (3.2), $|f(x_n) - L| < \varepsilon$ for all $n > N$. Thus $f(x_n) \to L$.

(2) Conversely, assume that whenever $\{x_n\} \subset A$, $x_n \neq a$ and $x_n \to a$ then $f(x_n) \to L$. We have to show $f(x) \to L$ as $x \to a$. Suppose not. Then for some $\varepsilon > 0$ there does not exist any $\delta > 0$ such that (3.1) implies (3.4). In other words, there exists an $\varepsilon > 0$ such that for every $\delta > 0$, there exists an x, depending on δ, with $|f(x) - L| \geq \varepsilon$. Choose such an ε, and for $\delta_n = \dfrac{1}{n}$ ($n = 1, 2, \ldots$) choose $x = x_n$. Then it follows that $0 < |x_n - a| < \dfrac{1}{n}$, and so $x_n \to a$ ($x_n \neq a$) but $|f(x_n) - L| \geq \varepsilon$, that is, $f(x_n)$ does not converge to L. This contradiction proves the required result. ∎

Theorem 3.2. *If a function f has a limit L at a point a, that is, $\lim\limits_{x \to a} f(x) = L$, then this limit is unique.*

☐ Suppose $f(x) \to L_1$ ($L \neq L_1$), $x \to a$. Let $\{x_n\} \subset A$ and $x_n \to a$ ($x_n \neq a$). Then it follows from Definition 3.2 that the numbers L and L_1 are limits of the sequence $\{f(x_n)\}$. But this contradicts Theorem 2.1. ∎

Theorem 3.3 (Cauchy's Criterion). *In order for the limit of a function f to exist at a point a, it is necessary and sufficient that for any $\varepsilon > 0$, there is a $\delta = \delta(\varepsilon)$ such that for all x', $x'' \in A$ satisfying $0 < |x' - a| < \delta(\varepsilon)$ and $0 < |x'' - a| < \delta(\varepsilon)$, we have $|f(x') - f(x'')| < \varepsilon$.*

☐ (*Necessity*). Suppose $\lim\limits_{x \to a} f(x) = L$. By Definition 3.1, for any $\varepsilon > 0$ there is a $\delta = \delta(\varepsilon) > 0$ such that for every x', $x'' \in A$ satisfying $0 < |x' - a| < \delta$ and $0 < |x'' - a| < \delta$, the inequalities $|f(x') - L| < \dfrac{\varepsilon}{2}$ and $|f(x'') - L| < \dfrac{\varepsilon}{2}$ hold. Then it follows that

$$|f(x') - f(x'')| = |f(x') - L| + |f(x'') - L| < \frac{\varepsilon}{2} + \frac{\varepsilon}{2} = \varepsilon.$$

(*Sufficiency*). Conversely, let, for any given $\varepsilon > 0$, $\delta = \delta(\varepsilon) > 0$ be such that for all x', $x'' \in A$ satisfying $0 < |x' - a| < \delta(\varepsilon)$ and $0 < |x'' - a| < \delta(\varepsilon)$, we have $|f(x') - f(x'')| < \varepsilon$. Let $\{x_n\} \subset A$ ($x_n \neq a$) be any sequence such that $x_n \to a$. We must show that $\{f(x_n)\}$ is a Cauchy sequence. Since $x_n \to a$, there is an integer N such that for all $n > N$,

$$0 < |x_n - a| < \delta(\varepsilon). \tag{3.3}$$

Then, taking into account (3.3), it follows from $|f(x') - f(x'')| < \varepsilon$ that for all $n, m > N$ we have $|f(x_n) - f(x_m)| < \varepsilon$. Thus $\{f(x_n)\}$ is a Cauchy sequence and consequently, by Theorem 2.12, converges. We now show that the limit of $\{f(x_n)\}$ does not depend on the choice of $\{x_n\}$. On the contrary, suppose there are two sequences $x_n \to a$ $(x_n \neq a, \{x_n \subset A\})$ and $\bar{x}_n \to a$ $(\bar{x}_n \neq a, \{\bar{x}_n\} \subset A)$ such that $f(x_n) \to L$ and $f(\bar{x}_n) \to \bar{L}$ $(L \neq \bar{L})$. Consider the sequence $x_1, \bar{x}_1, x_2, \bar{x}_2, \ldots, x_n, \bar{x}_n, \ldots$. It converges to a. Nevertheless the corresponding sequence of values, $f(x_1), f(\bar{x}_1), \ldots, f(x_n), f(\bar{x}_n), \ldots$, diverges, which is a contradiction. Hence, all sequences $\{f(x_n)\}$ have the same limit. Then by Definition 3.2, $\lim_{x \to a} f(x)$ exists. ∎

Let $f, g : A \to \mathbb{R}^1$. We define the sum, product, and quotient of functions as follows:

$$(f+g)(x) = f(x) + g(x), \quad (f \cdot g)(x) = f(x)g(x), \quad \left(\frac{f}{g}\right)(x) = \frac{f(x)}{g(x)}.$$

Theorem 3.4. *Let a be a limit point of a set A and suppose the functions f and g are defined on the same set A and the limits $\lim_{x \to a} f(x) = L_1$, $\lim_{x \to a} g(x) = L_2$ exist. Then the following limits $f+g$, $f \cdot g$, $\dfrac{f}{g}$ exist and have the stated values:*
(1) $\lim_{x \to a} (f+g)(x) = L_1 + L_2$;
(2) $\lim_{x \to a} (f \cdot g)(x) = L_1 \cdot L_2$;
(3) $\lim_{x \to a} \left(\dfrac{f}{g}\right)(x) = \dfrac{L_1}{L_2}$.
(provided in the last case that $g(x) \neq 0$, $x \in A$).

☐ Let $x_n \to a$ $(x_n \neq a, \{x_n\} \subset A)$. By Definition 3.2, $f(x_n) \to L_1$ and $g(x_n) \to L_2$ as $n \to \infty$, and so it follows by Theorem 2.3 that the limits of the sequences $\{f(x_n) + g(x_n)\}$, $\{f(x_n) \cdot g(x_n)\}$, $\left\{\dfrac{f(x_n)}{g(x_n)}\right\}$ are, $L_1 \pm L_2$, $L_1 \cdot L_2$, and $\dfrac{L_1}{L_2}$, respectively. Thus, in turn, as a result of Definition 3.2 these numbers are the limits of the functions $f+g$, $f \cdot g$, $\dfrac{f}{g}$, respectively, whenever $x \to a$. ∎

Remark 3.2. Let us investigate the domains of the sum, difference, product, and quotient functions $f \pm g$, $f \cdot g$, $\dfrac{f}{g}$. (We remind the reader that in the previous theorem, the set A was a common domain). It is easy to see that $\text{dom}(f \pm g) = \text{dom} f \cap \text{dom} g$ and $\text{dom}(f \cdot g) = \text{dom} f \cap \text{dom} g$. For the quotient, $\text{dom}\left(\dfrac{f}{g}\right) = (\text{dom} f \cap \text{dom} g) \setminus P_0$ where $P_0 = \{x : g(x) = 0\}$.

For example, let $f(x) = \sqrt{5-x}$ and $g(x) = \sqrt{x-3}$. Then since $\text{dom} f = (-\infty, 5]$, $\text{dom} g = [3, +\infty)$, we have $\text{dom}(f \pm g) = \text{dom}(f \cdot g) = [3, 5]$, and $\text{dom}\left(\dfrac{f}{g}\right) = [3, 5] \setminus \{3\} = (3, 5]$ $(P_0 = \{3\})$.

Moreover, for the composite function $(f \circ g)(x) = f(g(x))$,

$$\text{dom}(f \circ g) = \{x \in \text{dom}\, g \; : \; g(x) \in \text{dom}\, f\}.$$

If $f(x) = x^2 + 3$ and $g(x) = \sqrt{3}$, then $\text{dom}\, g = [0, +\infty)$ and $\text{dom}\, f = (-\infty, +\infty)$, and so $\text{dom}(f \circ g) = [0, +\infty)$.

Example 3.9. Show that $\lim\limits_{x \to a} x^n = a^n$ ($n \geqslant 1$).

By (2) of Theorem 3.4, setting $f(x) = g(x) = x$, we have

$$\lim_{x \to a} x^n = \lim_{x \to a} x \cdot \lim_{x \to a} x^{n-1} = \left(\lim_{x \to a} x\right)^2 \lim_{x \to a} x^{n-1} = \cdots = \left(\lim_{x \to a} x\right)^n = a^n.$$

Example 3.10. For polynomial functions $P(x) = a_0 x^n + \cdots + a_n$ and $Q(x) = b_0 x^m + \cdots + b_m$ of degrees n and m, respectively, the quotient law (3) of Theorem 3.4 implies

$$\lim_{x \to a} \frac{P(x)}{Q(x)} = \frac{P(a)}{Q(a)} \quad (Q(a) \neq 0).$$

Example 3.11. Compute the limit of the rational function $\dfrac{P(x)}{Q(x)}$ as $x \to \infty$, where $P(x)$, $Q(x)$ are the same polynomials considered in previous example. By Theorem 3.4,

$$\lim_{x \to \infty} \frac{P(x)}{Q(x)} = \lim_{x \to \infty} \frac{x^n \left(a_0 + \dfrac{a_1}{x} + \cdots + \dfrac{a_n}{x^n}\right)}{x^m \left(b_0 + \dfrac{b_1}{x} + \cdots + \dfrac{b_m}{x^m}\right)} = \lim_{x \to \infty} \frac{a_0}{b_0} x^{n-m} = \begin{cases} \dfrac{a_0}{b_0}, & \text{if } m = n, \\[2mm] \infty \cdot \text{sgn}\left(\dfrac{a_0}{b_0}\right), & \text{if } m < n, \\[2mm] 0, & \text{if } m > n, \end{cases}$$

Example 3.12. Compute the limit, $\lim\limits_{x \to a} \dfrac{P(x)}{Q(x)}$. If $\lim\limits_{x \to a} P(x) = 0 = \lim\limits_{x \to a} Q(x)$, then we say that the quotient $\dfrac{P(x)}{Q(x)}$ has the indeterminate form $\frac{0}{0}$ at $x = a$ (for more detail, see Chapter 7). Suppose that a is a repeated root of multiplicity α and β of the algebraic equations $P(x) = 0$ and $Q(x) = 0$, respectively. Then by the factor theorem, we can write

$$P(x) = (x-a)^\alpha P_1(x) \;\; (P_1(a) \neq 0), \quad Q(x) = (x-a)^\beta Q_1(x) \;\; (Q_1(a) \neq 0),$$

where $P_1(x)$ and $Q_1(x)$ are polynomials of degree $n - \alpha$ and $m - \beta$, respectively. Then by parts (1), (2), and (3) of Theorem 3.4, we have

$$\lim_{x \to a} \frac{P(x)}{Q(x)} = \lim_{x \to a} \frac{P_1(a)}{Q_1(a)} (x-a)^{\alpha-\beta} = \begin{cases} \dfrac{P_1(a)}{Q_1(a)}, & \text{if } \alpha = \beta, \\[2mm] 0, & \text{if } \alpha > \beta, \\[2mm] \infty, & \text{if } \alpha < \beta. \end{cases}$$

Theorem 3.5 (Comparison Test). *Let the functions f, g, and h be defined on the same set A. Then:*

(1) *(Limit inequality) If $f(x) \leqslant g(x)$ for all $x \in A$, and if the limits $\lim_{x \to a} f(x)$ and $\lim_{x \to a} g(x)$ exist, then*

$$\lim_{x \to a} f(x) \leqslant \lim_{x \to a} g(x).$$

(2) *(Sandwich or squeeze property) If $f(x) \leqslant h(x) \leqslant g(x)$ for all $x \in A$ (with the possible exception of the point a) and the functions f and g have the same limit as $x \to a$, then*

$$\lim_{x \to a} f(x) = \lim_{x \to a} g(x) = \lim_{x \to a} h(x).$$

☐ Let $x_n \to a$ ($x_n \neq a$, $\{x_n\} \subset A$). Then by hypothesis, we have

$$f(x_n) \leqslant g(x_n) \quad \text{and} \quad f(x_n) \leqslant h(x_n) \leqslant g(x_n) \quad (n = 1, 2, \ldots). \tag{3.4}$$

Now, taking into account Definition 3.2 and Theorem 2.4, by passing to the limit in inequalities (3.4), we have the desired result. ∎

Remark 3.3. Obviously it can be seen that Theorem 3.5 is true in the case when $x \to \pm\infty$.

∎

In the formulation of the next theorem we specialize to the case where the open set A is one of the intervals: (a, b), $(-\infty, a)$, $(a, +\infty)$, $(-\infty, +\infty)$.

Theorem 3.6 (One-sided and Two-sided Limits). *Suppose that the function f is defined on an open interval Ω. Then the limit $\lim_{x \to x_0} f(x)$ ($x_0 \in \Omega$) exists and is equal to the limit L if and only if the one-sided limits $\lim_{x \to x_0+0} f(x)$ and $\lim_{x \to x_0-0} f(x)$ both exist and are both equal to L.*

☐ If $\lim_{x \to x_0} f(x) = L$, then, by Definition 3.1, for every $\varepsilon > 0$ there is $\delta > 0$ such that for $x \in \Omega$, $0 < |x - x_0| < \delta$ implies $|f(x) - L| < \varepsilon$. In particular, for $x < x_0$ ($x \in \Omega$), $0 < |x - x_0| < \delta$, we have $|f(x) - L| < \varepsilon$. So, $f(x_0 - 0) = L$. Similarly, for $x > x_0$ ($x \in \Omega$), $0 < |x - x_0| < \delta$, it follows that $|f(x) - L| < \varepsilon$ and $f(x_0 + 0) = L$. Conversely, if $f(x_0 - 0) = L = f(x_0 + 0)$, then for any $\varepsilon > 0$ there is $\delta_1 > 0$ such that for $x < x_0$ ($x \in \Omega$), $0 < |x - x_0| < \delta_1$ we have $|f(x) - L| < \varepsilon$, and there is $\delta_2 > 0$ such that for $x > x_0$ ($x \in \Omega$), $0 < |x - x_0| < \delta_2$ it follows that $|f(x) - L| < \varepsilon$. Let δ denote the smaller of the two numbers δ_1, $\delta_2 > 0$. Then for $x \in \Omega$, satisfying $0 < |x - x_0| < \delta$, we have $|f(x) - L| < \varepsilon$, that is, $\lim_{x \to x_0} f(x) = L$. ∎

3.3 Local Properties of a Continuous Function

We first define the notion of continuity at a point in terms of limits, and then we give a few useful equivalent definitions.

Definition 3.6. Suppose that a function f is defined on a subset A of the set of real numbers, that a is limit point of A, and that $a \in A$. If $\lim_{x \to a} f(x)$ exists and if

$$\lim_{x \to a} f(x) = f(a), \tag{3.5}$$

then we say that f is *continuous* at a.

In terms of "$\varepsilon - \delta$," the function f is continuous at a if for any $\varepsilon > 0$ there is a $\delta = \delta(\varepsilon) > 0$ such that for all $x \in A$, $0 < |x - a| < \delta$, it follows that $|f(x) - f(a)| < \varepsilon$.

Briefly, the continuity of f at a means that the limit of f at the point a is equal to its value there. In other words, when x is close to a, then $f(x)$ is close to $f(a)$.

Then we see that in order to be continuous at the point a, the function f must satisfy the following three conditions:

(1) f must be defined at a; (2) the limit, $\lim_{x \to a} f(x)$, must exist; (3) $\lim_{x \to a} f(x)$ must equal $f(a)$. If any one of these conditions is not satisfied, then f is not continuous at a.

Definition 3.7. The function f is continuous on a set A if f is continuous at each point of the set A. If the function f is not continuous at a, we say that it is *discontinuous* there, or that a is a *discontinuity* of f.

Definition 3.8. If at a point $a \in A$ the limit $\lim_{x \to a} f(x)$ exists, but $f(a)$ either is not defined or $\lim_{x \to a} f(x) \neq f(a)$, then a is called a *removable point of discontinuity*.

Example 3.13. The function

$$f(x) = \begin{cases} x^3, & \text{if } x \neq 0, \\ 2, & \text{if } x = 0, \end{cases}$$

is discontinuous at $a = 0$ and $\lim_{x \to 0} f(x) = 0$ so that $\lim_{x \to 0} f(x) \neq f(0)$. Thus the point 0 is a removable point of discontinuity. If, instead of $f(0)$, we took the value $\lim_{x \to 0} f(x)$, then f would be continuous at $x = 0$.

Definition 3.9. If f is discontinuous at a point $a \in A$, and if $f(a - 0)$ and $f(a + 0)$ exist, but $f(a - 0) \neq f(a + 0)$, then f is said to have a *discontinuity of the first kind*.

Example 3.14. (a) The function

$$f(x) = \begin{cases} x \cdot \cos \dfrac{1}{x}, & \text{if } x < 0, \\ 0, & \text{if } x = 0, \\ \cos \dfrac{1}{x}, & \text{if } x > 0, \end{cases}$$

has limit from the left at $a = 0$, that is $\lim\limits_{x \to 0-0} f(x) = 0$. Indeed, if $\{x_n\}$ is a sequence that converges to 0 and satisfies $x_n < 0$, then

$$0 \leqslant |f(x_n)| = |x_n| \cdot \left| \cos \frac{1}{x_n} \right| \leqslant |x_n|.$$

Since $x_n \to 0$, as $n \to \infty$, then $\lim\limits_{n \to \infty} f(x_n) = 0$. We now show that this function has no limit at $a = 0$ from the right. For if we consider the two following subsequences, which consist of positive numbers:

$$x'_n = \frac{1}{\dfrac{\pi}{2} + \pi n}; \quad x''_n = \frac{1}{2\pi n}.$$

Then

$$\lim_{n \to \infty} f(x'_n) = \lim_{n \to \infty} \cos\left(\frac{\pi}{2} + \pi n\right) = 0, \quad \lim_{n \to \infty} f(x''_n) = \lim_{n \to \infty} \cos 2\pi n = 1.$$

Thus the function has no right-hand limit at $a = 0$. Therefore we have the second kind of discontinuity at $a = 0$.

(b) For the function $f(x) = \dfrac{1}{1 + 2^{\frac{1}{x-1}}}$, $\lim\limits_{x \to 1+0} f(x) = 0$ and $\lim\limits_{x \to 1-0} f(x) = 1$ and so $x = 1$ is a discontinuity of the first kind.

Definition 3.10. If at $a \in A$ at least one of the one-sided limits does not exist or is infinite, then we say that the point a is a *discontinuity of the second kind*.

A function f is called *continuous from the right* (*from the left*) at the point a if the conditions below are satisfied

$$\lim_{x \to a+0} f(x) = f(a+0) \ (\lim_{x \to a-0} f(x) = f(a-0)) \ \text{and} \ f(a+0) = f(a) \ (f(a+0) = f(a))$$

Example 3.15. (a) For

$$f(x) = \begin{cases} \sin \dfrac{1}{x}, & \text{if } x > 0, \\ 0, & \text{if } x \leqslant 0, \end{cases}$$

$\lim\limits_{x \to 0+0} f(x)$ does not exist, since

$$f(x_n) \to \begin{cases} 0, & \text{if } x_n = \dfrac{1}{\pi n}, \\ 1, & \text{if } x_n = \dfrac{1}{\dfrac{\pi}{2} + 2\pi n}. \end{cases}$$

On the other hand, since $\lim\limits_{x \to 0-0} f(x) = 0$, the function f is continuous from the left at the point $a = 0$ and so this point is a discontinuity of the second kind.

(b) The function

$$f(x) = \begin{cases} x, & \text{if } x \text{ is rational,} \\ 0, & \text{if } x \text{ is irrational,} \end{cases}$$

is continuous at $a = 0$ and has a discontinuity of the second kind at any other point.

(c) The Dirichlet function,

$$f(x) = \begin{cases} 1, & \text{if } x \text{ is rational,} \\ 0, & \text{if } x \text{ is irrational,} \end{cases}$$

has a discontinuity of the second kind at the point x, because neither $f(x+0)$ nor $f(x-0)$ exist.

Theorem 3.7. *If the functions f and g, defined on a set A, are continuous at $a \in A$, then the fuctions $f \pm g$, $f \cdot g$ and $\dfrac{f}{g}$ $(g(a) \neq 0)$ are continuous at a.*

☐ The proof of the theorem follows immediately from Theorem 3.4. ■

Definition 3.11. If there is a real number such that for all $x \in A$, $f(x) \leqslant M$ $(f(x) \geqslant m)$, then we say that f is *bounded above* (*bounded below*) on the set. The number $M(m)$ is called the *upper bound* (*lower bound*) of the function f on. If the set of values of f is bounded above (bounded below), then the least upper bound (greatest lower bound), that is $\sup\limits_{x \in A} f(x)$ $(\inf\limits_{x \in A} f(x))$ exists.

Theorem 3.8. *If a function f is defined on neighboring points of a and is continuous at this point, then there is a $\delta_1 > 0$ such that f is bounded on the set of points satisfying $|x - a| < \delta_1$. Moreover, if $f(a) \neq 0$ then there is a $\delta_2 > 0$ such that f does not change its sign on the interval $(a - \delta_2, a + \delta_2)$.*

☐ By the continuity of f at the point a (see (3.1) and (3.2)), for any fixed $\varepsilon > 0$ there is a $\delta_1 > 0$ such that $f(a) - \varepsilon < f(x) < f(a) + \varepsilon$ for all $|x - a| < \delta_1$, i.e., f is bounded on the interval $(a - \delta_1, a + \delta_1)$. Now, let $f(a) \neq 0$. For convenience, suppose $f(a) > 0$. Setting $\varepsilon = f(a)$, it follows from continuity that

$$-f(a) < f(x) - f(a) < f(a)$$

whenever $|x - a| < \delta_2$, that is, $f(x) > 0$ whenever $|x - a| < \delta_2$. ■

Theorem 3.9. *Let functions f and φ be continuous at the points x_0 and $t_0 = f(x_0)$, respectively. Then the composite function $\varphi \circ f$ is continuous at x_0.*

☐ By the continuity of φ at $t_0 = f(x_0)$, for every $\varepsilon > 0$ there is a $\delta_1 > 0$ such that the inequality

$$|t - t_0| < \delta_1 \tag{3.6}$$

implies

$$|\varphi(t) - \varphi(t_0)| < \varepsilon. \tag{3.7}$$

On the other hand, f is continuous at x_0, and so for any $\delta_1 > 0$ there is a $\delta > 0$ such that the inequality

$$|x - x_0| < \delta \tag{3.8}$$

implies

$$|f(x) - f(x_0)| < \delta_1. \tag{3.9}$$

Setting $t = f(x)$ and $t_0 = f(x_0)$ in (3.6) and (3.7) and taking into account (3.8) and (3.9), we have from $|x - x_0| < \delta$ that $|\phi(f(x)) - \phi(f(x_0))| < \varepsilon$, which means that $\varphi \circ f$ is continuous at the point x_0. ∎

3.4 Basic Properties of Continuous Functions on Closed Intervals

In this section we consider functions defined on a closed interval $\Omega = [a, b]$.

Theorem 3.10 (Intermediate value theorem). *If f is continuous on the closed interval $[a, b]$, $f(a) \neq f(b)$, and C is any number between $f(a)$ and $f(b)$, then there is at least one number c ($c \in (a, b)$) such that $f(c) = C$.*

☐ For definiteness, suppose, for example $f(a) < f(b)$ and $f(a) < C < f(b)$. Let us denote $I_1 = [a, b]$. Assume, that I_n has been defined. We describe (inductively) how to define I_{n+1}, and this shows in turn how to define I_2, I_3, I_4, ... Let a_n, b_n be the left-hand and right-hand endpoints of I_n, respectively. Moreover, let m_n be the midpoint of the line segment joining a_n and b_n. Clearly, if $f(m_n) > C$, then $f(a_n) < C < f(m_n)$; in this case, put $a_{n+1} = a_n$, $b_{n+1} = m_n$, and $I_{n+1} = [a_{n+1}, b_{n+1}]$. If $f(m_n) < C$, then we put $a_{n+1} = m_n$ and $b_{n+1} = b_n$. Thus at each step we bisect I_n and let I_{n+1} be the half of I_n on which f takes on values both above and below C. Observe, that if $f(m_n) = C$, then we simply take $c = m_n$. Obviously, the sequence of closed intervals $\{I_n\}$ satisfies the hypothesis of Cantor's Theorem (Theorem 1.9). Suppose, that c is the unique real number common to all the intervals I_n. It remains to prove, that $f(c) = C$. Since $\lim_{n \to \infty} b_n = c$, by continuity of f the sequence $\{f(b_n)\}$ has limit, that is $\lim_{n\to\infty} f(b_n) = f(c)$. On the other hand, since $f(b_n) > C$ for all n, by Comparison Test (Theorem 2.4) $f(c) \geq C$. By considering the sequence $\{a_n\}$, it follows, that $f(c) \leq C$. Consequently, $f(c) = C$. ∎

Geometrically, the intermediate value property implies that if we draw a horizontal line $y = C$ crossing the y-axis between $f(a)$ and $f(b)$, then this line will cross the curve $y = f(x)$ at least once over the interval $[a, b]$.

Theorem 3.11 (Weierstrass[2]). *If a function f is continuous on the closed interval $[a, b]$, then f is bounded on $[a, b]$.*

☐ Suppose the contrary. Then for any number $M > 0$, there is a point $x_0 \in [a, b]$ such that $|f(x_0)| > M$. Choose $M = n$ $(n = 1, 2, \ldots)$ and construct a sequence $\{x_n\} \subset [a, b]$ satisfying

$$|f(x_n)| > n.$$

The sequence $\{x_n\}$ is bounded and so, by Remark 2.1, there is a convergent subsequence, $\{x_{n_k}\}$. Let $x_{n_k} \to \bar{x}$ $(a \leq x_{n_k} \leq b)$. Then, by the comparison test (Theorem 2.4), $\bar{x} \in [a, b]$. Now, on the one hand the continuity of f implies that $f(x_{n_k}) \to f(\bar{x})$, and on the other hand, $|f(x_{n_k})| > n_k$ and so $f(x_{n_k}) \to \infty$. This contradiction ends the proof of theorem. ∎

Theorem 3.12 (Weierstrass). *Let the function f be continuous on the closed interval $[a, b]$. Then the supremum and infimum are attained there, that is, there exist points $x_1, x_2 \in [a, b]$ such that*

$$f(x_1) = \sup_{a \leq x \leq b} f(x) \quad and \quad f(x_2) = \inf_{a \leq x \leq b} f(x).$$

☐ Since $\inf_{a \leq x \leq b} f(x) = - \sup_{a \leq x \leq b} [-f(x)]$ we prove the theorem for the first case: $f(x_1) = \sup_{a \leq x \leq b} f(x)$. Let $M = \sup_{a \leq x \leq b} f(x)$. We must prove that there exists a point $\bar{x} \in [a, b]$ such that $f(\bar{x}) = M$. Suppose the contrary. Then by Theorem 3.7, the function $q(x) = \dfrac{1}{M - f(x)}$ is continuous on the interval $[a, b]$. Clearly, $q(x) > 0$ for all $x \in [a, b]$. Also, by Theorem 3.11, $q(x)$ is bounded above, i.e., $q(x) \leq M_1$ for all $x \in [a, b]$. It follows from the last inequality that $f(x) \leq M - \dfrac{1}{M_1} < M$, that is, $f(x) < M$. But this is impossible, because M is the least upper bound of the set of values $f(x)$. This contradiction proves the theorem. ∎

Remark 3.4. That the intervals in the hypotheses of Theorems 3.11 and 3.12 be closed is essential for the validity of the theorems. For example, the function $f(x) = \dfrac{1}{x}$ is continuous on a half-closed interval $(0, 1]$, but is not bounded.

---

[2]K. Weierstrass (1815-1897), German mathematician.

On the other hand, the function $y = \arctan x$ is continuous on the x-axis, which is a closed unbounded set, but there do not exist any points $\bar{x}$, such that $\arctan \bar{x} = M = \dfrac{\pi}{2}$ and $\arctan \underline{x} = m = -\dfrac{\pi}{2}$.

Moreover, consider the following function f on the closed interval $[0, 1]$.

$$f(x) = \begin{cases} x^3, & \text{if } 0 < x < 1, \\ \dfrac{1}{3}, & \text{if } x = 0, x = 1. \end{cases}$$

Obviously, this function is bounded on $[0, 1]$: its infimum is $m = 0$ and its supremum is $M = 1$. But points $x_1, x_2 \in [0, 1]$ such that $f(x_1) = M$ and $f(x_2) = m$ do not exist. The reason is that in this case f is not continuous.

Definition 3.12. A function f defined on A is *uniformly continuous* on A if for each $\varepsilon > 0$ there exists a positive $\delta = \delta(\varepsilon)$ such that for all $x', x'' \in A$,

$$|x' - x''| < \delta(\varepsilon) \text{ implies } |f(x') - f(x'')| < \varepsilon.$$

Obviously, a uniformly continuous function on A is continuous on A.

Theorem 3.13 (Cantor). *Let f be continuous on the closed interval $[a, b]$. Then f is uniformly continuous.*

☐ Suppose that f is not uniformly continuous on $[a, b]$, that is, for some $\varepsilon > 0$ and arbitrary $\delta > 0$ there are points $x, x' \in [a, b]$, such that $|x - x'| < \delta$ and $|f(x) - f(x')| \geqslant \varepsilon$. Choosing a sequence of positive numbers $\{\delta_n\}$ convergent to zero, construct two sequences $\{x_n\} \subset [a, b]$, $\{x'_n\} \subset [a, b]$, such that $|x_n - x'_n| < \delta_n$, but

$$|f(x_n) - f(x'_n)| \geqslant \varepsilon. \tag{3.10}$$

By Bolzano–Weierstrass theorem (Theorem 2.10) it can be choosen from bounded sequence $\{x_n\}$ a convergent subsequence $\{x_{n_k}\}$. Let $x_{n_k} \to x_0$ as $k \to \infty$. Clearly, $x_0 \in [a, b]$, because $\{x_{n_k}\} \subset [a, b]$. In turn the inequality $|x_n - x'_n| < \delta_n$ means that $x_{n_k} - \delta_{n_k} < x'_{n_k} < x_{n_k} + \delta_{n_k}$. By passing to the limit in the latter inequality, as $k \to \infty$, we have $x'_{n_k} \to x_0$. Thus, the sequences $\{x_n\}$, $\{x'_n\}$ tend to the same limit point $x_0 \in [a, b]$. Also by continuity of f the sequences $\{f(x_{n_k})\}$ and $\{f(x'_{n_k})\}$ tend to the same limit $f(x_0)$. However, this is impossible, because in view of the inequality (3.10) $|f(x_n) - f(x'_n)| \geqslant \varepsilon$. The obtained contradiction proves the theorem. ■

It should be pointed out, that this theorem is true, if we replace $[a, b]$ by arbitrary bounded, closed set A.

Example 3.16. Show that $f(x) = \dfrac{1}{x}, x \in [1, +\infty)$, is uniformly continuous.

Since for all $x', x'' \in [1, +\infty)$

$$|f(x') - f(x'')| = \left| \frac{1}{x'} - \frac{1}{x''} \right| = \frac{|x' - x''|}{x' \cdot x''} \leqslant |x' - x''|$$

for each $\varepsilon > 0$ we set $\delta = \varepsilon$. Then the inequality $|x' - x''| < \delta$ implies $|f(x') - f(x'')| < \varepsilon$, which in turn means that f is uniformly continuous.

We now show that the same function is not uniformly continuous on $(0, +\infty)$. For if not, then let, for each $\varepsilon > 0$, $\delta = \delta(\varepsilon) > 0$ be such that $x', x'' > 0$, $|x' - x''| < \delta(\varepsilon)$ implies the inequality $|f(x') - f(x'')| = \dfrac{|x' - x''|}{x' \cdot x''} < \varepsilon$. These inequalities are satisfied also for $0 < x' < \delta(\varepsilon)$ and $0 < x'' < \delta(\varepsilon)$. It follows from the inequality $\dfrac{|x' - x''|}{x' \cdot x''} < \varepsilon$ that for fixed x' ($0 < x' < \delta(\varepsilon)$), the function $\dfrac{|x' - x''|}{x' \cdot x''}$ of x'' ($0 < x'' < \delta(\varepsilon)$) is bounded. On the other hand, $\dfrac{|x' - x''|}{x' \cdot x''} \to +\infty, x'' \to 0 + 0$. This contradicts uniform continuity.

Example 3.17. Show that $f(x) = \sin \dfrac{1}{x}$ is not uniformly continuous on $(0, 1)$. In terms of "$\varepsilon - \delta$," this means that for some $\varepsilon > 0$ and arbitrary $\delta > 0$ there exist $x', x'' \in (0, 1)$ such that

$$|x' - x''| < \delta \text{ implies } |f(x') - f(x'')| \geqslant \varepsilon.$$

For each $n = 1, 2, \ldots$, take points of the form $x'_n = \dfrac{1}{\pi n}$ and $x''_n = \dfrac{1}{\frac{\pi}{2} + 2\pi n}$ (see Example 3.15). Clearly, $\{x'_n\}, \{x''_n\} \subset (0, 1)$. Then for any small $\delta > 0$, there is an integer n such that $|x'_n - x''_n| < \delta$. Besides, $|f(x'_n) - f(x''_n)| = 1$ and for $\varepsilon = \dfrac{1}{2}$ we have $|f(x'_n) - f(x''_n)| > \varepsilon$. Thus f is not a uniformly continuous function on $(0, 1)$.

Remark 3.5. In Definition 3.12, the point is that δ may depend on ε, but does not depend on x' or x''. We remind the reader that in Definition 3.6 of continuity, δ depended on both ε and $x = a$.

In the theory of functions there is also the notion of absolute continuity, given below.

If for any $\varepsilon > 0$ there is a $\delta > 0$ such that for any pairwise disjoint intervals $(a_i, b_i) \subset [a, b]$ $(i = 1, 2, 3, \ldots, n)$

$$\sum_{i=1}^{n} |b_i - a_i| < \delta \text{ implies } \sum_{i=1}^{n} |f(b_i) - f(a_i)| < \varepsilon,$$

then we say that f is absolutely continuous on $[a, b]$. In particular, if $i = 1$ it follows that f is uniformly continuous.

Here, pairwise disjointness means that

$$\bigcup_{i=1}^{n} [a_i, b_i] \subset [a, b] \text{ and } [a_i, b_i] \cap [a_j, b_j] = \varnothing \ (i \neq j).$$

3.5   Monotone and Inverse Functions

Definition 3.13. A function $f : \Omega \to \mathbb{R}$ is called *nondecreasing (nonincreasing)* if for every $x_1, x_2 \in \Omega$

$$x_1 < x_2 \text{ implies } f(x_1) \leqslant f(x_2) \ (f(x_1) \geqslant f(x_2)).$$

A function $f : \Omega \to \mathbb{R}$ is called *increasing (decreasing)* if for every $x_1, x_2 \in \Omega$

$$x_1 < x_2 \text{ implies } f(x_1) < f(x_2) \ (f(x_1) > f(x_2)).$$

Nondecreasing and nonincreasing functions are called *monotone*, and increasing or decreasing functions are called *strictly monotone functions*.

Theorem 3.14. *Let f be a monotone function defined on an open set Ω. Then at each point $x_0 \in \Omega$ there exist one-sided limits and for a nondecreasing (nonincreasing) function f,*

$$f(x_0 - 0) \leqslant f(x_0) \leqslant f(x_0 + 0) \quad (f(x_0 - 0) \geqslant f(x_0) \geqslant f(x_0 + 0)). \tag{3.11}$$

□   For definiteness let f be a nondecreasing function on Ω. Then the set of values $\{f(x) : x < x_0\}$ is bounded above by $f(x_0)$ and so the least upper bound $M = \sup_{x \in A} f(x)$ exists. Clearly $M \leqslant f(x_0)$. We show that $\lim_{x \to x_0 - 0} f(x) = f(x_0 - 0) = M$. Indeed, according to the definition of the least upper bound, for any $\varepsilon > 0$ there is $\delta > 0$ such that $M - \varepsilon < f(x_0 - \delta) \leqslant M$. Since f is nondecreasing for $x_0 - \delta < x < x_0$ it follows that $M - \varepsilon < f(x_0 - \delta) \leqslant f(x) \leqslant M$. Consequently, $\lim_{x \to x_0 - 0} f(x) = f(x_0 - 0) = M \leqslant f(x_0)$. Similarly, it can be proved that $\lim_{x \to x_0 + 0} f(x) = f(x_0 + 0) \geqslant f(x_0)$.   ∎

It follows from inequality (3.11) that if $f(x_0 - 0) = f(x_0 + 0)$, then x_0 is a point of continuity of f. And if $f(x_0 - 0) \neq f(x_0 + 0)$, then x_0 is a discontinuity of the first kind. In other words, each point $x_0 \in \Omega$ is either a point of continuity or a discontinuity of the first kind. Thus monotone functions don't have the second kind of discontinuity.

Theorem 3.15. *Suppose f is a monotone function defined on an open set Ω. Then the set of discontinuity points of f is at most denumerable.*

□   For definiteness let f be a nondecreasing function. Let $x_0 \in A$ be a point of discontinuity of f. Then, according to (3.11), $f(x_0 - 0) < f(x_0 + 0)$. By Theorem 1.12, there exists a rational number r such that $f(x_0 - 0) < r < f(x_0 + 0)$. Thus to every discontinuity point x_0 we have associated a rational number r. If $r_i, i = 1, 2$ are the numbers corresponding to the discontinuity points $x_1 < x_2$, then $r_1 < f(x_1 + 0) \leqslant f(x_2 - 0) < r_2$, that is, $r_1 < r_2$. Thus there is a one-to-one (injective) correspondence between the set of discontinuity points of f and a subset of the rational numbers, which is denumerable by Theorem 1.4. ∎

Remark 3.6. For non-monotone functions the set of discontinuity points can be uncountable. For example, for the Dirichlet function (see Example 3.2), every $x \in (-\infty, +\infty)$ is a point of discontinuity.

Definition 3.14. Let $f : [a, b] \to [\alpha, \beta]$ be a one-to-one function. Then to each element $y \in [\alpha, \beta]$, assign a unique element $x \in [a, b]$ by the rule $f(x) = y$. Such a function defined on $[\alpha, \beta]$ is called the *inverse function* of f and is denoted f^{-1}, that is, $x = f^{-1}(y)$.

Here, instead of closed intervals $[a, b]$ and $[\alpha, \beta]$, open intervals (a, b), (α, β), $(-\infty, b)$, $(a, +\infty)$, $(-\infty, \beta)$, $(\alpha, +\infty)$, $(-\infty, +\infty)$ can be taken.

In turn, it is obvious that $y = f(x)$ is the inverse function of $x = f^{-1}(y)$ and $f(f^{-1}(y)) = y$, $f^{-1}(f(x)) = x$.

If f and f^{-1} are inverses of each other, then the graphs of the two equations $y = f(x)$ and $x = f^{-1}(y)$ are identical. On the other hand, the graphs of the two equations $y = f^{-1}(x)$ and $y = f(x)$ are reflections of one another across the line $y = x$.

Example 3.18. For the function $y = 3x$ defined on $[a, b]$, its image is the closed interval $[3a, 3b]$.

Then the inverse function is defined by $x = f^{-1}(y) = \dfrac{y}{3}$, $y \in [3a, 3b]$.

Example 3.19. The function defined by $y = e^x$ is an increasing function on $(-\infty, +\infty)$. The inverse function of e^x is $x = \ln y$ defined on $(0, +\infty)$ (Figure 3.1).

Example 3.20. For the following function defined on $[0, 1]$, irrational if

$$y = \begin{cases} x, & \text{if } x \text{ is rational,} \\ 1 - x, & \text{if } x \text{ is irrational,} \end{cases}$$

the inverse function,

$$x = \begin{cases} y, & \text{if } y \text{ is rational,} \\ 1 - y, & \text{if } y \text{ is irrational,} \end{cases}$$

is defined on $[0, 1]$.

Theorem 3.16. *Let a function f defined on $[a, b]$ be a continuous and increasing (decreasing) function and put $\alpha = f(a)$ and $\beta = f(b)$. Then there exists an increasing and continuous inverse function $x = f^{-1}(y)$ defined on $[\alpha, \beta]$ $([\beta, \alpha])$.*

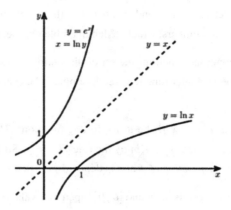

Fig. 3.1   The graphs of $y = \ln x$ and $x = \ln y$

☐   For definiteness, let f be an increasing function. Since by Theorem 3.10 a continuous function must assume every possible value between $f(a)$ and $f(b)$ as x varies from a to b, the range of f is a closed interval $[f(a), f(b)]$. By hypothesis, $f : [a, b] \to [\alpha, \beta]$ is one-to-one. Thus there exists an inverse function $x = f^{-1}(y)$ defined on the closed interval $[\alpha, \beta]$. Now, let us show that this inverse function is increasing.

Let $y_1 < y_2$ ($y_1, y_2 \in [\alpha, \beta]$). We show that $x_1 = f^{-1}(y_1) < x_2 = f^{-1}(y_2)$. If not, then if $x_1 \geqslant x_2$, then since $y = f(x)$ is increasing, it follows that $y_1 \geqslant y_2$. On the other hand we had $y_1 < y_2$. This contradiction proves that f^{-1} is an increasing function. To complete the proof of the theorem it remains to prove the continuity of this inverse function f^{-1} defined by $x = f^{-1}(y)$ on the interval $[\alpha, \beta]$. Suppose f^{-1} is not continuous and y_0 is a point of discontinuity. Then either $f^{-1}(y_0 - 0) < f^{-1}(y_0)$ or $f^{-1}(y_0) < f^{-1}(y_0 + 0)$ holds. In the first case $f^{-1}(y) \leqslant f^{-1}(y_0 - 0)$ if $y < y_0$ and $f^{-1}(y) \geqslant f^{-1}(y_0)$ if $y \geqslant y_0$, that is $f^{-1}(y)$ can not be the value of an inverse function f^{-1} belonging to $(f^{-1}(y_0 - 0), f^{-1}(y_0))$. In the second case, by analogy it can be shown that f^{-1} does not take a value in $(f^{-1}(y_0), f^{-1}(y_0 + 0))$. Therefore, we conclude that $[a, b]$ is not the range of f^{-1}. This contradiction ends the proof of theorem.   ■

3.6   The Main Elementary Functions

1. The Power Function.   The function $y = x^{\alpha}$, where the constant α is a real number, is called the power function. It has not yet been defined except when α is a natural number.

In this case, the function $y = x^n$ $(n = 1, 2, \ldots)$ is continuous on the real axis. It follows, by Theorem 3.7, that $y = x^{-n}$ $(x \neq 0)$ is continuous.

The function $y = \sqrt[n]{x}$ $(x \geq 0, n = 2, 3, \ldots)$ is defined as the inverse of the function $x = y^n$ $(y \geq 0)$. Since the latter is continuous and increasing, this is possible by Theorem 3.16.

If $r = \dfrac{p}{q}$ $(q \geq 2, p = \text{integer})$ is a rational number, then, by Theorem 3.9, the function $y = x^r = \left(\sqrt[q]{x}\right)^p$ $(x > 0)$ is continuous.

Now, let $\alpha \geq 0$ be an arbitrary positive number and put $\underline{\alpha}_m, \overline{\alpha}_m$ equal to its m-valued lower and upper decimal representations (Definition 1.11). By the definition of the power function, $x^{\underline{\alpha}_m} < x^\alpha < x^{\overline{\alpha}_m}$ for $x > 1$. By continuity, $\lim\limits_{x \to 1+0} x^{\underline{\alpha}_m} = \lim\limits_{x \to 1+0} x^{\overline{\alpha}_m} = 1$. It follows from Theorem 3.5 that $\lim\limits_{x \to 1+0} x^\alpha = 1$. Similarly, $\lim\limits_{x \to 1-0} x^\alpha = 1$. Thus the function x^α is continuous at $x = 1$. Then, for any positive $x_0 > 0$,

$$\lim_{x \to x_0} x^\alpha = \lim_{x \to x_0} x_0^\alpha \left(\frac{x}{x_0}\right)^\alpha = x_0^\alpha \lim_{t \to 1} t^\alpha = x_0^\alpha, \quad t = \frac{x}{x_0}.$$

Therefore, x^α is continuous on the infinite half-line $x > 0$ for every $\alpha > 0$. Since $x^\alpha = \dfrac{1}{x^{-\alpha}}$, the function x^α is continuous even if $\alpha < 0$.

On the other hand, for $x_2 > x_1$ and $\alpha > 0$, since $\dfrac{x_2^\alpha}{x_1^\alpha} = \left(\dfrac{x_2}{x_1}\right)^\alpha > 1$, the function x^α is increasing. If $\alpha < 0$, then the function $x^\alpha = \dfrac{1}{x^{-\alpha}}$ is decreasing. The domain of the function x^α is the open interval $(0, \infty)$.

2. Exponential and Logarithmic Functions.

An exponential function is one of the form $y = a^x$, where x is variable, and a is constant. In the case $a > 1$, if $x_2 > x_1$, then $\dfrac{a^{x_2}}{a^{x_1}} = a^{x_2 - x_1} > a^0 = 1$ and so the exponential function $y = a^x$ is an increasing function. Similarly, for $a < 1$ the function $y = a^x$ is decreasing. The domain of $y = a^x$ is the real line $(-\infty, +\infty)$ and the set of values of $y = a^x$ is $(0, +\infty)$. We show that the exponential function is continuous. If $a > 1$ and $-\dfrac{1}{n} < x < \dfrac{1}{n}$ we have $\dfrac{1}{a^{1/n}} = a^{-1/n} < a^x < a^{1/n}$. Since $\lim\limits_{n \to \infty} a^{1/n} = 1$, by the comparison theorem (3.5), $\lim\limits_{x \to 0} a^x = 1$. By analogous arguments, the same result can be proved if $a < 1$.

As for the continuity of $y = a^x$ at any point $x \neq x_0$, it is easy to see that

$$\lim_{x \to x_0} a^x = \lim_{x \to x_0} a^{x_0} a^{x - x_0} = a^{x_0} \lim_{t \to 0} a^t = a^{x_0} \quad (t = x - x_0).$$

By Theorem 3.16, the inverse function of the exponential function $y = a^x$ is continuous and strictly monotone. This function is denoted by $x = \log_a y$ and is called the *logarithm function*. The logarithm function is increasing if $a > 1$, and decreasing if $a < 1$. Its domain is $(0, +\infty)$ and its range is $(-\infty, +\infty)$. The constant a is referred to as the base of the logarithm function. The logarithm function with base $a = e$ is called the *natural logarithm function* and notated $\ln x$.

The following laws of logarithms are easy to derive from the laws of exponents:

(1) $\log_a xy = \log_a x + \log_a y$,

(2) $\log_a \dfrac{x}{y} = \log_a x - \log_a y$,

(3) $\log_a x^\alpha = \alpha \log_a x$ (α any real number).

(4) $\log_a x = \dfrac{\log_b x}{\log_b a}$.

These formulas hold for any positive x, y, a, b ($a \neq 1, b \neq 1$).

3. Hyperbolic Functions.   The four hyperbolic functions, hyperbolic cosine, hyperbolic sine, hyperbolic tangent, and hyperbolic cotangent are defined as follows:

$$\cosh x = \frac{e^x + e^{-x}}{2}, \qquad \sinh x = \frac{e^x - e^{-x}}{2},$$

$$\tanh x = \frac{\sinh x}{\cosh x}, \qquad \coth x = \frac{\cosh x}{\sinh x}.$$

These combinations of familiar exponentials are useful in certain applications of calculus and are helpful in evaluating certain integrals. Moreover, the trigonometric terminology and notation for hyperbolic functions stem from the fact that they satisfy a list of identities that, apart from an occasional difference of sign, much resemble the familiar trigonometric identites:

$$\sinh(x+y) = \sinh x \cdot \cosh y + \cosh x \cdot \sinh y),$$
$$\cosh(x+y) = \cosh x \cdot \cosh y + \sinh x \cdot \sinh y. \tag{3.12}$$

Setting $x = y$ in the first formula, we have the familiar trigonometric identity

$$\sinh 2x = 2\sinh x \cdot \cosh x.$$

Also, from their definitions, it follows that

$$\cosh^2 x - \sinh^2 x = 1,$$
$$\tanh(x_1 + x_2) = \frac{\tanh x_1 + \tanh x_2}{1 + \tanh x_1 \cdot \tanh x_2}, \qquad \coth(x_1 + x_2) = \frac{1 + \coth x_1 \cdot \coth x_2}{\coth x_1 + \coth x_2}. \tag{3.13}$$

The first identity tells us that the point $(r\cosh t, r\sinh t)$ lies on the hyperbola $x^2 - y^2 = r^2$ for all t and this is the reason for the name hyperbolic function.

The function $\cosh x$ is even, because $\cosh(-x) = \cosh x$. The other three functions, $\sinh x$, $\tanh x$, and $\coth x$, are odd:

$$\sinh(-x) = \sinh x, \quad \tanh(-x) = \tanh x, \quad \coth(-x) = \coth x.$$

As is seen from their definitions, the domain of the functions $\sinh x$, $\cosh x$, and $\tanh x$ is the real line $(-\infty, +\infty)$, and the domain of $\coth x$ is $(-\infty, +\infty) \smallsetminus \{0\}$.

The sets of values of the hyperbolic functions are given below:

for $\sinh x$ and $\cosh x$ it is $(-\infty, +\infty)$;

for $\tanh x$ it is $(-1, +1)$;

for $\coth x$ it is $(-\infty, -1) \cup (1, +\infty)$.

We note that none of the hyperbolic functions is periodic. The graphs of these four functions are shown in Figs. 3.2–3.5.

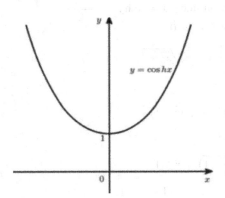

Figure 3.2 The hyperbolic cosine

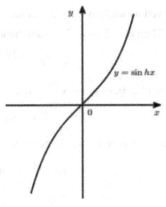

Figure 3.3 The hyperbolic sine

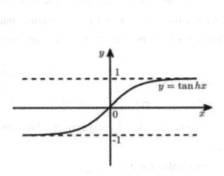

Figure 3.4 The hyperbolic tangent

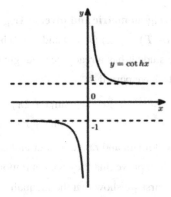

Figure 3.5 The hyperbolic cotangent

All hyperbolic functions are continuous at every point of their domains.

By formulas (3.12) and (3.13), it is clear that hyperbolic functions are strictly monotone functions if either $x > 0$ or $x < 0$. For example, if $x > 0$ (for $x < 0$ we use the evenness or the oddness of the functions to reduce to the case of positive x), then $\sinh x$, $\cosh x$ and $\tanh x$ are increasing and $\coth x$ is decreasing.

Now, let's investigate the inverse hyperbolic functions. The graphs of the four inverse hyperbolic functions may be obtained by reflecting through the line $y = x$ the graphs of the hyperbolic functions (restricted to the half-line $x \geqslant 0$ in the case of $\cosh x$). The hyperbolic sine, tangent and cotangent are one-to-one where they are defined. But the hyperbolic cosine is one-to-one on the half-line $x \geqslant 0$ (or $x \leqslant 0$). Let us derive an inverse function for one of them, say for $\sinh^{-1} x$ (the formulas for $\cosh^{-1} x$, $\tanh^{-1} x$ and $\coth^{-1} x$ can be derived by analogy). By definition of the function $\sinh y$, $x = \sinh y = \dfrac{e^y - e^{-y}}{2}$.

Then $e^y - 2x - e^{-y} = 0$ and hence $e^{2y} - 2xe^y - 1 = 0$.

$$e^y = \frac{2x \pm \sqrt{4x^2 + 4}}{2} = x \pm \sqrt{x^2 + 1}.$$

Since $e^y > 0$, $e^y = x + \sqrt{x^2 + 1}$, and so

$$y = \sinh^{-1} x = \ln\left(x + \sqrt{x^2 + 1}\right).$$

The other three inverse hyperbolic functions are:

$$\cosh^{-1} x = \ln\left(x + \sqrt{x^2 - 1}\right), \quad x \geqslant 1,$$

$$\tanh^{-1} x = \frac{1}{2} \ln \frac{1 + x}{1 - x}, \qquad -1 < x < 1,$$

$$\coth^{-1} x = \frac{1}{2} \ln \frac{x + 1}{x - 1}, \qquad |x| > 1.$$

4. Trigonometric and Inverse Trigonometric Functions.   If for a function f the equality $f(x + T) = f(x)$ ($T \neq 0$ and fixed) holds, then we say that f has period T. We assume that the student has already been taught that $\sin x$ and $\cos x$ have period 2π, and that $\tan x$ and $\cot x$ have period π:

$$\sin(x + 2\pi) = \sin x, \qquad \cos(x + 2\pi) = \cos x,$$
$$\tan(x + \pi) = \tan x, \qquad \cot(x + \pi) = \cot x.$$

The domain and range of $\sin x$ and $\cos x$ are $(-\infty, +\infty)$ and $[-1, +1]$, respectively. We now wish to prove that they are continuous functions.

First we show that the inequality $\sin x < x < \tan x$ holds for $0 < x < \dfrac{\pi}{2}$.

Figure 3.6 shows the triangles OMA and OAB and the circular sector OMA that contains the triangle OMA and is contained in the triangle OAB. The angle between OM and OA is x—and $|OA| = |OM| = 1$. Let S_1 and S_3 denote the areas of triangles OMA and OAB, respectively, and S_2, the area of the circular sector OMA. Obviously, $S_1 < S_2 < S_3$.

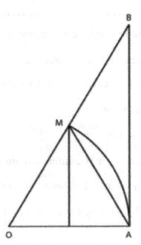

Fig. 3.6   The triangles OMA and OAB and the circular sector OMA that contains the triange OMA

Using the fact that the area of a circular sector with angle x and radius r is $\frac{1}{2}r^2 x$
(here $r = 1$), we have $S_2 = \frac{x}{2}$. On the other hand, taking into account that $S_1 = \frac{\sin x}{2}$
and $S_3 = \frac{\tan x}{2}$, we have the required inequality. Then by the Comparison Test (Theo-
rem 3.5), $\lim\limits_{x\to 0} \sin x = 0$. Moreover, the inequality $1 - \cos x = 2\sin^2\left(\frac{x}{2}\right) \leqslant \frac{x^2}{2}$ $(|x| < \pi)$
implies $\lim\limits_{x\to 0} \cos x = 1$. Thus,

$$\lim_{x\to x_0} \sin x = \lim_{t\to 0} \sin(x_0 + t) = \lim_{t\to 0}(\sin x_0 \cos t + \cos x_0 \sin t) = \sin x_0, \quad t = x - x_0,$$

and $\sin x$ is continuous. The continuity of $\cos x$ can be proved analogously.

It follows from its definition that is increasing on $\left[-\frac{\pi}{2}, \frac{\pi}{2}\right]$ and decreasing on $\left[\frac{\pi}{2}, \frac{3\pi}{2}\right]$.
The others intervals on which $\sin x$ is increasing or decreasing can be found by using its
periodicity. At the same time, by formula $\cos x = \sin\left(\frac{\pi}{2} - x\right)$ we can define the intervals
of strict monotonicity of $\cos x$.

Thus, in the intervals of monotonicity of $\sin x$ and $\cos x$, we may define the continuous
and strictly monotone inverse functions, $y = \arcsin x$ (or $\sin^{-1} x$) $(-1 \leqslant x \leqslant +1, -\frac{\pi}{2} \leqslant$
$y \leqslant \frac{\pi}{2})$ and $y = \arccos x$ (or $\cos^{-1} x$) $(-1 \leqslant x \leqslant +1, 0 \leqslant y \leqslant \pi)$. Similarly, we may define
$\arctan x$ $(-\infty < x < +\infty, -\frac{\pi}{2} < y < \frac{\pi}{2})$ and $y = \operatorname{arc cot} x$ $(-\infty < x < +\infty, 0 < y < \pi)$. It
follows from the definition of the inverse trigonometric functions that

$$\arcsin x + \arccos x = \frac{\pi}{2} \quad \text{and} \quad \arctan x + \operatorname{arc cot} x = \frac{\pi}{2}.$$

If we take into account the intervals of monotonicity of the the trigonometric functions, then we obtain a collection of one-to-one inverse trigonometric functions:

$$\arcsin x = (-1)^n \arcsin x + \pi n, \quad \arccos x = \pm \arccos x + 2\pi n,$$

$$\arctan x = \arctan x + \pi n, \quad \text{arccot}\, x = \text{arc cot}\, x + \pi n, \quad n = 0, \pm 1, \pm 2, \ldots.$$

Now we compute some important limits.

(1) The basic trigonometric limit is $\lim\limits_{x \to 0} \dfrac{\sin x}{x} = 1$.

Since the function $\dfrac{\sin x}{x}$ is even, we may confine our attention to the right-handed limit. The previously proved inequality $\sin x < x < \tan x$, for $0 < x < \dfrac{\pi}{2}$, implies that $\cos x < \dfrac{\sin x}{x} < 1$ for $0 < x < \dfrac{\pi}{2}$. Then, by passing to the limit and using Theorem 3.5, we obtain the desired result.

(2) Another very important limit is $\lim\limits_{x \to \infty} \left(1 + \dfrac{1}{x}\right)^x = e$.

It suffices to take the limit as $x \to +\infty$, because by substituting $t = -(1 + x)$, the case $x \to -\infty$ can be transformed into the case of $x \to +\infty$. As before, let $[\![x]\!]$ denote the integer part of the number x, so that $[\![x]\!] \leqslant x \leqslant [\![x]\!] + 1$.

Then

$$\left(1 + \frac{1}{[\![x]\!] + 1}\right)^{[\![x]\!]} \leqslant \left(1 + \frac{1}{x}\right)^x \leqslant \left(1 + \frac{1}{[\![x]\!]}\right)^{[\![x]\!] + 1}. \tag{3.14}$$

Let $n = [\![x]\!]$. Then, using the fact that $\lim\limits_{n \to \infty} \left(1 + \dfrac{1}{n}\right)^n = e$ (Remark 2.6), we obtain the desired formula by passing to the limit in (3.14) (Theorem 3.5).

(3) We next show that $\lim\limits_{x \to 0} \dfrac{a^x - 1}{x} = \ln a$.

Setting $a^x - 1 = t$ yields $x = \log_a(1 + t)$, and so

$$\lim\limits_{x \to 0} \frac{a^x - 1}{x} = \lim\limits_{t \to 0} \frac{t}{\log_a(1 + t)} = \frac{1}{\log_a e} = \ln a$$

(here we used the fact that $\lim\limits_{x \to \infty} \dfrac{\log_a(1 + x)}{x} = \log_a e$). In particular, if $x = \dfrac{1}{n}$, then it follows from (3) that $\lim\limits_{n \to +\infty} n\left(\sqrt[n]{a} - 1\right) = \ln a$.

(4) Next, we show that $\lim\limits_{x \to 0} \dfrac{(1 + x)^\mu - 1}{x} = \mu$.

Using the substitution $(1 + x)^\mu - 1 = t$, we have $\mu \ln(1 + x) = \ln(1 + t)$ and so

$$\lim\limits_{x \to 0} \frac{(1 + x)^\mu - 1}{x} = \lim\limits_{t \to 0} \frac{t}{x} = \lim\limits_{t \to 0} \left[\frac{t}{\ln(1 + t)} \cdot \mu \cdot \frac{\ln(1 + x)}{x}\right] = \mu.$$

3.7 Comparison of Infinitely Small and Infinitely Large Functions

Definition 3.15. A function f defined in a neighborhood of a point a is called *infinitely small* (or *infinitesimal*) at a, if $\lim_{x \to a} f(x) = 0$.

For example, the functions x, $\sin x$, $\tan x$, and $1 - \cos x$ at $x = 0$ are infinitely small functions.

$$\text{If } \lim_{x \to a} \phi(x) = b, \text{ then } f(x) = \phi(x) - b \text{ is infinitely small at } a.$$

Definition 3.16. For a function F with $\lim_{x \to a+0} F(x) = \pm\infty$ ($\lim_{x \to a-0} F(x) = \pm\infty$), then we say that $F(x)$ is *infinitely large from the right (left) hand side* of a.

Definition 3.17. Let f and g be infinitely small functions at a point a. Then, if $\lim_{x \to a} \dfrac{f(x)}{g(x)} = 0$, we say that the function f is *infinitely small of a higher order at a than g*.

Definition 3.18. Let f and g be infinitely small functions at a point a. Then if $\lim_{x \to a} \dfrac{f(x)}{g(x)} = k$ ($k \neq 0$), then the functions f and g *have the same infinite smallness order* at point a.

Definition 3.19. If $\lim_{x \to a} \dfrac{f(x)}{g(x)} = 1$, then the functions $f(x)$ and $g(x)$ are *equivalently infinitely small* at a.

If f is higher order infinitely small at a with respect to g, then we write $f = o(g)$.

The following properties follow from this definition:

(1) We have $o(g) + o(g) = o(g)$ and $o(g) - o(g) = o(g)$.
(2) If $\gamma = o(g)$, then $o(g) + o(\gamma) = o(g)$.
(3) If f and g are infinitely small at a point a, then either $f \cdot g = o(f)$ or $f \cdot g = o(g)$.

Definition 3.20. Let $\lim_{x \to a+0} F(x) = +\infty$ and $\lim_{x \to a+0} G(x) = +\infty$. Then if $\lim_{x \to a+0} \dfrac{f(x)}{g(x)} = +\infty$, we say that the function $F(x)$ is *higher order large with respect to $G(x)$*.

If $\lim_{x \to a+0} \dfrac{f(x)}{g(x)} = k \neq 0$, then the functions $F(x)$ and $G(x)$ have the same order of largeness at a.

Example 3.21. The functions $f(x) = x^3 - x^5$ and $g(x) = 5x^3 + x^4$ have the same smallness order at a point $x = 0$, because

$$\lim_{x \to 0} \frac{f(x)}{g(x)} = \lim_{x \to 0} \frac{x^3 - x^5}{5x^3 + x^4} = \lim_{x \to 0} \frac{1 - x^2}{5 + x} = \frac{1}{5}.$$

Example 3.22. The functions $f(x) = \sin x$ and $g(x) = x$ have the same order of smallness at $x = 0$.

Indeed,

$$\lim_{x \to 0} \frac{\sin x}{x} = 1.$$

Example 3.23. The functions $F(x) = \frac{2+x}{x}$ and $G(x) = \frac{1}{x}$ are the same infinite largeness order at $x = 0$:

$$\lim_{x \to 0} \frac{f(x)}{g(x)} = \lim_{x \to 0} (2 + x) = 2.$$

By analogy infinitely small and infinitely large functions are defined, when $x \to \pm\infty$.

3.8 Sequences and Series of Functions

Let $f_n : A \to \mathbb{R}$ $(n = 1, 2, 3, \ldots)$, be functions defined on the same set A. Then by $\{f_n(x)\}$, we denote the sequence of functions f_n. An infinite series of functions $u_k(x)$ $(k = 1, 2, \ldots)$, defined on A is denoted by $\sum\limits_{k=1}^{\infty} u_k(x)$.

Definition 3.21. We say that a *sequence of functions* $f_n : A \to \mathbb{R}$ *converges pointwise* on a set A to a function $f : A \to \mathbb{R}$ if for every fixed $x \in A$, $\lim\limits_{n \to \infty} f_n(x) = f(x)$, i.e., for every $\varepsilon > 0$, there exists an integer N depending on x and $\varepsilon > 0$, such that $|f_n(x) - f(x)| < \varepsilon$ for all $n > N$. This kind of convergence is denoted $f_n(x) \to f(x), x \in A$.

Definition 3.22. A *series of functions* $\sum\limits_{k=1}^{\infty} u_k(x)$ *converges pointwise* on A to a function $S : A \to \mathbb{R}$ if $S_n(x) \to S(x), x \in A$, where $\{S_n(x)\}$ denotes the sequence of partial sums.

Definition 3.23. If for any $\varepsilon > 0$ there exists an integer $N(\varepsilon)$ such that $|f_n(x) - f(x)| < \varepsilon$ for all $n > N(\varepsilon)$ and any $x \in A$, then we say that $\{f_n(x)\}$ *converges uniformly* to $f(x)$. In this case we write $f_n \to f, x \in A$.

Definition 3.24. If $S_n(x) \to S(x), x \in A$, then we say that the *series of functions* $\sum\limits_{k=1}^{\infty} u_k(x)$ *converges uniformly* on A and its sum is $S(x), x \in A$.

In Definition 3.21, the integer N depends on both $\varepsilon > 0$ and $x \in A$, i.e., $N = N(\varepsilon, x)$, while in Definition 3.23, the integer $N(\varepsilon)$ does not depend on x. Therefore, it is clear that if $f_n(x) \rightrightarrows f(x)$, then $f_n(x) \to f(x), x \in A$, and so uniform convergence implies pointwise convergence.

Example 3.24. Consider the sequence of functions $\{f_n(x)\} = \{x^n\}$ on $A = [0, 1)$. Obviously, $\lim\limits_{n\to\infty} f_n(x) = \lim\limits_{n\to\infty} x^n = 0$ for any fixed $x \in A$, that is, $f_n(x) \to 0$ converges pointwise. But this convergence is not uniform: for suppose that $f_n(x)$ did converge uniformly on A. Then for $\varepsilon = \dfrac{1}{2}$, there would exist an integer N such that for all $A = [0, 1)$, $|x^n - 0| < \dfrac{1}{2}$ whenever $n > N$. On the other hand, $|x^n| > \dfrac{1}{2}$ for points sufficiently close to 1. This contradiction implies that $\{f_n\}$ does not converge uniformly on A.

Example 3.25. The series $\sum\limits_{k=0}^{\infty} x^k$, $x \in (-1, 1)$ converges as a geometric series and its sum is $\dfrac{1}{1-x}$. But we will see that the series does not converge uniformly on $(-1, 1)$. For suppose the contrary, i.e., that $\dfrac{1 - x^{n+1}}{1 - x} = S_n \to S(x) = \dfrac{1}{1-x}$ for $x \in (-1, 1)$. Then for any $\varepsilon > 0$, there exists an $N = N(\varepsilon)$ such that $|S_n(x) - S(x)| = \dfrac{|x^{n+1}|}{1-x} < \varepsilon$ for all $n > N$ and $x \in (-1, 1)$. But, clearly, $\lim\limits_{x\to 1-0} \dfrac{|x^{n+1}|}{1-x} = +\infty$. This contradiction proves that the series does not converge uniformly on $(-1, 1)$.

Theorem 3.17 (Cauchy's Test). (1) *For the uniform convergence of a sequence* $\{f_n(x)\}$ *on A, it is necessary and sufficient that for every $\varepsilon > 0$ there exists an integer $N = N(\varepsilon)$ such that*

$$|f_n(x) - f_m(x)| < \varepsilon \tag{3.15}$$

for all $n, m > N(\varepsilon)$ and every $x \in A$.

(2) *For the uniform convergence of the series of functions* $\sum\limits_{k=1}^{\infty} u_k(x)$ *on a set A, it is necessary and sufficient that for any $\varepsilon > 0$, there is an integer $N = N(\varepsilon)$ such that*

$$\left| \sum_{k=n+1}^{n+p} u_k(x) \right| < \varepsilon$$

for all $n > N(\varepsilon)$, all $p \geqslant 1$, and all $x \in A$.

☐ It suffices to prove the theorem for a sequence of functions.

First, suppose $f_n(x) \to f(x)$ on A. By Definition 3.23, then, for any $\varepsilon > 0$ there exists an N such that

$$|f_n(x) - f(x)| < \frac{\varepsilon}{2} \text{ for all } n > N(\varepsilon) \text{ and } x \in A.$$

If $n, m > N = N(\varepsilon)$, then for every $x \in A$,

$$|f_n(x) - f_m(x)| \leqslant |f_n(x) - f(x)| + |f(x) - f_m(x)| < \frac{\varepsilon}{2} + \frac{\varepsilon}{2} = \varepsilon.$$

Thus, the necessity of the stated condition is proved.

In order to prove the sufficiency, let us suppose that, for any $\varepsilon > 0$, there is a number N such that the inequality (3.15) is satisfied for all $n, m > N$ and $x \in A$. Then, for every fixed x ($x \in A$), $\{f_n(x)\}$ is a Cauchy sequence, and so by Theorem 2.12, $f_n \to f$. By passing to the limit in (3.15) as $m \to \infty$, we obtain that for any $n > N(\varepsilon)$ and any $x \in A$ $|f_n(x) - f(x)| < \varepsilon$; that is, $f_n(x) \to f(x)$ on A. ∎

Definition 3.25. Let $\sum\limits_{k=1}^{\infty} u_k(x)$ be series of functions on A. If for all natural numbers k and every $x \in A$ the inequality $|u_k(x)| \leqslant \alpha_k$ holds and the numerical series $\sum\limits_{k=1}^{\infty} \alpha_k$ converges, then the series $\sum\limits_{k=1}^{\infty} \alpha_k$ is called a majorizing series for $\sum\limits_{k=1}^{\infty} u_k(x)$, and the series $\sum\limits_{k=1}^{\infty} u_k(x)$ is said to be *majorized*.

Theorem 3.18 (Weierstrass). *If the series of functions $\sum\limits_{k=1}^{\infty} u_k(x)$ is majorized on A, then it converges uniformly and absolutely on A.*

☐ If $\sum\limits_{k=1}^{\infty} \alpha_k$ is a majorizing series for the given series of functions $\sum\limits_{k=1}^{\infty} u_k(x)$, then by Cauchy's criterion, for any $\varepsilon > 0$ there exists a natural number $N = N(\varepsilon)$ such that for all $n > N$ and $p \geqslant 1$ we have $\sum\limits_{k=n+1}^{n+p} \alpha_k < \varepsilon$. Then for $n > N$, $p \geqslant 1$, and every $x \in A$,

$$\left| \sum_{k=n+1}^{n+p} u_k(x) \right| \leqslant \sum_{k=n+1}^{n+p} |u_k(x)| \leqslant \sum_{k=n+1}^{n+p} \alpha_k < \varepsilon.$$

By Theorem 3.17, the result follows from this inequality. ∎

Theorem 3.19. (1) *Let $f_n(x) \to f(x)$ on A. If a is a limit point of A and the limit $\lim\limits_{x \to a} f_n(x)$ exists, then*

$$\lim_{x \to a} \lim_{n \to \infty} f_n(x) = \lim_{n \to \infty} \lim_{x \to a} f_n(x). \tag{3.16}$$

(2) *Let $\sum\limits_{k=1}^{\infty} u_k(x)$ be uniformly convergent on A and suppose further that the limit $\lim\limits_{x \to a} u_k(x)$ exists for all $x \in A$. Then*

$$\lim_{x \to a} \sum_{k=1}^{\infty} u_k(x) = \sum_{k=1}^{\infty} \lim_{x \to a} u_k(x). \tag{3.17}$$

☐ It suffices to prove (1). Let $\lim\limits_{x \to a} f_n(x) = \alpha_n$. We show that $\{\alpha_n\}$ converges. Since $f_n(x) \to f(x)$ on A, then, by Theorem 3.17, for every $\varepsilon > 0$ there exists an integer $N = N(\varepsilon)$ such that for all $m, n > N$ and every $x \in A$, $|f_n(x) - f_m(x)| < \varepsilon$. Hence, by passing to the limit, we have $|\alpha_n - \alpha_m| < \varepsilon$. This implies that $\{\alpha_n\}$ is a Cauchy sequence.

Let us prove (3.16). Since $f_n(x) \to f(x)$ on A, for any $\varepsilon > 0$ there exists $N_1(\varepsilon)$ such that for all $m, n > N_1(\varepsilon)$ and every $x \in A$,

$$|f_n(x) - f(x)| < \frac{\varepsilon}{3}. \tag{3.18}$$

Let now $\lim_{n \to \infty} \alpha_n = \beta$. Then there is an $N_2(\varepsilon)$ such that for all $n > N_2(\varepsilon)$,

$$|\alpha_n - \beta| < \frac{\varepsilon}{3}. \tag{3.19}$$

Denote by N the greater of the two numbers $N_1(\varepsilon)$ and $N_2(\varepsilon)$. Clearly, for $n > N$, (3.18) and (3.19) are satisfied.

On the other hand, it follows from $\lim_{x \to a} f_n(x) = \alpha_n$ that there is a $\delta = \delta(\varepsilon) > 0$ such that for all $x \in A$ satisfying $|x - a| < \delta(\varepsilon)$,

$$|f_n(x) - \alpha_n| < \frac{\varepsilon}{3}. \tag{3.20}$$

Thus, for every $x \in A$ such that $|x - a| < \delta$, we have from (3.18)–(3.20) that

$$|f(x) - \beta| \leqslant |f(x) - f_n(x)| + |f_n(x) - \alpha_n| + |\alpha_n - \beta| < \varepsilon,$$

or

$$\lim_{x \to a} f(x) = \beta = \lim_{n \to \infty} \alpha_n.$$

∎

Theorem 3.20. (1) *If the sequence of continuous functions $\{f_n(x)\}$ converges to f uniformly on a closed and bounded interval $[a,b]$, i.e., $f_n \to f$, then $f(x)$ is continuous on $[a,b]$.*

(2) *If each term $u_k(x)$ of a uniformly convergent series $\sum_{k=1}^{\infty} u_k(x)$ is continuous on $[a,b]$, then its sum, $S(x)$, is continuous on $[a,b]$.*

☐ For simplicity, we carry out the proof of the theorem only for the case (1). Let x_0 be an arbitrary point of $[a,b]$. By Theorem 3.19, $\lim_{x \to x_0} f(x) = \lim_{n \to \infty} f_n(x_0) = f(x_0)$. Thus, f is continuous at every point $x_0 \in [a,b]$ and, consequently, f is continuous on $[a,b]$. ∎

Example 3.26. (a) By Theorem 3.18, the series of functions $\sum_{k=1}^{\infty} \frac{\cos kx}{k^2}$ converges absolutely and uniformly on $\mathbb{R}$ because $\left| \frac{\cos kx}{k^2} \right| \leqslant \frac{1}{k^2}$ and $\sum_{k=1}^{\infty} \frac{1}{k^2}$ converges.

(b) Show that $f_n(x) = \frac{\sin nx}{\sqrt{n}}$ converges uniformly on $\mathbb{R}$.

Note that the pointwise limit is $f(x) = \lim_{n \to \infty} \frac{\sin nx}{\sqrt{n}} = 0$, and $|f_n(x) - f(x)| = |f_n(x)| \leqslant \frac{1}{\sqrt{n}}$, for any $x \in \mathbb{R}$. Now, if for every $\varepsilon > 0$ we choose $N(\varepsilon) = \left[\left[\frac{1}{\varepsilon^2} \right]\right] + 1$, then $|f_n(x) - f(x)| < \varepsilon$ for all $n > N(\varepsilon)$.

Example 3.27. Let $\{f_n(x)\} = \{x^n\}$ $(n = 1, 2, 3, \ldots)$. It is clear that f_n is continuous on $[0, 1]$ for each n. Obviously, the limit function

$$f(x) = \lim_{n \to \infty} x^n = \begin{cases} 0, & \text{if } 0 \leqslant x < 1, \\ 1, & \text{if } x = 1 \end{cases}$$

is discontinuous and the convergence of $\{f_n(x)\}$ on $[0, 1]$ is not uniform. (See Example 3.24.)

Example 3.28. The series $\sum\limits_{k=1}^{\infty} \dfrac{x^k}{k}$ is absolutely convergent on $(-1, 1)$, because by D'Alembert's test, $\lim\limits_{n \to \infty} \left| \dfrac{u_{n+1}(x)}{u_n(x)} \right| = \lim\limits_{n \to \infty} \dfrac{n \cdot |x|}{n+1} = |x| < 1$. Let $[a, b] \subset (-1, 1)$. Since for every $x \in [a, b]$,

$$\left| \frac{x^k}{k} \right| \leqslant \frac{x_0^k}{k}, \quad x_0 = \max\{|a|, |b|\} < 1,$$

and $\sum\limits_{k=1}^{\infty} \dfrac{x_0^k}{k}$ converges, then, by Theorem 3.18, the series converges uniformly on $[a, b]$. Also, by Theorem 3.20, $S(x) = \sum\limits_{k=1}^{\infty} \dfrac{x^k}{k}$ is continuous on $(-1, 1)$.

Example 3.29. Consider the series $\sum\limits_{k=0}^{\infty} \dfrac{x^2}{(1+x^2)^n}$. In the present case, $u_k(x) = \dfrac{x^2}{(1+x^2)^k}$ and is differentiable. Let $S(x)$ be the sum of this series. Obviously, $u_k(0) = 0$ and its sum is 0. For $x \neq 0$, the sum of the geometric series

$$S(x) = x^2 \left(1 + \frac{1}{1+x^2} + \frac{1}{(1+x^2)^2} + \frac{1}{(1+x^2)^3} + \cdots \right) = x^2 \frac{1}{1 - \dfrac{1}{1+x^2}} = 1 + x^2.$$

Thus the series converges pointwise on the real line to

$$S(x) = \begin{cases} 0, & \text{if } x = 0, \\ 1 + x^2, & \text{if } x \neq 0, \end{cases}$$

which is discontinuous.

Example 3.30. Let $f_m(x) = \lim\limits_{n \to \infty} (\cos m! \pi x)^{2n}$, $m = 1, 2, 3, \ldots$. If $m! x$ is an integer, then $f_m(x) = 1$. For the all other numbers x, $f_m(x) = 0$.

Let now $f(x) = \lim\limits_{m \to \infty} f_m(x)$. If x is an irrational number, then $f_m(x) = 0 (m = 1, 2, 3, \ldots)$, and so $f(x) = 0$. If $x = \dfrac{p}{q}$ is a rational number, then for $m \geqslant q$, the number $m! x$ is an integer, and so $f(x) = 1$. Finally,

$$f(x) = \lim_{m \to \infty} f_m(x) = \lim_{m \to \infty} \lim_{n \to \infty} (\cos m! \pi x)^{2n} = \begin{cases} 0, & \text{if } x \text{ is irrational}, \\ 1, & \text{if } x \text{ is rational}. \end{cases}$$

Thus the limit function f is discontinuous everywhere.

Definition 3.26. If for a sequence of functions $f_n : [a,b] \to \mathbb{R}$ ($n = 1, 2, \ldots$) there exists a real number $C > 0$ such that $|f_n(x)| \leqslant C$ for each n and every $x \in [a,b]$, then we say that the sequence of functions $\{f_n(x)\}$ is *uniformly bounded* on $[a,b]$.

Definition 3.27. Let $f_n : [a,b] \to \mathbb{R}$ ($n = 1, 2, \ldots$). If for every $\varepsilon > 0$ there exists $\delta = \delta(\varepsilon) > 0$ such that

$$|x_1 - x_2| < \delta(\varepsilon) \text{ implies } |f_n(x_1) - f_n(x_2)| < \varepsilon$$

for all $x_1, x_2 \in [a,b]$ and all n, we say that $\{f_n(x)\}$ is *equicontinuous* on $[a,b]$.

Theorem 3.21. *If a sequence of continuous functions $\{f_n(x)\}$ converges uniformly, i.e., $f_n(x) \to f(x)$ on $[a,b]$, then the sequence is uniformly bounded and equicontinuous on $[a,b]$.*

$\square$ By the uniform continuity of $f_n(x) \to f(x)$, $x \in [a,b]$, for any $\varepsilon > 0$ there exists an integer $N = N(\varepsilon)$ such that

$$|f_n(x) - f(x)| < \frac{\varepsilon}{3} \tag{3.21}$$

for all $n > N(\varepsilon)$ and for all $x \in [a,b]$.

By Theorem 3.20, $f(x) = \lim\limits_{n \to \infty} f_n(x)$ is continuous and also, by Theorem 3.11,

$$|f(x)| \leqslant C_0.$$

It follows from (3.21) that

$$|f_n(x)| \leqslant |f(x)| + |f(x) - f_n(x)| < C_0 + \frac{\varepsilon}{3}. \tag{3.22}$$

It is also clear, by Theorem 3.11, that for arbitrarily fixed ε and n ($n > N(\varepsilon)$), the functions f_i, $i = 1, \ldots, n-1$, are bounded on $[a,b]$:

$$|f_i(x)| < C_i, \quad i = 1, \ldots, n-1. \tag{3.23}$$

Lef $C = \max\{C_1, C_2, \ldots, C_{n-1}, C_0 + \frac{\varepsilon}{3}\}$. Then for all $n = 1, 2, 3, \ldots$ and all $x \in [a,b]$, from (3.22) and (3.23) we have

$$|f_n(x)| \leqslant C.$$

Thus, $\{f_n(x)\}$ is uniformly bounded.

Let's prove the equicontinuity of $\{f_n(x)\}$. It is clear that

$$|f_n(x_1) - f_n(x_2)| \leqslant |f_n(x_1) - f(x_1)| + |f(x_1) - f(x_2)| + |f(x_2) - f_n(x_2)|. \tag{3.24}$$

Since $f(x)$ is continuous on $[a,b]$, it is uniformly continuous by Cantor's Theorem (Theorem 3.13). Hence there exists a $\delta_1 = \delta_1(\varepsilon) > 0$ such that for all $x_1, x_2 \in [a,b]$ satisfying $|x_1 - x_2| < \delta_1(\varepsilon)$, we have

$$|f(x_1) - f(x_2)| < \frac{\varepsilon}{3}. \tag{3.25}$$

Now, taking into account (3.21), (3.24), and (3.25), for all $n > N(\varepsilon)$ and $x_1, x_2 \in [a,b]$ such that $|x_1 - x_2| < \delta_1(\varepsilon)$, we have

$$|f_n(x_1) - f_n(x_2)| < \varepsilon. \tag{3.26}$$

Fix an $n > N(\varepsilon)$. By Cantor's Theorem, the functions $\{f_i\}$ $(i = 1, 2, \ldots, n-1)$ are uniformly continuous on $[a,b]$. This means that there exists a $\delta_2 = \delta_2(\varepsilon) > 0$ such that for all $x_1, x_2 \in [a,b]$ satisfying $|x_1 - x_2| < \delta_2(\varepsilon)$,

$$|f_i(x_1) - f_i(x_2)| < \varepsilon, \quad i = 1, \ldots, n-1. \tag{3.27}$$

Then, by choosing $\delta(\varepsilon) = \min\{\delta_1(\varepsilon), \delta_2(\varepsilon)\}$, it follows from (3.26) and (3.27) that for all $n = 1, 2, \ldots$ and $x_1, x_2 \in [a,b]$ such that $|x_1 - x_2| < \delta(\varepsilon)$, the inequality $|f_n(x_1) - f_n(x_2)| < \varepsilon$ holds. Consequently, $\{f_n(x)\}$ is equicontinuous. ∎

Theorem 3.22 (Arzela[3]). *Suppose the sequence $\{f_n\}$ is uniformly bounded and equicontinuous on $[a,b]$. Then it has a subsequence that converges uniformly on $[a,b]$.*

☐   The set of rational numbers in $[a,b]$ is countable (Theorem 1.4) and can be put in a sequence $\{x_n\}$. By hypothesis, $\{f_n(x)\}$ is bounded and so, by the Bolzano–Weierstrass theorem (Theorem 2.10), $\{f_n(x)\}$ has a subsequence $\{f_n^{(1)}(x)\}$ that converges at x_1. In turn, $\{f_n^{(1)}(x)\}$ has a subsequence $\{f_n^{(2)}(x)\}$ that converges at point x_2. Thus the subsequence $\{f_n^{(2)}(x)\}$ is convergent simultaneously at points x_1 and x_2. Continuing this process, we obtain a countable set of sequences $\{f_n^{(1)}(x)\}, \ldots, \{f_n^{(k)}(x)\}, \ldots$, such that:

(1) $\{f_n^{(k+1)}(x)\}$ is a subsequence of $\{f_n^{(k)}(x)\}$ $(k = 1, 2, \ldots)$ for each k,
(2) $\{f_n^{(k)}(x)\}$ converges at the points $x_1, x_2, \ldots, x_k$.

Consider the sequence $\{f_n^{(n)}(x)\}$. Since, by construction, $\{f_n^{(n)}(x)\} \subset \{f_n^{(k)}(x)\}$ when $n > k$, the sequence $\{f_n^{(n)}(x)\}$ converges at every point x_k. In other words, $\{f_n^{(n)}(x)\}$ $(n > k)$ is a subsequence of $\{f_n^{(k)}(x)\}$ and so converges at x_k since $\{f_n^{(k)}(x)\}$ does.

We will next use Theorem 3.17 to show that $\{f_n^{(n)}(x)\}$ converges uniformly on $[a,b]$. Let $\varepsilon > 0$ be any real number. By the equicontinuity of the sequence $\{f_n(x)\}$ on $[a,b]$,

---
[3]C. Arzela (1847–1912), Italian mathematician.

there exists a $\delta = \delta(\varepsilon)$ such that for all $x_1, x_2 \in [a,b]$ satisfying $|x_1 - x_2| < \delta$, and for all $n = 1, 2, 3, \ldots,$

$$|f_n(x_1) - f_n(x_2)| < \frac{\varepsilon}{3}. \tag{3.28}$$

Since $\{x_k\}$ contain all the rational numbers within $[a,b]$, a finite number of points $x_{i_1}, x_{i_2}, \ldots, x_{i_q}$ can be chosen such that for any $x \in [a,b]$, there exists a point (say x_j) satisfying the inequality $|x_j - x| < \delta(\varepsilon)$ $(j = 1, 2, 3, \ldots, q)$.

Moreover, $\{f_n^{(n)}(x)\}$ converges and so is a Cauchy sequence at every point x_{i_k}, $k = 1, 2, 3, \ldots, q$. Hence there exists an integer $N = N(\varepsilon)$ such that for all $m, n > N$ and $k = 1, 2, 3, \ldots, q$

$$\left| f_n^{(n)}(x_{i_k}) - f_m^{(m)}(x_{i_k}) \right| < \frac{\varepsilon}{3}. \tag{3.29}$$

We have the following inequality:

$$\left| f_n^{(n)}(x) - f_m^{(m)}(x) \right| \leqslant \left| f_n^{(n)}(x) - f_n^{(n)}(x_j) \right| + \left| f_n^{(n)}(x_j) - f_m^{(m)}(x_j) \right|$$
$$+ \left| f_m^{(m)}(x_j) - f_m^{(m)}(x) \right|. \tag{3.30}$$

Since $|x - x_j| < \delta(\varepsilon)$, the inequality (3.28) implies

$$\left| f_n^{(n)}(x) - f_n^{(n)}(x_j) \right| < \frac{\varepsilon}{3} \quad \text{and} \quad \left| f_m^{(m)}(x_j) - f_m^{(m)}(x) \right| < \frac{\varepsilon}{3}. \tag{3.31}$$

If $m, n > N(\varepsilon)$, then from (3.29) we have

$$\left| f_n^{(n)}(x_j) - f_m^{(m)}(x_j) \right| < \frac{\varepsilon}{3}.$$

Then, from (3.29)–(3.31), we obtain, for all $m, n > N(\varepsilon)$ and all $x \in [a,b]$,

$$\left| f_n^{(n)}(x) - f_m^{(m)}(x) \right| < \varepsilon.$$

Theorem 3.17 implies that the sequence $\{f_n^{(n)}(x)\}$ converges uniformly on $[a,b]$. ∎

The symbol $C[a,b]$ denotes the set of continuous functions on $[a,b]$. For functions $f_1 = f_1(x)$ and $f_2 = f_2(x)$, if $f_1, f_2 \in C[a,b]$, we will use the notation

$$\rho(f_1, f_2) = \max_{x \in [a,b]} |f_1(x) - f_2(x)|.$$

Then it can be easily seen that this is a metric, and so the set of functions $C[a,b]$ is a metric space.

Definition 3.28. Let $\{f_n\} \subset C[a,b]$ be a sequence of continuous functions. If the numerical sequences $\rho(f_n, f)$, $f \in C[a,b]$ converge, i.e., if $\rho(f_n, f) \to 0$ as $n \to \infty$, then we say that $\{f_n\}$ *converges to f in the metric ρ.*

It follows from the definition that for any $\varepsilon > 0$ there exists an integer $N = N(\varepsilon)$ such that for all $n > N(\varepsilon)$,

$$\max_{x \in [a,b]} |f_n(x) - f(x)| < \varepsilon,$$

i.e.,

$$|f_n(x) - f(x)| < \varepsilon \text{ for every } x \in [a,b].$$

Consequently, the convergence of $\{f_n\}$ as elements in the metric space $C[a,b]$ (in terms of the metric ρ) is equivalent to uniform convergence as functions on $[a,b]$.

Definition 3.29. A sequence $\{f_n\} \subset C[a,b]$ is said to be a *Cauchy sequence with respect to the metric ρ* if for any $\varepsilon > 0$ there exists a natural number $N(\varepsilon)$ such that $\rho(f_n, f_m) < \varepsilon$ for every $n, m > N(\varepsilon)$.

Since $\rho(f_n, f_m) < \varepsilon$ implies $|f_n(x) - f_m(x)| < \varepsilon$, the hypotheses of Theorem 3.17 are satisfied. Thus, $\{f_n\} \subset C[a,b]$ converges to f in the metric ρ if and only if it is a Cauchy sequence in this metric.

Definition 3.30. If there exists a real number $M > 0$ such that $\rho(0, f_n) \leqslant M$ for all $n = 1, 2, 3, \ldots$, then the sequence $\{f_n\} \subset C[a,b]$ is bounded in the metric space $C[a,b]$.

In other words, a bounded sequence in the metric space $C[a,b]$ is uniformly bounded as functions on $[a,b]$.

Example 3.31. Consider a sequence of continuous functions in $C[0,1]$:

$$f_n(x) = \begin{cases} 1, & 0 \leqslant x < \dfrac{1}{n+1}, \\ (n+1)(1-nx), & \dfrac{1}{n+1} \leqslant x \leqslant \dfrac{1}{n}, \\ 0, & \dfrac{1}{n} < x \leqslant 1. \end{cases}$$

This sequence is bounded in the metric, because $\rho(0, f_n) = 1$. But since $\rho(f_n, f_m) = 1$, $n \neq m$, the sequence $\{f_n\}$ is not a Cauchy sequence with respect to the metric. (Compare with the Corollary 2.1.)

Remark 3.7. Let $f_n(x) = u_n(x) + iv_n(x)$ $(n = 1, 2, \ldots)$, where $u_n(x)$ and $v_n(x)$ are real-valued functions. Then $\{f_n\}$ is a sequence of complex-valued functions. By using the absolute value of complex-valued functions, the concept of uniform convergence for the sequence $\{f_n\}$ can be formulated. For the convergence of $\{f_n\}$ to $f(x) = u(x) + iv(x)$ on a set A it is necessary and sufficient that $\{u_n\}$ and $\{v_n\}$ converge to u and v, respectively, on A. This can be seen from the following relations:

$$|f_n(x) - f(x)| = \sqrt{[u_n(x) - u(x)]^2 + [v_n(x) - v(x)]^2},$$

$$|u_n(x) - u(x)| \leqslant |f_n(x) - f(x)|, \text{ and } |v_n(x) - v(x)| \leqslant |f_n(x) - f(x)|.$$

Finally, the convergence of an infinite series of complex-valued functions reduces to the convergence of the sequence of partial sums. Furthermore, the analogues of Cauchy's criterion, of the comparison tests for convergence, and of Arzela's theorem, are all true.

3.9 Problems

In problems (1)–(7), describe the domains of the functions:

(1) $f(x) = \sqrt{\sin(\sqrt{x})}$ *Answer:* $4k^2\pi^2 \leqslant x \leqslant (2k+1)^2\pi^2$ $(k = 0, 1, 2, \ldots)$.

(2) $f(x) = \ln\left(\sin\dfrac{\pi}{x}\right)$.   *Answer:* $\dfrac{1}{2k+1} < x < \dfrac{1}{2k}$ ve

$$-\frac{1}{2k+1} < x < -\frac{1}{2k+2} \ (k = 0, 1, 2, \ldots).$$

(3) $f(x) = \dfrac{\sqrt{x}}{\sin \pi x}$ *Answer:* $x > 0$, $x \neq n$ $(n = 1, 2, \ldots)$.

(4) $f(x) = (2x)!$ *Answer:* $x = \dfrac{1}{2}, 1, \dfrac{3}{2}, 2, \ldots$

(5) $f(x) = \sqrt{3x - x^2}$ *Answer:* $-\infty < x \leqslant -\sqrt{3}$ and $0 \leqslant x \leqslant \sqrt{3}$.

(6) $f(x) = \cos^{-1}(2\sin x)$ *Answer:* $|x - k\pi| \leqslant \dfrac{\pi}{6}$ $(k = 0, \pm1, \pm2, \ldots)$.

(7) $f(x) = (x + |x|)\sqrt{x\sin^2 \pi x}$ *Answer:* $x = -1, -2, -3, \ldots$ and $x \geqslant 0$.

In each of the problems (8)–(14), describe the range of the given function.

(8) $f(x) = \sqrt{2 + x - x^2}$ *Answer:* $0 \leqslant f(x) \leqslant 1\dfrac{1}{2}$.

(9) $f(x) = (-1)^x$ *Answer:* $f(x) = \pm1$.

(10) $f(x) = \cos^{-1}\dfrac{2x}{1 + x^2}$ *Answer:* $0 \leqslant f(x) \leqslant \pi$.

(11) $f(x) = \ln(1 - 2\cos x)$ *Answer:* $-\infty < f(x) \leqslant \ln 3$.

(12) $f(x) = |x|$, $1 \leqslant |x| \leqslant 2$ *Answer:* $1 \leqslant f(x) \leqslant 2$.

(13) $f(x) = \dfrac{1}{\pi}\tan^{-1} x$, $-\infty < x < +\infty$ *Answer:* $1 \leqslant f(x) \leqslant 2$.

(14) $f(x) = x + [\![2x]\!]$ *Answer:* $0 < f(x) < \dfrac{1}{2}$ and $\dfrac{3}{2} \leqslant f(x) < 2$.

In problems (15) and (16), find the domain of the functions $f(u)$ defined on the interval $0 < u < 1$.

(15) $f(\sin x)$ *Answer:* $2k\pi < x < \pi + 2k\pi\ (k = 0, \pm 1, \pm 2, \ldots).$

(16) $f(\ln x)$ *Answer:* $1 < x < e.$

(17) Let $f(x) = \dfrac{1}{2}(a^x + a^{-x})\ (a > 0)$. Prove that $f(x+y) + f(x-y) = 2f(x)f(y).$

(18) Let $f_n(x) = \underbrace{f(f(\ldots f(x)))}_{n}.$ *Answer:* $f_n(x) = \dfrac{x}{\sqrt{1 + nx^2}}.$

   Find $f_n(x)$, if $f(x) = \dfrac{x}{\sqrt{1 + x^2}}$

In problems (19) and (20), determine whether the function is periodic or nonperiodic. If it
is periodic, find its period.

(19) $f(x) = \sin^2 x$ *Answer:* $T = \pi.$

(20) $f(x) = \sin x^2$ *Answer:* nonperiodic.

(21) Show that any rational number is a period of the Dirichlet function (Example 3.2).

In problems (22)–(24), show that the given function is increasing.

(22) $f(x) = 2x + \sin x,\ -\infty < x < +\infty.$

(23) $f(x) = \tan x,\ -\dfrac{\pi}{2} < x < \dfrac{\pi}{2}.$

(24) $f(x) = x^2,\ 0 \leqslant x < +\infty.$

(25) Suppose the functions $\varphi(x)$, $\psi(x)$ and $f(x)$ satisfying $\varphi(x) \leqslant f(x) \leqslant \psi(x)$ are
   increasing. Then prove that $\varphi(\varphi(x)) \leqslant f(f(x)) \leqslant \psi(\psi(x)).$

(26) Determine an inverse function of   *Answer:* $x = f(x) - k,$
   $f(x) = x + [\![x]\!]$ $2k \leqslant f(x) < 2k + 1\ (k = 0, \pm 1, \pm 2, \ldots).$

(27) When do the formulas $y = f(x)$ and   *Answer:* $f(f(x)) \equiv x.$
   $x = f^{-1}(y)$ determine the same function?

(28) Prove that $\tan^{-1} x + \tan^{-1} \dfrac{1}{x} = \dfrac{\pi}{2}\operatorname{sgn} x\ (x \neq 0)$ (see Example 3.3).

(29) Show that the range of the function $f(x) = 1 + \sin x\ (0 < x < 2\pi)$ is closed.

(30) Show that the inverse function of $f(x) = (1 + x^2)\operatorname{sgn} x$ is continuous even though the
   function itself is discontinuous.

In problems (31)–(41), find the limits.

(31) $\displaystyle\lim_{n\to\infty} \dfrac{1^2 + 2^2 + \cdots + n^2}{n^2}.$ *Answer:* $\dfrac{1}{3}.$

(32) $\displaystyle\lim_{x\to\infty} \dfrac{3x^2 - 2x - 1}{x^3 + 4}$ *Answer:* 0.

(33) $\displaystyle\lim_{h\to\infty} \dfrac{(x + h^3) - x^3}{h}.$ *Answer:* $3x^2.$

(34) $\displaystyle\lim_{x\to 4} \dfrac{\sqrt{x+1} - 3}{\sqrt{x-2} - \sqrt{2}}.$ *Answer:* $\dfrac{2\sqrt{2}}{3}.$

(35) $\lim\limits_{x\to 0} \dfrac{\sqrt{1+x}-1}{x}$.   *Answer:* $\dfrac{1}{2}$.

(36) $\lim\limits_{x\to 1} \dfrac{\sqrt[3]{x}-1}{\sqrt{x}-1}$.   *Answer:* $\dfrac{2}{3}$.

(37) $\lim\limits_{x\to \frac{\pi}{3}} \dfrac{1-2\cos x}{\sin\left(x-\frac{\pi}{3}\right)}$.   *Answer:* $\sqrt{3}$.

(38) $\lim\limits_{x\to \infty} \left(1-\dfrac{1}{x}\right)^x$.   *Answer:* $\dfrac{1}{e}$.

(39) $\lim\limits_{x\to \infty} \dfrac{a^x-1}{x}$ $(a>1)$.   *Answer:* $+\infty$.

(40) $\lim\limits_{x\to 0} x\cot x$.   *Answer:* 1.

(41) $\lim\limits_{x\to 0} \dfrac{\sin^2(x/3)}{x^2}$.   *Answer:* $\dfrac{1}{9}$.

In problems (42)–(46), in terms of "$\varepsilon - \delta$", prove the assertion made.

(42) $\lim\limits_{x\to 5} 3x = 15$.

(43) $\lim\limits_{x\to 0} \dfrac{x^2+x}{x} = 1$.

(44) $\lim\limits_{x\to 9} \sqrt{x} = 3$.

(45) $\lim\limits_{x\to 2} (x^2-3) = 1$.

(46) $\lim\limits_{x\to 4} \dfrac{x^2-9}{x+3} = 1$.

In problems (47)–(54), determine whether the function is continuous or discontinuous.

(47) $f(x) = \begin{cases} x\sin\dfrac{1}{x}, & x\neq 0, \\ 0, & x=0 \end{cases}$ *Answer:* Continuous.

(48) $f(x) = \sqrt{x} - [\![\sqrt{x}]\!]$.   *Answer:* At $x = k^2$ $(k=1,2,\ldots)$ there are discontinuities.

(49) $f(x) = \begin{cases} \sin\dfrac{1}{x}, & x\neq 0, \\ k, & x=0, \\ \text{for a } k \text{ constant.} \end{cases}$ *Answer:* Discontinuous at $x = 0$.

(50) $f(x) = x[\![x]\!]$.   *Answer:* Discontinuities of the first kind at $x = k$ $(k = \pm 1, \pm 2, \ldots)$.

(51) $f(x) = e^{-\frac{1}{x}}$.   *Answer:* At $x = 0$, a discontinuity of the second kind.

(52) $f(x) = \sqrt{x}\tan^{-1}\dfrac{1}{x}$.   *Answer:* At $x = 0$, removable discontinuity.

(53) $f(x) = x^2 - [\![x^2]\!]$.

Answer: Discontinuities of the first kind at $x = \pm\sqrt{n}$ $(n = 1, 2, \ldots)$.

(54) $f(x) = \tan^{-1}\dfrac{1}{x}$.

Answer: A discontinuity of the first kind at $x = 0$.

(55) Determine whether the functions $f(g(x))$ and $g(f(x))$ are continuous or discontinuous, where $f(x) = \operatorname{sgn} x$ and $g(x) = 1 + x^2$.

Answer: $f(g(x))$ is continuous, $g(f(x))$ is discontinuous at $x = 0$.

(56) For $f(x)$ and $g(x)$ continuous functions, prove that $\psi(x) = \min[f(x), g(x)]$ and $\psi(x) = \max[f(x), g(x)]$ are continuous.

(57) Show that $o(o(f(x))) = o(f(x))$.

(58) Let $x \to 0$. Prove that $\ln x = o\left(\dfrac{1}{x^\varepsilon}\right)$ $(\varepsilon > 0)$.

(59) Prove that the infinitely large functions $\sqrt{x + \sqrt{x + \sqrt{x}}}$ and $\sqrt{x}$ have the same order as x goes to infinity.

In problems (60)–(64), determine whether the functions are uniformly continuous or not.

(60) $f(x) = \dfrac{x}{4 - x^2}$, $-1 \leqslant x \leqslant 1$.

Answer: Uniformly continuous.

(61) $f(x) = \ln x$, $0 < x < 1$.

Answer: Is not uniformly continuous.

(62) $f(x) = \dfrac{\sin x}{x}$, $0 < x < \pi$.

Answer: Uniformly continuous.

(63) $f(x) = \tan^{-1} x$, $-\infty < x < +\infty$.

Answer: Uniformly continuous.

(64) $f(x) = e^x \cos\dfrac{1}{x}$, $0 < x < 1$.

Answer: Is not uniformly continuous.

In problems (65)–(70), determine whether the functions are uniformly convergent or not.

(65) $f_n(x) = x^n$, $0 \leqslant x \leqslant \dfrac{1}{2}$.

Answer: Uniformly convergent.

(66) $f_n(x) = x^n - x^{n+1}$, $0 \leqslant x \leqslant 1$.

Answer: Uniformly convergent.

(67) $f_n(x) = \dfrac{1}{x + n}$, $0 \leqslant x \leqslant +\infty$.

Answer: Uniformly convergent.

(68) $f_n(x) = \dfrac{\sin nx}{n}$, $-\infty < x < +\infty$.

Answer: Uniformly convergent.

(69) $f_n(x) = \sin\dfrac{x}{n}$, $-\infty < x < +\infty$.

Answer: Not uniformly convergent.

(70) $\sum\limits_{n=1}^{\infty}\left(\dfrac{x^n}{n} - \dfrac{x^{n+1}}{n+1}\right)$, $-1 \leqslant x \leqslant 1$.

Answer: Uniformly convergent.

In problems (71)–(74) use the Weierstrass test to determine whether the convergence of series is uniform or not.

(71) $\sum\limits_{n=1}^{\infty} \left(\dfrac{1}{x^2+n^2} \right)$, $-\infty \leqslant x \leqslant +\infty$.

(72) $\sum\limits_{n=1}^{\infty} \left(\dfrac{\cos nx}{n^2} \right)$, $|x| < +\infty$.

(73) $\sum\limits_{n=1}^{\infty} \left(\dfrac{(-1)^n}{x+2^n} \right)$, $-2 < x < +\infty$.

(74) $\sum\limits_{n=1}^{\infty} x^2 e^{-nx}$, $0 \leqslant x < +\infty$.

Chapter 4

Differential Calculus

In this chapter, important conceptions of mathematical analysis such as derivatives and differentials, and their mechanical and geometrical meanings, are given. Then the concept of the derivative is applied to study the free fall of a mass, linear motion at velocities near the speed of light, and applications in numerical analysis. There are important formulas for derivatives, such as Leibniz's formula for the derivatives of products of functions. The concept of the derivative is also applied to curvature, conics, and evolvents. These topics are treated both in Cartesian and in polar coordinates. We apply these techniques to geometric objects given explicitly or implicitly, or by parametric equations in either coordinate system.

4.1 Derivatives: Their Definition and Their Meaning for Geometry and Mechanics

Let $f : A \to \mathbb{R}$ be a function defined on $A \subseteq R$. For a given $x \in A$, taking $x + \Delta x \in A$, we denote

$$\Delta y = f(x + \Delta x) - f(x), \qquad (4.1)$$

where Δx is an increment of the independent variable and $\Delta y = \Delta f$ $(y = f(x))$ is the resulting increment defined by (4.1) and corresponding to Δx. For example, let $f(x) = \sin x$. Then

$$\Delta f = \sin(x + \Delta x) - \sin x = 2 \cos\left(x + \frac{\Delta x}{2}\right) \cdot \sin \frac{\Delta x}{2}. \qquad (4.2)$$

It follows from the definition of continuity that at a point x the given function f is continuous if and only if

$$\lim_{\Delta x \to 0} \Delta y = \lim_{\Delta x \to 0} [f(x + \Delta x) - f(x)] = 0. \qquad (4.3)$$

Since $\lim_{\Delta x \to 0} \sin \frac{\Delta x}{2} = 0$, we have, from (4.2), that $\lim_{\Delta x \to 0} \Delta y = 0$.

Definition 4.1. The derivative of the function f at a point x $(x \in A)$ is the following limit (if exists) as $\Delta x \to 0$ and denoted $f'(x)$ (Lagrange form) or $\frac{df}{dx}$ (Leibniz form):

$$f'(x) = \frac{df(x)}{dx} = \lim_{\Delta x \to 0} \frac{f(x + \Delta x) - f(x)}{\Delta x}.$$

E. Mahmudov, *Single Variable Differential and Integral Calculus*,
DOI: 10.2991/978-94-91216-86-2_4, © Atlantis Press 2013

If the function f has a derivative at every point $x \in A$, then we say that this function is differentiable on A. The process of finding a derivative is called differentiation. Clearly, f' is itself a function of x.

Definition 4.2. The function f has a derivative from the right (left) at x if there exists the following limit from the right (left):

$$\lim_{\Delta x \to 0+0} \frac{f(x+\Delta x) - f(x)}{\Delta x} \qquad \left(\lim_{\Delta x \to 0-0} \frac{f(x+\Delta x) - f(x)}{\Delta x} \right)$$

The derivatives from the right and left at a point x are denoted by $f'(x+0)$ and $f'(x-0)$, respectively. By the properties of one-sided limits, if at x both one-sided derivatives $f'(x+0)$, $f'(x-0)$ exist and are equal, then the derivative f' at x exists:

$$f'(x) = f'(x+0) = f'(x-0).$$

Example 4.1. Find the derivative of $f(x) = ax + b$. By Definition 4.1,

$$f'(x) = \lim_{\Delta x \to 0} \frac{f(x+\Delta x) - f(x)}{\Delta x} = \lim_{\Delta x \to 0} \frac{a(x+\Delta x) + b - (ax+b)}{\Delta x} = a.$$

Consequently,

$$f'(x) = a.$$

In particular, if $f(x) = b$ $(a = 0)$, then $f'(x) = 0$.

Example 4.2. Find the derivative of $f(x) = |x|$ at $x = 0$. By Definition 4.2,

$$f'(0+0) = \lim_{\Delta x \to 0+0} \frac{f(0+\Delta x) - f(0)}{\Delta x} = \lim_{\Delta x \to 0+0} \frac{\Delta x}{\Delta x} = 1,$$

$$f'(0-0) = \lim_{\Delta x \to 0-0} \frac{f(0+\Delta x) - f(0)}{\Delta x} = \lim_{\Delta x \to 0-0} \frac{-\Delta x}{\Delta x} = -1,$$

and so $f'(0+0) \neq f'(0-0)$. This means that the derivative $f'(0)$ does not exist.

Let a particle move along a straight line with position at time t given by the function $s = f(t)$. The particle moves from $f(t)$ to $f(t+\Delta t)$ in the interval of time $[t, t+\Delta t]$. Its displacement is then the increment

$$\Delta s = f(t+\Delta t) - f(t).$$

The average velocity of the particle during the time interval $[t, t+\Delta t]$ is

$$\frac{\Delta s}{\Delta t} = \frac{f(t+\Delta t) - f(t)}{\Delta t}.$$

We define the instantaneous velocity of the particle at the time t to be the limit of this average velocity as $\Delta t \to 0$:

$$f'(t) = \lim_{\Delta t \to 0} \frac{f(t+\Delta t) - f(t)}{\Delta t}.$$

Thus, velocity is the instantaneous rate of change of position.

Example 4.3. The case of vertical motion under the influence of gravity is of special interest. By the mechanics of Galilei[1], the particle's height s above the ground at time t is $s = \dfrac{gt^2}{2}$ during free fall. Here g denotes the acceleration due to gravity. Near the surface of the earth $g \approx -9.8 \text{m} \cdot \text{s}^{-2}$. We find the velocity v of the particle at time t,

$$\Delta s = \frac{g(t^2 + 2t\Delta t + \Delta t^2)}{2} - \frac{gt^2}{2} = gt\Delta t + g\frac{\Delta t^2}{2}.$$

Then we have

$$v = \lim_{\Delta t \to 0} \frac{\Delta s}{\Delta t} = \lim_{\Delta t \to 0} \left(gt + \frac{1}{2}g\Delta t \right) = gt.$$

We next find the geometric meaning of the derivative $f'(x)$ at a point x. First we write the equation of the line which passes through the two points $(x, f(x))$ and $(x + \Delta x, f(x + \Delta x))$,

$$Y = f(x) + \frac{f(x + \Delta x) - f(x)}{\Delta x}(X - x),$$

where (X, Y) denotes the points of line. Passing to the limit as $\Delta x \to 0$ gives

$$Y = f(x) + f'(x)(X - x). \tag{4.4}$$

Thus, the derivative has an important geometric interpretation: $f'(x) = \tan \alpha$ is the slope of the line tangent to the curve $y = f(x)$ at the point $(x, f(x))$, where α is its angle of inclination.

Remark 4.1. If x is a point of continuity of $y = f(x)$ and $f'(x) = \pm\infty$, then it follows from (4.4) that $X = x$. Indeed, if we rewrite the previous equation as follows,

$$X = x + \frac{\Delta x}{f(x + \Delta x) - f(x)}(Y - f(x)),$$

then by passing to the limit as $\Delta x \to 0$, we have $X = x$.

4.2   Differentials

Definition 4.3. Let f be a function on (a, b) and let Δx be an increment of the independent variable x with $x + \Delta x \in (a, b)$. Then f is said to be differentiable at x if there exists a number A such that the increment Δy of f can be represented as

$$\Delta y = A\Delta x + o(\Delta x). \tag{4.5}$$

[1]G. Galilei (1564–1642), Italian mathematician and physicist.

Here, the number A does not depend on Δx and $o(\Delta x)$ has a higher order of smallness than Δx.

Theorem 4.1. *In order that a function f be differentiable at the point x it is necessary and sufficient that there exists a finite derivative $f'(x)$.*

☐ Suppose that f is differentiable at a point x. Let Δy be the increment of f corresponding to $\Delta x \neq 0$ and suppose that (4.5) is true. Dividing both sides of (4.5) by Δx and passing to the limit as $\Delta x \to 0$, we have

$$\lim_{\Delta x \to 0} \frac{\Delta y}{\Delta x} = f'(x) = A.$$

Conversely, suppose $f'(x)$ exists, that is, $\lim\limits_{\Delta x \to 0} \dfrac{\Delta y}{\Delta x} = f'(x)$. Let $o(\Delta x)$ be given by

$$o(\Delta x) = \Delta y - f'(x)\Delta x. \tag{4.6}$$

Obviously, it follows from (4.6) that $o(\Delta x)$ has a higher order of smallness than Δx. Then (4.6) and (4.5) agree provided that $A = f'(x)$. ∎

Theorem 4.2. *If f has a differential at x, then f is continuous at x.*

☐ In fact, the existence of the differential of f at x implies the representation (4.5). And hence that $\lim\limits_{\Delta x \to 0} \Delta y = 0$, that is, that f is continuous at x. ∎

Remark 4.2. The converse assertion is not true, that is, the continuity of a function at some point x need not imply its differentiability at that point. For example, the function $f(x) = |x|$ is continuous everywhere (including $x = 0$). But, as we have seen (Example 4.2), this function has no derivative at $x = 0$.

On the other hand one of the important historical achievments of Mathematical Analysis was to investigate the Weierstrass function. This function is an everywhere continuous but nowhere differentiable function and is defined as follows:

$$f(x) = \sum_{n=0}^{\infty} b^n \cos(a^n \pi x).$$

Here a is an odd integer satisfying $ab \geq 1 + \dfrac{3\pi}{2}$, where $0 < b < 1$.

At once we see that by using Theorems 3.18 and 3.20, this series is majorized by the convergent series $\sum\limits_{n=0}^{\infty} b^n$, and so $f(x)$ is continuous at every point $x \in \mathbb{R}$.

Example 4.4. Let $f(x) = \begin{cases} x\sin\dfrac{1}{x}, & \text{if } x \neq 0, \\ 0, & \text{if } x = 0. \end{cases}$

This function is continuous everywhere. We show that this function is not differentiable at $x = 0$. By the definition of derivative, we have that

$$f'(0) = \lim_{\Delta x \to 0} \frac{f(0+\Delta x) - f(0)}{\Delta x} = \lim_{\Delta x \to 0}\left[\sin\frac{1}{\Delta x}\right].$$

But this limit does not exist, because if k is positive and odd, and if $\Delta x = \dfrac{2}{\pi k}$, then the values of $\sin\dfrac{1}{\Delta x}$ are equal to -1 and $+1$.

Let us return to Definition 4.3 of the differentiability. By (4.5), the increment, Δy, of a differentiable function f at a point x consists of two summands. The first is its linear homogeneous part $A\Delta x = f'(x)\Delta x$ (linear in Δx). The second part has a higher order of smallness than Δx. In Definition 4.1, we have been viewing the expression $\dfrac{df}{dx}$ as a single symbol for the derivative. We shall now show how to interpret the symbols df and dx so that $\dfrac{df}{dx}$ can be regarded as the ratio of df and dx.

Definition 4.4. The first summand in the right hand side of (4.5), $f'(x)\Delta x$, is called the *differential* of f at x and is denoted by

$$dy = f'(x)\Delta x. \tag{4.7}$$

If $f'(x) = 0$ then the differential of the function is zero. The distinction between the increment Δy and the differential dy for the same Δx is pictured in Figure 4.1. Think of x as fixed. Then (4.7) shows that the differential dy is a linear function of the increment Δx.

A special case is if $f(x) = x$: then $df = dx = 1 \cdot \Delta x = \Delta x$, that is, the differential of the independent variable is equal to its increment. Then (4.7) can be rewritten as

$$dy = f'(x)dx$$

and $f'(x) = \dfrac{dy}{dx}$. Whenever $f'(x) \neq 0$, we have

$$\lim_{\Delta x \to 0} \frac{\Delta y}{dy} = 1 + \lim_{\Delta x \to 0} \frac{o(\Delta x)}{f'(x)\Delta x} = 1.$$

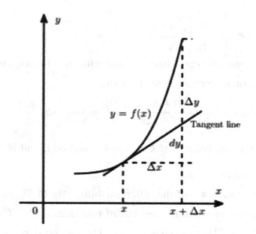

Fig. 4.1   The differential and increment of f with respect to Δx.

This observation motivates the tentative approximation $\Delta y \approx dy$ or

$$f(x+\Delta x) \approx f(x) + f'(x)\Delta x. \tag{4.8}$$

For this reason, dy is called the linear approximation to the exact increment Δy.

Example 4.5. Use the linear approximation formula, (4.8), to estimate $\sqrt{1.1}$.
Let $f(x) = \sqrt{x}$. Then we must compute $f(1.1)$.
By the definition of the derivative,

$$f'(x) = \lim_{\Delta x \to 0} \frac{\Delta y}{\Delta x} = \lim_{\Delta x \to 0} \frac{\sqrt{x+\Delta x} - \sqrt{x}}{\Delta x} = \lim_{\Delta x \to 0} \frac{\Delta x}{\Delta x(\sqrt{x+\Delta x} + \sqrt{x})} = \frac{1}{2\sqrt{x}}.$$

Setting $x = 1, x+\Delta x = 1,1$ it follows from (4.8) that

$$f(1.1) \approx f(1) + f'(1)(0.1)$$

or

$$\sqrt{1.1} \approx \sqrt{1} + \frac{1}{2\sqrt{1}}(0.1) = 1.05$$

Consequently, $\sqrt{1.1} \approx 1.05$.

   More generally, if $x = 1, \Delta x = \alpha$ then for $f(x) = \sqrt{x}$, Formula (4.8) gives us that

$$\sqrt{1+\alpha} \approx 1 + \frac{1}{2}\alpha.$$

Example 4.6. We can compute df and Δf for $f(x) = x^2$ if $x = 10$ and $\Delta x = 0.1$.

$$f'(x) = \lim_{\Delta x \to 0} \frac{\Delta f}{\Delta x} = \lim_{\Delta x \to 0} \frac{(x+\Delta x)^2 - x^2}{\Delta x} = \lim_{\Delta x \to 0} (2x + \Delta x) = 2x,$$

$$df = f'(x)\Delta x = 2x\Delta x,$$

and

$$\Delta f = (x+\Delta x)^2 - x^2 = 2x\Delta x + \Delta x^2.$$

Taking $x = 10$ and $\Delta x = 0.1$, we have $df = 2 \cdot 10 \cdot (0.1) = 2$ and $\Delta y = 2 \cdot 10 \cdot (0.1) + (0.1)^2 = 2.01$.

Example 4.7. Let $f(x) = \sin x$. Then by using (4.2), we can write

$$f'(x) = \lim_{\Delta x \to 0} \frac{2\cos\left(x + \dfrac{\Delta x}{2}\right)\sin\dfrac{\Delta x}{2}}{\Delta x} = \lim_{\Delta x \to 0} \cos\left(x + \frac{\Delta x}{2}\right) \cdot \lim_{\Delta x \to 0} \frac{\sin\dfrac{\Delta x}{2}}{\dfrac{\Delta x}{2}} = \cos x.$$

On the other hand, by the linear approximation formula (4.8),

$$\sin(x + \Delta x) \approx \sin x + \cos x \Delta x. \tag{4.9}$$

Let's estimate the value of $\sin 46°$. Since $x + \Delta x = \dfrac{\pi}{4} + \dfrac{\pi}{180}$ $(1° = \dfrac{\pi}{180}$ radians) and $x = \dfrac{\pi}{4}$, by (4.9),

$$\sin 46° = \sin\left(\frac{\pi}{4} + \frac{\pi}{180}\right) \approx \sin\frac{\pi}{4} + \frac{\pi}{180}\cos\frac{\pi}{4},$$

or

$$\sin 46° \approx \frac{\sqrt{2}}{2} + \frac{\sqrt{2}}{2}\frac{\pi}{180} = 0.7071 + 0.7071 \cdot 0.0175 = 0.7194.$$

4.3   Derivatives of Composite and Inverse Functions

Theorem 4.3 (The Chain rule). *Suppose that φ is differentiable at t and f is differentiable at $x = \varphi(t)$. Then the composition $f(\varphi)$ is differentiable at t and its derivative is*

$$[f(\varphi(t))]' = f'(x) \cdot \varphi'(t) = f'(\varphi(t)) \cdot \varphi'(t). \tag{4.10}$$

□ Let $\Delta t \neq 0$ be the increment of the independent variable at t. Then $\Delta x = \varphi(t + \Delta t) - \varphi(t)$ is the corresponding increment of the function φ at t. The increment of f corresponding to Δx is, in turn, $\Delta y = f(x + \Delta x) - f(x)$. On the other hand, by Theorem 4.1,

$$\Delta y = f'(x)\Delta x + o(\Delta x). \tag{4.11}$$

and $o(\Delta x)$ has a higher order of smallness than Δx. Obviously, (4.11) is valid for $\Delta x = 0$.

Dividing both sides of (4.11) by $\Delta t \neq 0$, we have

$$\frac{\Delta y}{\Delta t} = f'(x)\frac{\Delta x}{\Delta t} + \frac{o(\Delta x)}{\Delta x}\frac{\Delta x}{\Delta t}. \tag{4.12}$$

where $\Delta x \neq 0$. By Theorem 4.2, the function φ is continuous at t and $\lim\limits_{\Delta t \to 0} \Delta x = 0$.

Furthermore, the following limits exist:

$$\lim_{\Delta t \to 0} \frac{\Delta x}{\Delta t} = \varphi'(t), \quad \lim_{\Delta x \to 0} \frac{o(\Delta x)}{\Delta x} = 0.$$

Then by passing to the limit in (4.12) as $\Delta t \to 0$, we see that the $\lim\limits_{\Delta t \to 0} \frac{\Delta y}{\Delta t} = [f(\varphi(t))]'$ exists, and the required formula, (4.10), is true. ∎

Remark 4.3. It is not hard to see that by adding a new differentiable function F at the point $y = f(x)$, by analogy we can write:

$$\left[F\left(f(\varphi(t))\right)\right]' = F'\left(f(\varphi(t))\right) \cdot f'\left(\varphi(t)\right) \cdot \varphi'(t).$$

Example 4.8. Compute the derivatives of the composite functions $y = \sin\sqrt{t}$ and $x = \sin^2 t$. For the first example, let $f(x) = \sin x$, $x = \varphi(t) = \sqrt{t}$. By Formula (4.10),

$$(\sin\sqrt{t})' = (\sin x)' \cdot (\sqrt{t})' = \frac{\cos x}{2\sqrt{t}} = \frac{\cos\sqrt{t}}{2\sqrt{t}} \quad \text{(see Examples 4.5 and 4.7)}.$$

For the second composite, let $y = f(x) = x^2$ and $x = \varphi(t) = \sin t$.

Then $[\sin^2 t]' = (x^2)' \cdot (\sin t)' = 2x \cdot \cos t = 2\sin t \cos t = \sin 2t$ (see Examples 4.6 and 4.7.).

Remark 4.4. If f is a function of the independent variable x, then for a differentiable function at x,

$$dy = f'(x)dx. \tag{4.13}$$

It can easily be seen that if the function φ is differentiable at t, then the differential of the composite function $y = f(\varphi)$ has the form (4.13). Indeed, for the function $f(\varphi)$,

$$dy = [f(\varphi(t))]'dt. \tag{4.14}$$

But by Theorem 4.3, (4.14) has the form $dy = f'(x)\varphi'(t)dt$, and so, taking into account $dx = \varphi'(t)dt$, we see that $dy = f'(x)dx$. This property is called the invariance of first order differentials.

Theorem 4.4. *If the function f is an increasing (decreasing) monotone function and continuous in a neighborhood of x and if f is differentiable at x, with $f'(x) \neq 0$, then there exists a differentiable inverse function f^{-1} defined in a neighborhood of $y = f(x)$ such that*

$$\left[f^{-1}(y)\right]' = \frac{1}{f'(x)}. \tag{4.15}$$

☐   By Theorem 3.16, there exists a strictly monotone inverse function f^{-1} defined in a neighborhood of $y = f(x)$. Then for sufficiently small $\Delta y \neq 0$, there exists a nonzero $\Delta x = f^{-1}(y + \Delta y) - f^{-1}(y)$ and

$$\frac{\Delta x}{\Delta y} = \frac{1}{\dfrac{\Delta y}{\Delta x}}. \tag{4.16}$$

Since the function f^{-1} is continuous, then $\Delta y \to 0$ implies $\Delta x \to 0$. It can be seen easily that in the right hand side of (4.16), the limit exists as $\Delta x \to 0$. Since $x = f^{-1}(y)$, it follows from the formula for Δx that $x + \Delta x = f^{-1}(y + \Delta y)$, or, $y + \Delta y = f(x + \Delta x)$. Then $\Delta y = f(x + \Delta x) - y = f(x + \Delta x) - f(x)$, and by hypothesis, the limit of the quotient

$$\frac{\Delta y}{\Delta x} = \frac{f(x + \Delta x) - f(x)}{\Delta x}$$

exists as $\Delta x \to 0$. Then, since $f'(x) \neq 0$, the left hand side of (4.16) has the limit $\dfrac{1}{f'(x)}$.   ∎

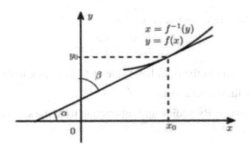

Fig. 4.2   The geometric interpretation of Formula (4.15): $\tan \alpha \cdot \tan \beta = 1$.

Remark 4.5. The geometric interpretation of Formula (4.15) is shown in Figure 4.2. If $f'(x_0) = \tan \alpha$ is the slope of the tangent line to the curve $y = f(x)$ at (x_0, y_0), then since $y = f(x)$ and $x = f^{-1}(y)$ have the same curve, we can write $[f^{-1}(y)]' = \tan \beta$. In this case $\alpha + \beta = \dfrac{\pi}{2}$ and it follows from (4.15) that $\tan \alpha \cdot \tan \beta = 1$.

Example 4.9. If $y = f(x) = \sqrt{x}$ $(x > 0)$, use (4.15) to compute the derivative of the inverse function. The inverse function is $x = f^{-1}(y) = y^2$ $(y < 0)$. Then, by (4.15),

$$[y^2]' = \frac{1}{(\sqrt{x})'} = 2\sqrt{x} = 2y.$$

4.4   Derivatives of Sums, Products, and Quotients

Theorem 4.5. *Let u and v be functions, differentiable at x. Then their sum, product, and quotient are differentiable at x, and*

$$[u+v]'(x) = u'(x) + v'(x),$$
$$(u \cdot v)'(x) = u'(x) \cdot v(x) + u(x) \cdot v'(x), \quad and \qquad (4.17)$$
$$\left(\frac{u}{v}\right)'(x) = \frac{u'(x)v(x) - u(x)v'(x)}{v^2(x)},$$

provided in the last case that $v(x) \neq 0$.

☐ (1) Let $y(x) = (u+v)(x)$ and Δu, Δv, Δy be the increments at x corresponding to Δx, of the functions u, v, y, respectively. Then we can write

$$\Delta y = y(x+\Delta x) - y(x) = [u(x+\Delta x) + v(x+\Delta x] - [u(x) + v(x)]$$
$$= [u(x+\Delta x) - u(x)] + [v(x+\Delta x) - v(x)] = \Delta u + \Delta v.$$

Dividing both sides of this relation by Δx, we have

$$\frac{\Delta y}{\Delta x} = \frac{\Delta u}{\Delta x} + \frac{\Delta v}{\Delta x}. \qquad (4.18)$$

Moreover, since the sum in the right hand side of (4.18) has a limit as $\Delta x \to 0$, the left hand side has the same limit, and so $y'(x) = u'(x) + v'(x)$.

(2) Let $y(x) = u(x) \cdot v(x)$. By adding and subtracting to Δy the same quantity, $u(x+\Delta x) \cdot v(x)$, we obtain

$$\Delta y = y(x+\Delta x) - y(x) = u(x+\Delta x) \cdot v(x+\Delta x) - u(x) \cdot v(x)$$
$$= [u(x+\Delta x) \cdot v(x+\Delta x) - u(x+\Delta x) \cdot v(x)] + [u(x+\Delta x) \cdot v(x) - u(x) \cdot v(x)]$$
$$= u(x+\Delta x)[v(x+\Delta x) - v(x)] + v(x)[u(x+\Delta x) - u(x)]$$
$$= u(x+\Delta x)\Delta v + v(x)\Delta u.$$

Again, dividing both sides of this equation by Δx and passing to the limit as $\Delta x \to 0$, we have

$$\lim_{\Delta x \to 0} \frac{\Delta y}{\Delta x} = \lim_{\Delta x \to 0} \left[u(x+\Delta x)\frac{\Delta v}{\Delta x} + v(x)\frac{\Delta y}{\Delta x} \right]. \qquad (4.19)$$

By Theorem 4.2, $\lim_{\Delta x \to 0} u(x+\Delta x) = u(x)$, and so the right hand side of (4.19) exists. Then the indicated limit on the left hand side exists, and so the second line of (4.17) is proved.

(3) Let now $y(x) = \dfrac{u(x)}{v(x)}$. The value of v is nonzero at x and so for sufficiently small $\Delta x \neq 0$, we have $v(x + \Delta x) \neq 0$ (see Theorem 3.8). Then

$$\Delta y = y(x + \Delta x) - y(x) = \frac{u(x + \Delta x)}{v(x + \Delta x)} - \frac{u(x)}{v(x)}$$

$$= \frac{[u(x + \Delta x)v(x) - u(x)v(x)] - [v(x + \Delta x)u(x) - u(x)v(x)]}{v(x)v(x + \Delta x)}$$

$$= \frac{v(x)[u(x + \Delta x) - u(x)] - u(x)[v(x + \Delta x) - v(x)]}{v(x)v(x + \Delta x)}.$$

Expressing $\Delta u = u(x + \Delta x) - u(x)$, $\Delta v = v(x + \Delta x) - v(x)$, and dividing both sides of this equality by Δx, we have

$$\frac{\Delta y}{\Delta x} = \frac{v(x)\dfrac{\Delta u}{\Delta x} - u(x)\dfrac{\Delta v}{\Delta x}}{v(x)v(x + \Delta x)}. \tag{4.20}$$

The right hand side of (4.20) has the limit

$$\frac{v(x)\,u'(x) - u(x)\,v'(x)}{v^2(x)}$$

as $\Delta x \to 0$. Then the limit, as $\Delta x \to 0$, of the left hand side of (4.20) exists and is equal to $y'(x) = \left[\dfrac{u(x)}{v(x)}\right]'$. ∎

Corollary 4.1. *If for functions u and v at a point x the hypotheses of Theorem 4.5 are satisfied, then the following formulas for differentials are true:*

$$d(u + v) = du + dv,$$
$$d(u \cdot v) = v\,du + u\,dv, \tag{4.21}$$
$$d\left(\frac{u}{v}\right) = \frac{v\,du - u\,dv}{v^2}.$$

☐ In order to prove (4.21) it is sufficient to multiply both sides of (4.17) by dx and take into account the formulas $du = u'(x)dx$ and $dv = v'(x)dx$. ∎

Newton's Second Law of Motion[2] states that when an object with constant mass m is subjected to a force F, it undergoes an acceleration a satisfying

$$ma = F. \tag{4.22}$$

The concept of acceleration is that a is the derivative of the velocity with respect to time t, i.e., $a = v'(t)$. The gravitational force, i.e., weight W, is given by

$$mg = W.$$

---

[2]I. Newton (1643-1727), English mathematician and physicist.

The gravitational constant g (see Example 4.3) is independent of the particular object.

At the beginning of the twentieth century Einstein[3] showed that, for large velocities, Newton's law is not applicable.

In Einstein's Theory of Relativity, this law becomes instead:

$$m_0 \frac{d}{dt} \frac{v}{\sqrt{1 - \left(\frac{v^2}{c^2}\right)}} = F. \tag{4.23}$$

Here m_0 is the rest-mass of the object and c is the velocity of light in vacuum, ($c = 3 \cdot 10^{10}$ cm$\cdot$s^{-1}).

If $\frac{v}{c}$ is negligible, it follows from (4.23) that $\sqrt{1 - \left(\frac{v^2}{c^2}\right)} \approx 1$ and instead of (4.23), we have (4.22) as a useful, practical, approximation.

By using Theorems 4.3 and 4.5, it is easy to see that

$$\frac{d}{dt} \frac{v}{\sqrt{1 - \left(\frac{v^2}{c^2}\right)}} = \left[\frac{d}{dv} \frac{v}{\sqrt{1 - \left(\frac{v^2}{c^2}\right)}}\right] \frac{dv}{dt}$$

$$= a \left[\frac{\sqrt{1 - \left(\frac{v^2}{c^2}\right)} + \frac{v^2}{c^2 \sqrt{1 - \left(\frac{v^2}{c^2}\right)}}}{1 - \left(\frac{v^2}{c^2}\right)}\right] = \frac{a}{\left(1 - \frac{v^2}{c^2}\right)^{3/2}}.$$

Then (4.23) has the form:

$$\frac{m_0 a}{\sqrt{\left(1 - \frac{v^2}{c^2}\right)^3}} = F. \tag{4.24}$$

Clearly, in (4.24), $\frac{F}{a}$ depends on both the mass and the velocity v of the object. If the velocity increases, then a given acceleration requires more and more force.

4.5 Derivatives of Basic Elementary Functions

In this section we will compute the derivatives of the elementary functions considered in Section 3.6.

[3] A. Einstein (1879–1955), German physicist.

1. The Exponential and Logarithmic Functions. The exponential, $y = a^x$ $(0 < a \neq 1)$, and logarithmic, $x = \log_a y$ $(0 < y < +\infty)$, functions are inverses of each other. By the definition of the derivative:

$$(a^x)' = \lim_{\Delta x \to 0} \frac{a^{x+\Delta x} - a^x}{\Delta x} = a^x \lim_{\Delta x \to 0} \frac{a^{\Delta x} - 1}{\Delta x} = a^x \ln a.$$

Here we use (3) of Section 3.6.

In particular, if $a = e$,

$$(e^x)' = e^x \ln e = e^x.$$

For the logarithmic function $y = \log_a x$, $0 < a \neq 1$, consider a point $x > 0$. By the definition of the derivative,

$$(\log_a)' = \lim_{\Delta x \to 0} \frac{\log_a(x+\Delta x) - \log_a x}{\Delta x} = \frac{1}{x} \lim_{\Delta x \to 0} \frac{\log_a\left(1 + \dfrac{\Delta x}{x}\right)}{\dfrac{\Delta x}{x}}.$$

Since $\lim\limits_{t \to 0} \dfrac{\log_a(1+t)}{t} = \log_a e$, by substituting $t = \dfrac{\Delta x}{x}$, it follows from the last relation that

$$(\log_a x)' = \frac{\log_a e}{x}. \tag{4.25}$$

Note that (4.25) can be deduced by using Theorem 4.4:

$$(\log_a y)' = \frac{1}{(a^x)'} = \frac{1}{a^x \ln a} = \frac{1}{y \ln a} = \frac{\log_a e}{y}.$$

By setting $a = e$ in (4.25), we find

$$(\ln x)' = \frac{1}{x} \quad (x > 0). \tag{4.26}$$

For the case $x < 0$ in (4.26) we derive, by substituting $u = -x$, for $y = \ln u$ that $y'_x = y'_u \cdot u'_x = \frac{1}{x}$. Consequently, $(\ln|x|)' = \frac{1}{x}$.

2. Power Function. By using Theorem 4.3, we compute the derivative of the power function. $y = x^\alpha$ ($x > 0$, α are any real number):

$$(x^\alpha)' = \left(e^{\alpha \ln x}\right)' = e^{\alpha \ln x} \alpha \frac{1}{x} = \alpha x^{\alpha-1}. \tag{4.27}$$

Generally, let $f(x) = u(x)^{v(x)}$ ($u(x) > 0$). As before, suppose that the functions u, v are differentiable at x. By the corresponding laws of logarithms,

$$\ln f(x) = v(x) \cdot \ln u(x).$$

By Theorem 4.5,

$$[\ln f(x)]' = v'(x) \ln u(x) + v(x) \frac{u'(x)}{u(x)},$$

or

$$\frac{f'(x)}{f(x)} = v'(x)\ln u(x) + v(x)\frac{u'(x)}{u(x)}.$$

Therefore,

$$[u(x)^{v(x)}]' = u(x)^{v(x)}\left[v'(x)\ln u(x) + v(x)\frac{u'(x)}{u(x)}\right]. \qquad (4.28)$$

In particular, if $u(x) = x$ and $v(x) = \alpha$ (α is a constant), then $v'(x) = 0$ and (4.28) implies (4.27).

3. Derivatives of Hyperbolic and Inverse Hyperbolic Functions.

$$(\cosh x)' = \left(\frac{e^x + e^{-x}}{2}\right)' = \frac{e^x - e^{-x}}{2} = \sinh x.$$

Similarly,

$$(\sinh x)' = \cosh x.$$

By the third formula of (4.17), we can compute the derivative of $\tanh x$:

$$(\tanh x)' = \left[\frac{\sinh x}{\cosh x}\right]' = \frac{\cosh^2 x - \sinh^2 x}{\cosh^2 x} = \frac{1}{\cosh^2 x} = \text{sech}^2 x.$$

Likewise,

$$(\coth x)' = -\text{csch}^2 x.$$

Now we compute the derivatives of the inverse hyperbolic functions of Section 3.6:

$$[\cosh^{-1} x]' = \left[\ln\left(x + \sqrt{x^2 - 1}\right)\right]' = \frac{\sqrt{x^2 - 1} + x}{\left(x + \sqrt{x^2 - 1}\right)\sqrt{x^2 - 1}} = \frac{1}{\sqrt{x^2 - 1}}, \quad x > 1,$$

$$[\sinh^{-1} x]' = \left[\ln\left(x + \sqrt{x^2 + 1}\right)\right]' = \frac{1}{\sqrt{x^2 + 1}},$$

$$[\tanh^{-1} x]' = \left[\frac{1}{2}\ln\frac{1+x}{1-x}\right]' = \frac{1}{2}\left[\frac{1-x}{1+x}\cdot\frac{2}{(1-x)^2}\right] = \frac{1}{1-x^2}, \quad |x| < 1,$$

and

$$[\coth^{-1} x]' = \frac{1}{1-x^2}, \quad |x| > 1.$$

4.6   Derivatives of Trigonometric and Inverse Trigonometric functions

In Example 4.7 we proved that $(\sin x)' = \cos x$. By using Theorem 4.3, we can obtain the formula for the derivative of $\cos x$:

$$(\cos x)' = \left[\sin\left(\frac{\pi}{2} - x\right)\right]' = \cos\left(\frac{\pi}{2} - x\right)\left(\frac{\pi}{2}x\right)' = \sin x.$$

By using Theorem 4.5 and these formulas for $(\sin x)'$ and $(\cos x)'$ it is easy to see that

$$(\tan x)' = \left[\frac{\sin x}{\cos x}\right]' = \frac{(\sin x)'\cos x - (\cos x)'\sin x}{\cos^2 x} = \frac{1}{\cos^2 x} = \sec^2 x \quad (\cos x \neq 0)$$

$$\left(x \neq \frac{\pi}{2} + k\pi, \ k = 0, \pm 1, \pm 2, \ldots\right).$$

In the same way,

$$(\cot x)' = \left(\frac{\cos x}{\sin x}\right)' = \frac{(\cos x)'\sin x - (\sin x)'\cos x}{\sin^2 x} = -\frac{1}{\sin^2 x} = -\csc^2 x \quad (\sin x \neq 0)$$

$$(x \neq \pi k, \ k = 0, \pm 1, \pm 2, \ldots).$$

Now, consider the inverse trigonometric function $y = \sin^{-1} x \ (-1 < x < 1)$. This function is the inverse of $x = \sin y \ (-\frac{\pi}{2} < y < \frac{\pi}{2})$. By Theorem 4.4, we have

$$(\sin^{-1} x)' = \frac{1}{(\sin y)'} = \frac{1}{\cos y} = \frac{1}{\sqrt{1 - \sin^2 y}} = \frac{1}{\sqrt{1 - x^2}} \quad (-1 < x < 1).$$

The functions $y = \cos^{-1} x \ (-1 < x < 1)$ and $x = \cos y \ (0 < y < \pi)$ are inverses of each other and so,

$$(\cos^{-1} x)' = \frac{1}{(\cos y)'} = \frac{1}{-\sin y} = -\frac{1}{\sqrt{1 - x^2}} \quad (-1 < x < 1).$$

The function $y = \tan^{-1} x \ (-\infty < x < +\infty)$ is the inverse of $x = \tan y$, $-\frac{\pi}{2} < y < \frac{\pi}{2}$. Hence,

$$(\tan^{-1} x)' = \frac{1}{(\tan y)'} = \cos^2 y = \frac{1}{1 + \tan^2 y} = \frac{1}{1 + x^2}.$$

The functions $y = \cot^{-1} x \ (-\infty < x < +\infty)$ and $x = \cot y \ (0 < y < \pi)$ are inverses of each other. Then,

$$(\cot^{-1} x)' = \frac{1}{(\cot y)'} = -\frac{1}{1 + x^2}.$$

Thus, we can give the following derivative table:

1. $(x^a)' = \alpha x^{a-1}$.

   In particular, $\left(\frac{1}{x}\right)' = -\frac{1}{x^2}; \ (\sqrt{x})' = \frac{1}{2\sqrt{x}}$.

2. $(a^x)' = a^x \ln a \ (0 < a \neq 1)$. In particular, $(e^x)' = e^x$.

3. $(\log_a x)' = \dfrac{1}{x} \log_a e \; (0 < a \neq 1, x > 0).$

    If $a = e$, then $(\ln x)' = \dfrac{1}{x}.$

4. $(\sin x)' = \cos x.$

5. $(\sin x)' = -\sin x.$

6. $(\tan x)' = \dfrac{1}{\cos^2 x} = \sec^2 x, \; x \neq \dfrac{\pi}{2} + k\pi, \; k = 0, \pm 1, \pm 2, \ldots.$

7. $(\cot x)' = -\dfrac{1}{\sin^2 x} = -(1 + \cot^2 x), \; x \neq k\pi, \; k = 0, \pm 1, \pm 2, \ldots.$

8. $(\sin^{-1} x)' = \dfrac{1}{\sqrt{1-x^2}}, \; -1 < x < 1.$

9. $(\cos^{-1} x)' = -\dfrac{1}{\sqrt{1-x^2}}, \; -1 < x < 1.$

10. $(\tan^{-1} x)' = \dfrac{1}{1+x^2}.$

11. $(\cot^{-1} x)' = -\dfrac{1}{1+x^2}.$

12. $(\sinh x)' = \cosh x.$

13. $(\cosh x)' = \sinh x.$

14. $(\tanh x) = \operatorname{sech}^2 x.$

15. $(\coth x)' = -\operatorname{csch}^2 x.$

16. $(\sinh^{-1} x)' = \dfrac{1}{\sqrt{1+x^2}}.$

17. $(\cosh^{-1} x)' = \dfrac{1}{\sqrt{x^2-1}}, \; x > 1.$

18. $(\tanh^{-1} x)' = \dfrac{1}{1-x^2}, \; |x| < 1.$

19. $(\coth^{-1} x)' = \dfrac{1}{1-x^2}, \; |x| > 1.$

Example 4.10. Compute the derivative of $f(x) = \dfrac{\sin x}{1 + \cos x}.$

$$f'(x) = \frac{(1 + \cos x)\cos x - \sin x(-\sin x)}{(1 + \cos x)^2} = \frac{1}{1 + \cos x} \quad \text{(see Theorem 4.5).}$$

Example 4.11. Let $f(x) = \sin\left(\dfrac{x}{x+2}\right).$ Then $f'(x)$ is given by

$$f'(x) = \left[\sin\left(\frac{x}{x+2}\right)\right]' = \cos\left(\frac{x}{x+2}\right)\left(\frac{x}{x+2}\right)' = \frac{2}{(x+2)^2}\cos\left(\frac{x}{x+2}\right) \quad \text{(see Theorem 4.3)}$$

Example 4.12. Find the derivative of the function $f(x) = x^x \; (x < 0).$

By Formula (4.28),

$$f'(x) = x^x(\ln x + 1).$$

Example 4.13. Find the derivative of $f(x) = x^{\sin x}$. Setting $u = x$ and $v = \sin x$, it follows from (4.28) that

$$f'(x) = e^{\sin x}\left(\cos x \ln x + \frac{\sin x}{x}\right).$$

Example 4.14. Let $f(x) = (1 + x^3)^{\frac{4}{5}}$. Obviously, this function is defined everywhere. Its derivative is $f'(x) = \frac{4}{5}(1 + x^3)^{-\frac{1}{5}} \cdot 3x^2$. The derivative function is defined if $x \neq 1$.

Example 4.15. The function

$$f(x) = \begin{cases} x^2 \cos\dfrac{1}{x}, & \text{if } x \neq 0, \\ 0, & \text{if } x = 0, \end{cases}$$

given on $(-1, +1)$, is differentiable on $(-1, +1)$.

Indeed, for points $x \neq 0$,

$$f'(x) = 2x\cos\frac{1}{x} + \sin\frac{1}{x}.$$

And, by the definition of the derivative, at we can write

$$\lim_{\Delta x \to 0} \frac{f(0 + \Delta x) - f(0)}{\Delta x} = \lim_{\Delta x \to 0} \Delta x \cos\frac{1}{\Delta x} = 0.$$

Note that $f'(x)$ has no one-sided limits at $x = 0$. This is because the function $\sin\dfrac{1}{x}$ has no such limits at $x = 0$, which is a discontinuity of the second type (Example 3.15).

4.7 Higher Order Derivatives and Differentials

If the derivative f' of a function f defined on (a, b) is itself differentiable at x, then its derivative is called the second derivative of f at this point and is denoted by $y''(x)$, $f''(x)$, or $\dfrac{d^2 f(x)}{dx^2}$. The third order derivative is denoted by y''', f''', or $\dfrac{d^3 f}{dx^3}$. As long as the result is still differentiable, we can continue the process of taking derivatives, to obtain still higher derivatives of f. In general, the n-th derivative of f with respect to x is denoted by $y^{(n)}(x) = \left[y^{(n-1)}(x)\right]'$ $f^{(n)}(x)$, or $\dfrac{d^n f(x)}{dx^n}$.

Example 4.16. We compute the n-th derivative of $f(x) = x^\alpha$ ($x > 0$, α a real number):

$$y' = \alpha x^{\alpha-1}, \quad y'' = \alpha(\alpha - 1)x^{\alpha-2}, \quad y''' = \alpha(\alpha - 1)(\alpha - 2)x^{\alpha-3}, \ldots$$

By mathematical induction, it is easy to see that

$$(x^\alpha)^{(n)} = \alpha(\alpha - 1)\cdots(\alpha - 1 + n)x^{\alpha-n}.$$

In particular, if $\alpha = k$ (k natural number), then

$$(x^k)^{(k)} = k!, \quad (x^k)^{(n)} = 0, \quad n > k.$$

Example 4.17. Find the n-th derivative of $f(x) = a^x$.

$$y' = a^x \ln a, \quad y'' = a^x \ln^2 a, \dots$$

Then by mathematical induction,

$$(a^x)^{(n)} = a^x \ln^n a.$$

In the case of $a = e$, it follows that $(e^x)^{(n)} = e^x$.

Example 4.18. For the function $y = \cos x$,

$$y' = -\sin x = \cos\left(x + \frac{\pi}{2}\right) \quad \text{and} \quad y'' = -\cos x = \cos\left(x + 2\frac{\pi}{2}\right).$$

Consequently,

$$(\cos x)^{(n)} = \cos\left(x + n\frac{\pi}{2}\right).$$

Now, by mathematical induction, we will prove that for the function $y = \tan^{-1} x$,

$$y^n = (n-1)! \cos^n y \sin\left[n\left(y + \frac{\pi}{2}\right)\right]. \tag{4.29}$$

Indeed, if $n = 1$, then Formula 4.29 signifies that

$$y' = \cos y \sin\left(y + \frac{\pi}{2}\right) = \cos^2 y.$$

On the other hand, since

$$y' = (\tan^{-1} x)' = \frac{1}{1+x^2} = \cos^2 y,$$

the formula is true for $n = 1$. Suppose the formula holds for n. We prove that it also holds for $n+1$, as follows:

$$y^{(n+1)} = (n-1)! \frac{d}{dx}\left\{\cos^n y \sin\left[y + \frac{\pi}{2}\right]\right\}$$

$$= (n-1)! \frac{d}{dy}\left\{\cos^n y \sin\left[n\left(y + \frac{\pi}{2}\right)\right]\right\} y'$$

$$= (n-1)! \frac{d}{dy}\left\{\cos^n y \sin\left[n\left(y + \frac{\pi}{2}\right)\right] + \cos^n y \frac{d}{dy}\left[\sin\left(y + \frac{\pi}{2}\right)\right]\right\} y'$$

$$= n! \cos^{n+1} y\left\{-\sin y \sin\left[n\left(y + \frac{\pi}{2}\right)\right] + \cos y \cos\left[n\left(y + \frac{\pi}{2}\right)\right]\right\}$$

$$= n! \cos^{n+1} y \cos\left[(n+1)y + h\frac{\pi}{2}\right]$$

$$= n! \cos^{n+1} y \sin\left[(n+1)\left(y + \frac{\pi}{2}\right)\right].$$

Now, suppose that the n-th order derivatives of the functions u and v exist. Then it is clear that

$$[cu(x)]^{(n)} = cu^{(n)}(x), \quad u^{(0)}(x) = u(x), \quad c \text{ being a constant},$$

and

$$[u(x) + v(x)]^{(n)} = u^{(n)}(x) + v^{(n)}(x).$$

By mathematical induction, we prove the Leibniz formula for the n-th derivative of a product of two functions:

$$(uv)^{(n)} = u^{(n)}v + C_n^1 u^{(n-1)}v' + C_n^2 u^{(n-2)}v'' + \cdots + uv^{(n)}; \quad C_n^k = \frac{n!}{k!(n-k)!}. \tag{4.30}$$

For $n = 1$ the formula holds. We must prove that if (4.30) holds for n, it holds for $n+1$:

$$(uv)^{(n+1)} = u^{(n+1)}v + \left[C_n^0 u^{(n)}v' + C_n^1 u^{(n)}v' \right] + \left[C_n^1 u^{(n-1)}v'' + C_n^2 u^{(n-1)}v'' \right]$$
$$+ \left[C_n^2 u^{(n-2)}v''' + C_n^3 u^{(n-2)}v''' \right] + \cdots + uv^{(n+1)}. \tag{4.31}$$

We show that, for $n \geqslant k$,

$$C_n^k + C_n^{k-1} = C_{n+1}^k. \tag{4.32}$$

Indeed,

$$C_{n+1}^k - C_n^{k-1} = \frac{(n+1)!}{k!(n+1-k)!} - \frac{n!}{(k-1)!(n+1-k)!} = \frac{(n+1)! - n!k}{k!(n+1-k)!} = \frac{n!}{k!(n-k)!} = C_n^k.$$

Then, with the use of (4.32), from (4.31) we have that

$$(uv)^{(n+1)} = u^{(n+1)}v + C_{n+1}^1 u^{(n)}v' + C_{n+1}^2 u^{(n-1)}v'' + \cdots + uv^{(n+1)}.$$

This ends the proof of (4.30).

Example 4.19. Find the n-th order derivative of $f(x) = x^3 e^x$. Denoting $u = e^x$, $v = x^3$ for every n, we have $u^{(n)} = e^x$ and $v' = 3x^2$, $v'' = 6x$, $v''' = 6$, $v^{(4)} = v^{(5)} = \cdots = 0$.

Then, by Leibniz's formula (4.30),

$$(x^3 e^x)^{(n)} = e^x \cdot x^3 + C_n^1 e^x \cdot 3x^2 + C_n^2 e^x \cdot 6x + C_n^3 e^x \cdot 6$$
$$= e^x \left[x^3 + 3nx^2 + 3n(n-1)x + n(n-1)(n-2) \right].$$

Now, by using the n-th derivative, we introduce the concept of the n-th differential.

Recall that if the function f is differentiable at x, then there is the following connection between differentials dx and dy,

$$dy = f'(x)dx. \tag{4.33}$$

For higher order differentials, we will consider the case where x is the independent variable. In this case we consider that dx does not depend on x, and that $dx = \Delta x$ for all x and $d(dx) = 0$. Then by using Leibniz's formula for the differential of a product function (4.33),

$$d^2 y = d[f'(x)]dx + f'(x)d(dx) = f''(x)(dx)^2. \tag{4.34}$$

Denoting $(dx)^2 = dx^2$, we can write

$$d(dy) = d^2y = f''(x)dx^2.$$

In an analogous way, the n-th order differential, denoted as $d^n y$, is the differential of the $(n-1)$-th order differential, and is:

$$d^n y = d(d^{n-1}y) = \left[f^{(n-1)}(x)dx^{n-1} \right]' dx, \quad d^n y = f^{(n)}(x)dx^n. \tag{4.35}$$

Thus we can write:

$$f'(x) = \frac{dy}{dx}, \ldots, f^{(n)}(x) = \frac{d^n y}{dx^n}.$$

Remark 4.6. When x is n-th order $(n \geqslant 2)$ differentiable function of some independent variable t, the formula (4.35) is not true.

Indeed, consider the composition $y = f(x)$ where $x : A \to \mathbb{R}$ and $f : x(A) \to \mathbb{R}$ are functions differentiable to n-th order.

It follows from Leibniz's formula that $y(t)$ is an n-times differentiable function, and

$$y^{(n)}(t) = (y'(t))^{(n-1)} = (f'(x)x')^{(n-1)}$$

$$= \sum_{k=0}^{n-1} C_{n-1}^k \frac{d^k}{dt^k} f'(x) \frac{d^{n-k-1}}{dt^{n-k-1}} x' = \sum_{k=0}^{n-1} C_{n-1}^k x^{(n-k)} \frac{d^k}{dt^k} f'(x).$$

Therefore

$$d^n y = y^{(n)}(t)dt^n = \sum_{k=0}^{n-1} C_{n-1}^k d^k f'(x) d^{n-k} x$$

In particular, for a twice-differentiable $(n = 2)$ function, it follows that

$$d^2 y = f''(x)(dx)^2 + f'(x)d^2 x \tag{4.36}$$

As we see from (4.36), for higher order differentials, the invariance property is not true.

Example 4.20. Let $y = f(x) = \cos x$ and $x = \sqrt{t}$. Compute $d^2 y$.

$$dy = -\sin x \cdot \frac{1}{2\sqrt{t}} dt = -\sin x dx.$$

By Formula (4.36),

$$d^2 y = -\cos x dx^2 - \sin x d^2 x = -\cos x dx^2 - \sin x \cdot x'' dt^2$$

$$= -\cos x \left(\frac{1}{2\sqrt{t}} \right)^2 dt^2 - \sin x \left(-\frac{1}{4t^{3/2}} \right) dt^2,$$

$$d^2 y = \sin x \cdot \frac{1}{4t^{3/2}} dt^2 - \cos x \cdot \frac{1}{4t} dt^2.$$

4.8   Derivatives of Functions Given by Parametric Equations

Suppose the function f with $y = f(x)$ is given by parametric equations $x = \varphi(t)$ and $y = \psi(t)$. Here φ and ψ have the necessary order of differentiability with respect to t on the interval where these functions are defined. In order to obtain an explicit equation $y = f(x)$, suppose that an inverse function $t = \varphi^{-1}(x)$ exists. Conversely, any explicit curve $y = f(x)$ can be viewed as a parametric curve by writing $x = t$, $y = f(t)$, with t running through the values in the original domain of f.

Now, we try to compute the derivative of $y = y(x)$ with respect to x. By invariance of first order differentials,

$$dy = \psi'(t)dt, \quad dx = \varphi'(t)dt. \tag{4.37}$$

Hence,

$$y'(x) = \frac{dy}{dx} = \frac{\psi'(t)}{\varphi'(t)}. \tag{4.38}$$

Then, from formulas (4.37) and (4.38), we have

$$y''(x) = \frac{d[y'(x)]}{dx} = \frac{\left[\dfrac{\psi'(t)}{\varphi'(t)}\right]' dt}{\varphi'(t)} = \frac{\psi''(t)\varphi'(t) - \varphi''(t)\psi'(t)}{[\varphi'(t)]^2}. \tag{4.39}$$

By the definition of third order derivatives,

$$y'''(x) = \frac{d[y''(x)]}{dx}.$$

Then, taking into account (4.39) and the second formula in (4.37), we can define $y'''(x)$. By continuing this process, the higher order derivatives can be found.

Example 4.21. A function is given by parametric equations

$$\begin{aligned} x &= t - \sin t, \\ y &= 1 - \cos t, \end{aligned} \qquad -\infty < t < +\infty.$$

Find the derivatives $y'(x)$ and $y''(x)$.

This curve is known as a cycloid. The curve traced by some point (say M) on the edge of a rolling circle is called a cycloid. The circle is to roll along a straight line without slipping. As parameter t, we take the angle through which the circle has turned since it began with M at the origin. We remind the reader that the solution of the brachistochrone problem proposed by J. Bernoulli in 1696 is a cycloid curve.

With the use of (4.38) and (4.39), we obtain

$$y'(x) = \frac{\sin t}{1 - \cos t} = \cot \frac{t}{2},$$

$$y''(x) = \frac{\left(\cot \frac{t}{2}\right)'}{1-\cos t} = -\frac{1}{4\sin^4 \frac{t}{2}}.$$

Example 4.22. For the function given parametrically by $x = a\cos t$ and $y = b\sin t$, find $\frac{d^2 y}{dx^2}$.

$$y'(x) = \frac{b\cos t}{-a\sin t} = -\frac{b}{a}\cot t,$$

$$y''(x) = \frac{\left(-\frac{b}{a}\cot t\right)'}{-a\sin t} = -\frac{\frac{b}{a}\csc^2 t}{-a\sin t} = -\frac{b}{a^2 \sin^3 t}.$$

Example 4.23. Suppose that a mass is thrown out from an airplane which is flying with velocity v_0 at altitude y_0. Determine the trajectory of the mass and the place where it will fall to earth (the force of air resistance is to be neglected).

As is seen in Figure 4.3, the point from which the mass is thrown is $(0, y_0)$.
The velocity of the airplane is constant, and so $x = v_0 t$.
On the other hand, by the free fall formula,

$$S = \frac{gt^2}{2} \quad (g \approx 9,8\,\mathrm{m}\cdot\mathrm{s}^{-2}).$$

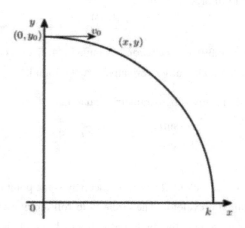

Fig. 4.3   The trajectory of the mass and the place where the mass falls to earth.

Therefore, for any given time t, the distance between the mass and the earth's surface is

$$y = y_0 - \frac{gt^2}{2}. \tag{4.40}$$

The pair of free fall formula's function (4.40) is a parametric description of our trajectory. Then,

$$y = y_0 - \frac{g}{2v_0^2} x^2. \tag{4.41}$$

The equation (4.41) is an explicit equation for a parabola. Let the abscissa of a point K be x_0. Then since at K the ordinate $y = 0$, from formula (4.41) we then have

$$0 = y_0 - \frac{g}{2v_0^2} x_0^2 \text{ or } x_0 = v_0 \sqrt{\frac{2y_0}{g}}.$$

4.9   Curvature and Evolvents

1. Equations of Tangent and Normal Lines.   Consider a point $M(x_0, y_0)$ on a curve $y = f(x)$. We determine the tangent line to the curve $y = f(x)$ at $M(x_0, y_0)$, supposing the function f is differentiable at x_0. An equation for the straight line through the point M with slope α is

$$y - y_0 = \alpha(x - x_0). \tag{4.42}$$

If, in (4.42), $\alpha = f'(x_0)$, then the equation of the tangent line to the curve $y = f(x)$ at M is

$$y - y_0 = f'(x_0)(x - x_0). \tag{4.43}$$

The line through the point $M(x_0, y_0)$ on the curve $y = f(x)$ that is perpendicular to the tangent line (4.43) is called the normal line at M. Since the slope of a perpendicular line is the negative reciprocal of the slope of the other, the slope of the normal is $\beta = -\frac{1}{\alpha}$. Then

$$y - y_0 = -\frac{1}{f'(x_0)} (x - x_0)$$

is the required equation.

Example 4.24. Determine the tangent and normal lines to the parabola $y = x^2$ at $M(1.1)$.

We have $y' = 2x$ and $y'(1) = 2$. Then $y - 1 = 2(x - 1)$ or $y = 2x - 1$ is the equation of tangent line. Thus, the equation of the normal line has the form

$$y - 1 = -\frac{1}{2}(x - 1) \text{ or } y = -\frac{1}{2}x + \frac{3}{2}.$$

Definition 4.5. The length of the segment T between M and the point of intersection of the tangent line with the x-axis, is called the *tangent length* and is denoted by $|T|$.

The length of the projection of T on Ox is called the *subtangent* and is denoted by S_T.

The normal length $|N|$ and the length of its projection S_N are defined similarly. The S_N is called the *subnormal*.

It follows from the definition that

$$S_T = \left| \frac{y_0}{y_0'} \right| \quad (y_0' = y'(x_0)),$$

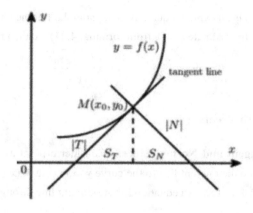

Fig. 4.4   The subnormal and the subtangent at a point $M(x_0, y_0)$.

$$|T| = \sqrt{y_0^2 + \frac{y_0^2}{y_0'^2}} = \left| \frac{y_0}{y_0'} \right| \sqrt{y_0'^2 + 1},$$

and

$$S_N = |y_0 y_0'|,$$

$$|N| = \sqrt{y_0^2 + (y_0 y_0')^2} = \left| y_0 \sqrt{1 + y_0'^2} \right|.$$

Example 4.25. Consider the ellipse given by the parametric equations $x = a\cos t$ and $y = b\sin t$. If $t = \frac{\pi}{4}$, define the tangent and normal lines at point $M(x_0, y_0)$ and compute the values of subtangent S_T, $|T|$, subnormal S_N, and $|N|$. By formula (4.38),

$$\frac{dy}{dx} = -\frac{b}{a}\cot t, \quad \text{and so,} \quad \left(\frac{dy}{dx} \right)_{t = \frac{\pi}{4}} = -\frac{b}{a}.$$

Since the coordinates of M are

$$x_0 = \frac{a}{\sqrt{2}} \quad \text{and} \quad y_0 = \frac{b}{\sqrt{2}},$$

it follows that the equation of the tangent line has the form

$$y - \frac{b}{\sqrt{2}} = -\frac{b}{a}\left(x - \frac{a}{\sqrt{2}} \right), \quad \text{or} \quad bx + ay - ab\sqrt{2} = 0.$$

It is easy to see that this equation can be rewritten as
$$\frac{xx_0}{a^2} + \frac{yy_0}{b^2} = 1.$$
Now, we can write the equation of the normal line as
$$y - \frac{b}{\sqrt{2}} = \frac{a}{b}\left(x - \frac{a}{\sqrt{2}}\right), \quad \text{or} \quad (ax - by)\sqrt{2} - a^2 + b^2 = 0.$$
On the other hand, it is not hard to see that
$$S_T = \left|\frac{b/\sqrt{2}}{-b/a}\right| = \frac{a}{\sqrt{2}} \text{ and } S_N = \left|\frac{b}{\sqrt{2}}\left(-\frac{b}{a}\right)\right| = \frac{b^2}{a\sqrt{2}}.$$
Further, the tangent and normal lengths are given by
$$|T| = \frac{1}{\sqrt{2}}\sqrt{a^2 + b^2} \text{ and } |N| = \frac{b}{a\sqrt{2}}\sqrt{a^2 + b^2}.$$

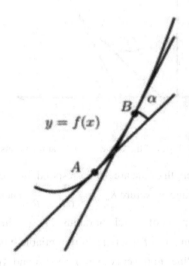

Fig. 4.5   The curvature of curve AB.

2. Circle of Curvature and Evolvents.   Suppose we have a curve that does not intersect itself, and that has a unique tangent line at every point. Let A and B be two points on this curve. The angle α formed by the tangent lines at these points is called the curvature of the arc AB. (Figure 4.5)

Definition 4.6. The ratio of the curvature α to the length of the arc AB is called the *average curvature* of the arc.
$$K_{\text{aver.}} = \frac{\alpha}{|AB|}. \tag{4.44}$$

Definition 4.7. The *curvature* of the arc AB at A is the limit of the average curvature as the length $|AB|$ approaches zero (B approaches A) and is denoted by K_A:

$$K_A = \lim_{AB \to 0} \frac{\alpha}{|AB|}. \tag{4.45}$$

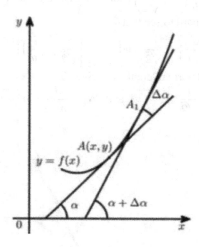

Fig. 4.6   The slopes of the curve $y = f(x)$ at the points A and A_1.

For example, since the length of the arc AB corresponding to central angle α of a circle with radius R is αR, the average curvature $K_{\text{aver.}} = \dfrac{\alpha}{\alpha R} = \dfrac{1}{R}$. Therefore, $K = \lim\limits_{a \to 0} \dfrac{\alpha}{\alpha R} = \dfrac{1}{R}$, that is, the curvature at each point of a circle of radius R is $\dfrac{1}{R}$. The curve $y = f(x)$ is shown in Figure 4.5. Suppose the function f is a twice differentiable function. Let α and $\alpha + \Delta\alpha$ be the slopes of the tangent lines to the curve $y = f(x)$ at A and A_1, respectively. By formula (4.44),

$$K_{\text{aver.}} = \left| \frac{\Delta\alpha}{\Delta s} \right| \quad (\Delta s = |AA_1|).$$

Then, by formula (4.45), the curvature at A is

$$K = \lim_{\Delta s \to 0} \left| \frac{\Delta\alpha}{\Delta s} \right|.$$

Since the values α and s depend on x, we consider α as a function of s. Then

$$\lim_{\Delta s \to 0} \frac{\Delta\alpha}{\Delta s} = \frac{d\alpha}{ds} \quad \text{and} \quad K = \left| \frac{d\alpha}{ds} \right|.$$

Thus K is the absolute value of the rate of change of the angle α with respect to arc length s.

If we think the variable x as a parameter, then

$$\frac{d\alpha}{ds} = \frac{d\alpha/dx}{ds/dx}. \tag{4.46}$$

On the other hand, $\tan \alpha = \dfrac{dy}{dx}$ or $\alpha = \tan^{-1} \dfrac{dy}{dx}$, and so

$$\frac{d\alpha}{dx} = \frac{y''}{1+y'^2}. \tag{4.47}$$

We will show later (see Remark 10.4), that there is the following formula for $\dfrac{ds}{dx}$:

$$\frac{ds}{dx} = \sqrt{1+y'^2}. \tag{4.48}$$

By substituting (4.47) and (4.48) into formula (4.46), we have

$$\frac{d\alpha}{ds} = \frac{\dfrac{y''}{1+y'^2}}{\sqrt{1+y'^2}} = \frac{y''}{[1+y'^2]^{\frac{3}{2}}},$$

or

$$K = \frac{|y''|}{[1+y'^2]^{\frac{3}{2}}}. \tag{4.49}$$

Alternatively, we could have used (4.49) for calculating the curvature K of a circle of radius R: our point of departure would have been the equation $x^2 + y^2 = R^2$ of the circle, and we would have computed y' and y'' by implicit differentiation.

Example 4.26. Find the curvature of the parabola $y^2 = 2x$ at the point $A(x,y)$.

We compute the derivatives of first and second order:

$$y' = \frac{1}{\sqrt{2x}} \quad \text{and} \quad y'' = -\frac{1}{(2x)^{\frac{3}{2}}}.$$

Then, by substituting the expressions y' and y'' into (4.49), we have

$$K = \frac{1}{(2x+1)^{\frac{3}{2}}}.$$

In particular, at the points $x = 0$, $y = 0$ and $x = \dfrac{1}{2}$, $y = 1$, we obtain $K = 1$ and $K = \dfrac{1}{2\sqrt{2}}$, respectively.

Remark 4.7. If we need to derive a formula that is effective in computing the curvature of a smooth parametric curve $x = \varphi(t)$ and $y = \psi(t)$, we should use the formulae for y' and y'', (4.38) and (4.39). We easily obtain

$$K = \frac{|\psi''\varphi' - \psi'\varphi''|}{[\varphi'^2 + \psi'^2]^{\frac{3}{2}}}. \tag{4.50}$$

Example 4.27. Determine the curvature of the cycloid

$$x = t - \sin t,$$
$$y = 1 - \cos t,$$
$$-\infty < + < \infty,$$

at the point (x, y). Clearly,

$$x'(t) = \varphi'(t) = 1 - \cos t, \qquad x''(t) = \varphi''(t) = \sin t,$$
$$y'(t) = \psi'(t) = \sin t, \qquad y''(t) = \psi''(t) = \cos t.$$

By substituting these expressions for the derivatives into (4.50), we see that

$$K = \frac{|\cos t - 1|}{2^{\frac{3}{2}}(1 - \cos t)^{\frac{3}{2}}} = \frac{1}{2^{\frac{3}{2}}(1 - \cos t)^{\frac{3}{2}}} = \frac{1}{4\left|\sin \dfrac{t}{2}\right|}.$$

Definition 4.8. Consider the circle tangent to a curve at A which has the same curvature there, the center of the circle lying on the concave side of the curve. This circle is called *circle of curvature* of the curve at the given point A. Correspondingly, its radius is called the *radius of curvature*.

It follows from the example above that the radius of curvature of the circle with radius R is $\dfrac{1}{K} = R$, that is,

$$R = \frac{1}{K} = \frac{(1 + y'^2)^{\frac{3}{2}}}{|y''|}, \tag{4.51}$$

and in parametric form (see (4.50)),

$$R = \frac{(\varphi'^2 + \psi'^2)^{\frac{3}{2}}}{\psi''\varphi' - \psi'\varphi''}. \tag{4.52}$$

Note that if the slope of the tangent line to the curve $y = f(x)$ at A is y', then the equation of the normal line at the same point is

$$Y - y = -\frac{1}{y'}(X - x).$$

Here (X, Y) denotes a point of the curve. Now, if we denote by $C(\xi, \eta)$ the center of the circle of curvature with radius R (Figure 4.6), then it is clear that

$$\eta - y = -\frac{1}{y'}(\xi - x) \tag{4.53}$$

On the other hand, the distance between the points $A(x, y)$ and $C(\xi, \eta)$ is R, and so

$$(\xi - x)^2 + (\eta - y)^2 = R^2. \tag{4.54}$$

Then we derive, from (4.53) and (4.54), that

$$\xi = x \pm \frac{y'}{\sqrt{1 + y'^2}} R, \quad \eta = y \mp \frac{1}{\sqrt{1 + y'^2}} R,$$

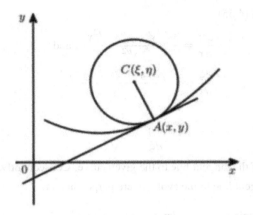

Fig. 4.7 The circle of curvature at a given point A.

Then (4.51) gives that

$$\xi = x \pm \frac{y'(1+y'^2)}{|y''|} \quad \text{and} \quad \eta = y \mp \frac{1+y'^2}{|y''|}.$$

Here, if $y'' > 0$ ($y'' < 0$) then $\eta > y$ ($\eta < y$), and consequently

$$\xi = x - \frac{y'(1+y'^2)}{y''} \quad \text{and} \quad \eta = y + \frac{1+y'^2}{y''}. \tag{4.55}$$

It follows from (4.55) that to each point $A_1(x,y)$ with nonzero curvature of the curve, there is assigned some center of curvature $C_1(\xi,\eta)$. The locus of the centers of curvature is called the evolvent of the given curve. By eliminating x in (4.55) we obtain the equation of the evolvent in cartesian coordinates.

For a smooth parametric curve $x = \varphi(t)$, $y = \psi(t)$, by substituting (4.38) and (4.39) into (4.55), we have the equation of the evolvent in parametric equations:

$$\xi = x - \frac{y'(x'^2+y'^2)}{x'y'' - x''y'},$$
$$\eta = y + \frac{x'(x'^2+y'^2)}{x'y'' - x''y'}. \tag{4.56}$$

Theorem 4.6. *The normal line to a curve is the tangent line to its evolvent.*

☐ Obviously, we can write

$$\frac{d\eta}{d\xi} = \frac{\dfrac{d\eta}{dx}}{\dfrac{d\xi}{dx}}.$$

On the other hand, by (4.55),

$$\frac{d\eta}{dx} = \frac{3y''^2 y' - y''' - y'^2 y'''}{y''^2} \quad \text{and}$$

$$\frac{d\xi}{dx} = -y' \frac{3y'y''^2 - y''' - y'^2 y'''}{y''^2}.$$

Thus

$$\frac{d\eta}{d\xi} = -\frac{1}{y'}.$$

Here y' is the slope of the tangent line to the given curve. Consequently, the normal line to the curve and the tangent line to the evolvent are perpendicular. ∎

Example 4.28. Determine the evolvent of the parabola $y^2 = 2x$.

By substituting $y' = \dfrac{1}{\sqrt{2x}}$ and $y'' = -\dfrac{1}{(2x)^{3/2}}$ into (4.55), we obtain

$$\xi = 3x + 1, \quad \eta = -(2x)^{3/2}.$$

Then, by eliminating the variable x, we have the equation of the evolvent

$$\eta^2 = \frac{8}{27} (\xi - 1)^3.$$

Example 4.29. Find the evolvent of the cycloid:

$$\begin{aligned} x &= t - \sin t, \\ y &= 1 - \cos t, \end{aligned} \quad -\infty < t < +\infty.$$

We calculate the derivatives x', x'' and y', y'':

$$\begin{aligned} x' &= 1 - \cos t, & x'' &= \sin t, \\ y' &= \sin t, \text{ and} & y'' &= \cos t. \end{aligned}$$

Then it follows from (4.56) that

$$\xi = t + \sin t,$$
$$\eta = \cos t - 1.$$

By substituting $\xi = \xi' - \pi$, $\eta = \eta' - 2$, and $t = \tau - \pi$, we verify that the evolvent

$$\xi' = \tau - \sin \tau,$$
$$\eta' = 1 - \cos \tau,$$

is a cycloid.

3. Conic sections in polar coordinates. The phrase conic sections stems from the fact that these are the curves in which a plane intersects a cone. If the cutting plane is parallel to some generator of the cone, then the curve of intersection is a parabola. Otherwise, it is either an ellipse or a hyperbola with two branches. In order to investigate the orbits of planets or comets around the sun, or moons orbiting a planet, we need the equations of the conic sections in polar coordinates. From elementary geometry, the equation of any conic section in Cartesian coordinates has the form

$$Ax^2 + Bxy + Cy^2 + Dx + Ey + F = 0, \tag{4.57}$$

where A, B, C, D, E, F are real numbers.

Clearly, if $A = B = C = 0$, then (4.57) is the equation of a straight line. If $B = 0$ and $A = C = 1$, then we get a circle's equation.

In general, the nature of the conic section described by (4.57) depends on the sign of $B^2 - 4AC$, that is, (4.57) describes a:

   (1) parabola if $B^2 - 4AC = 0$,
   (2) ellipse if $B^2 - 4AC < 0$,
   (3) hyperbola if $B^2 - 4AC > 0$.

Between Cartesian (x, y) and polar (r, θ) coordinates there exist the connections $x = r\cos\theta$ and $y = r\sin\theta$. Then we get the equation in polar coordinates of the straight line:

$$r(D\cos\theta + E\sin\theta) + F = 0.$$

It follows from the more general equation in Cartesian coordinates of a circle, $x^2 + y^2 + Dx + Ey + F = 0$, that its equation in polar coordinates has the form

$$r^2 + r(D\cos\theta + E\sin\theta) + F = 0.$$

Note that the equation in polar coordinates of a conic section,

$$r^2(A\cos^2\theta + B\sin\theta\cos\theta + C\sin^2\theta) + r(D\cos\theta + E\sin\theta) + F = 0,$$

is not very useful for applications.

To derive the polar equation of a conic section, suppose that its focus is at the pole and that its directrix is the vertical line $x = p(p > 0)$. In the notation of Figure 4.7, the fact that $|OP| = e|PQ|$ means that

$$r = e(p + r\cos\theta),$$

so that

$$r = \frac{pe}{1 - e\cos\theta}.$$

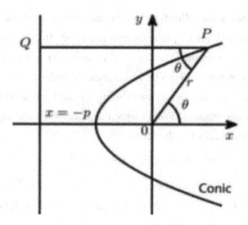

Fig. 4.8   Conic section: $|OP| = e|PQ|$.

If the directrix is the line $x = +p$, then a similar calculation gives the same formula except that the denominator $1 - e\cos\theta$ is replaced by $1 + e\cos\theta$.

Thus the polar equation of the conic section with eccentricity e, focus O, and directrix $x = \pm p$, is

$$r = \frac{pe}{1 \pm e\cos\theta}. \tag{4.58}$$

This equation is an ellipse if $e < 1$, a parabola if $e = 1$, and a hyperbola if $e > 1$. Note that since $\cos\theta = \cos(-\theta)$, its graph is symmetric about the polar axis.

It is easy to see that if the directrix of the conic section is the line $y = \pm p$, then a similar calculation yields its equation in polar coordinates,

$$r = \frac{pe}{1 \pm e\sin\theta}.$$

Example 4.30. The hyperbola $xy = 1$ is rotated counterclockwise through an angle of $\alpha = \dfrac{\pi}{4}$ about the origin. In the rotated system with new Cartesian coordinates (x', y'), determine its equation.

By using the connection between the old and new coordinates,

$$x = x'\cos\frac{\pi}{4} - y'\sin\frac{\pi}{4}$$

and

$$y = x'\sin\frac{\pi}{4} + y'\cos\frac{\pi}{4},$$

we can write

$$x = \frac{x' - y'}{\sqrt{2}} \quad \text{and} \quad y = \frac{x' + y'}{\sqrt{2}}.$$

The original equation $xy = 1$ thus becomes

$$\frac{x'^2}{(\sqrt{2})^2} - \frac{y'^2}{(\sqrt{2})^2} = 1.$$

So, in the $x'y'$-coordinate system, we have a hyperbola with $a = b = \sqrt{2}$, $c = \sqrt{a^2 + b^2} = 2$, and foci $(\pm 2, 0)$.

4.10   Problems

Consider the secant line to the parabola $y = x^2$ passing through $A(2,4)$ and $B(2 + \Delta x,\, 4 + \Delta y)$. Determine its slope for the given values of the increment.

(1) $\Delta x = 1.$ Answer: 5.

(2) $\Delta x = 0.1.$ Answer: 4.1.

(3) $\Delta x = 0.01.$ Answer: 4.01.

In problems (4)–(6) use the definition to find the derivatives.

(4) $f(x) = x^3.$ Answer: $3x^2.$

(5) $f(x) = \dfrac{1}{x}.$ Answer: $-\dfrac{1}{x^2}.$

(6) $f(x) = \sqrt[3]{x}.$ Answer: $\dfrac{1}{3\sqrt[3]{x^2}}\ (x \neq 0).$

(7)   If a function f is differentiable and n is a natural number, prove that

$$\lim_{n \to \infty} n \left[f\left(x + \frac{1}{n}\right) - f(x) \right] = f'(x).$$

Conversely, if this limit exists, determine whether the function is differentiable or not. Investigate functions similar to the Dirichlet function of Example 3.2.

(8)   Show that $f(x) = \begin{cases} x^2 & \text{if } x \text{ is rational,} \\ 0, & \text{if } x \text{ is irrational,} \end{cases}$

is differentiable only at the point $x = 0$.

(9)   (a) Show that the derivative of

$$f(x) = \begin{cases} x^2 \sin \dfrac{1}{x}, & \text{if } x \neq 0, \\ 0, & \text{if } x = 0, \end{cases}$$

is discontinuous at $x = 0$.

(b) Show that the derivative of $f(x) = \begin{cases} x^2 \sin \dfrac{1}{x^2}, & \text{if } x \neq 0, \\ 0, & \text{if } x = 0, \end{cases}$

is unbounded on $[-1, +1]$.

(c) Prove that $f(x) = \begin{cases} x^{2n} \sin \dfrac{1}{x}, & \text{if } x \neq 0, \\ 0, & \text{if } x = 0. \end{cases}$

has an n-th order derivative and does not have an $(n+1)$-st order derivative.

In problems (10)–(12), determine the positive integer n such that the function

$$f(x) = \begin{cases} x^n \sin \dfrac{1}{x}, & \text{if } x \neq 0, \\ 0, & \text{if } x = 0. \end{cases}$$

(10) is continuous at $x = 0$;   *Answer:* $n > 0$;

(11) is differentiable at $x = 0$;   *Answer:* $n > 1$;

(12) has a continuous derivative at $x = 0$;   *Answer:* $n > 2$.

(13) Let $\eta(x)$ be continuous at $x = a$. Find the   *Answer:* $f'(a) = \eta(a)$.
derivative of $f(x) = (x - a)\eta(x)$ at the point
$x = a$.

(14) If $f_{ij}(x)$ $(i, j = 1, \ldots, n)$ are differentiable functions, show that

$$\frac{d}{dx} \begin{vmatrix} f_{11}(x) & \ldots\ldots & f_{1n}(x) \\ \vdots & \ldots\ldots & \vdots \\ f_{k1} & \ldots\ldots & f_{kn}(x) \\ f_{n1}(x) & \ldots\ldots & f_{nn}(x) \end{vmatrix} = \sum_{k=1}^{n} \begin{vmatrix} f_{11}(x) & \ldots\ldots & f_{1n}(x) \\ \vdots & \ldots\ldots & \vdots \\ f'_{k1} & \ldots\ldots & f'_{kn}(x) \\ f_{n1}(x) & \ldots\ldots & f_{nn}(x) \end{vmatrix}.$$

(15) Find the derivative $f'(x)$ of   *Answer:* $3x^2 + 15$.

$$f(x) = \begin{vmatrix} x-1 & 1 & 2 \\ -3 & x & 3 \\ -2 & -3 & x+1 \end{vmatrix}.$$

(16) Prove that $f(x) = \begin{cases} e^{-\frac{1}{x^2}}, & x \neq 0, \\ 0, & x = 0, \end{cases}$

is a function possessing derivatives of all orders at $x = 0$.

(17) Determine conditions on a and b so that   *Answer:* $a = 2x_0$, $b = -x_0^2$

$$f(x) = \begin{cases} x^2, & \text{if } x \leqslant x_0, \\ ax+b, & \text{if } x > x_0, \end{cases}$$

will be differentiable at the point $x = x_0$.

In (18) and (19) can we say that a function $F(x) = f(x) + g(x)$ necessarily is nondifferentiable at $x = x_0$:

(18) if $f(x)$ is differentiable at $x = x_0$, but $g(x)$ is nondifferentiable   *Answer:* Yes.
at $x = x_0$?

(19) If both $f(x)$ and $g(x)$ are nondifferentiable at $x = x_0$?   *Answer:* No.

In problems (20) and (21), determine whether the functions are differentiable or not.

(20) $f(x) = |\cos x|$.   *Answer:* At points $x = \dfrac{2k-1}{2}\pi$ (k an integer).

(21) $f(x) = |\pi^2 - x^2| \sin^2 x$.   *Answer:* Differentiable everywhere.

In problems (22) and (23), find the one-sided derivatives $f'(x+0)$ and $f'(x-0)$.

(22) $f(x) = [x] \sin \pi x$.   *Answer:* $f'(x+0) = f'(x-0) = \pi[x] \cos \pi x$
(if x is non-integral);
$f'(k-0) = \pi(k-1)(-1)^k$,
$f'(k+0) = \pi k(-1)^k$
(if k is an integer).

(23) $f(x) = x \left| \cos \dfrac{\pi}{x} \right|$ ($x \neq 0$), $f(0) = 0$. *Answer:* $f'(x+0) = f'(x-0)$.

$$= \left(\cos \frac{\pi}{x} + \frac{\pi}{x} \sin \frac{\pi}{x} \right) \operatorname{sgn} \left(\cos \frac{\pi}{x} \right), \quad \text{if } x \neq \frac{2}{2k+1} \text{ (k an integer)};$$

$$f\left(\frac{2}{2k+1} - 0 \right) = -(2k+1)\frac{\pi}{2}, \quad f'\left(\frac{2}{2k+1} + 0 \right) = (2k+1)\frac{\pi}{2}.$$

(24) Prove that $\left(\dfrac{ax+b}{cx+d}\right)' = \dfrac{\begin{vmatrix} a & b \\ c & d \end{vmatrix}}{(cx+d)^2}$.

In problems (25)–(39), find the derivatives.

(25) $f(x) = \dfrac{2x}{1-x^2}$.

Answer: $\dfrac{2(1+x^2)}{(1-x^2)^2}$ $(|x| \neq 1)$.

(26) $f(x) = \dfrac{(1-x)^p}{(1+x)^q}$.

Answer: $-\dfrac{(1-x)^{p-1}[(p+q)+(p-q)x]}{(1+x)^{q+1}}$

$(|x| \neq -1)$.

(27) $f(x) = \dfrac{x}{\sqrt{a^2-x^2}}$.

Answer: $\dfrac{a^2}{(a^2-x^2)^{\frac{3}{2}}}$ $(|x| < |a|)$.

(28) $f(x) = \sqrt{x + \sqrt{x + \sqrt{x}}}$.

Answer: $\dfrac{1 + 2\sqrt{x} + 4\sqrt{x}\sqrt{x+\sqrt{x}}}{8\sqrt{x}\sqrt{x+\sqrt{x}} + \sqrt{x+\sqrt{x+\sqrt{x}}}}$

$(x > 0)$.

(29) $f(x) = (2-x^2)\cos x + 2x\sin x$.

Answer: $x^2 \sin x$.

(30) $f(x) = \sin[\sin(\sin x)]$.

Answer: $\cos x \cdot \cos(\sin x) \cdot \cos[\sin(\sin x)]$.

(31) $f(x) = \dfrac{1}{4}\ln\dfrac{x^2-1}{x^2+1}$.

Answer: $\dfrac{x}{x^4-1}$ $(|x| > 1)$.

(32) $f(x) = \ln(x + \sqrt{x^2+1})$.

Answer: $\dfrac{1}{\sqrt{x^2+1}}$.

(33) $f(x) = \cos^{-1}\dfrac{1}{x}$.

Answer: $\dfrac{1}{|x|\sqrt{x^2-1}}$ $(|x| > 1)$.

(34) $f(x) = \tan^{-1}\dfrac{x^2}{a}$.

Answer: $\dfrac{2ax}{x^4+a^2}$.

(35) $f(x) = \sin^{-1}\dfrac{1-x^2}{1+x^2}$.

Answer: $-\dfrac{2\,\mathrm{sgn}\,x}{1+x^2}$ $(x \neq 0)$.

(36) $f(x) = \tan^{-1}(\tan^2 x)$.

Answer: $\dfrac{\sin 2x}{\sin^4 x + \cos^4 x}$

$\left(x \neq \dfrac{2k-1}{2}\pi,\, k \text{ an integer}\right)$.

(37) $f(x) = x + x^2 + x^{x^x}$ $(x > 0)$.

Answer:

$1 + x^x(1 + \ln x) + x^x x^{x^x}\left(\dfrac{1}{x} + \ln x + \ln^2 x\right)$

$(x > 0)$.

(38) $f(x) = x|x|$.

Answer: $2|x|$.

(39) $f(x) = |\sin^3 x|$.

Answer: $\dfrac{3}{2}\sin 2x|\sin x|$.

In problems (40)–(43), determine the inverse f^{-1} of the given function f.

(40) $f(x) = 13x - 5.$

Answer: $f^{-1}(x) = \dfrac{x+5}{13}$, dom $f^{-1} = \mathbb{R}^1.$

(41) $f(x) = 3x^3 + 1.$

Answer: $f^{-1}(x) = \left(\dfrac{x-1}{3}\right)^{1/3}$,

dom $f^{-1} = \mathbb{R}^1.$

(42) $f(x) = (x^3 - 1)^{17}, 0 \leqslant x \leqslant 1.$

Answer: $f^{-1}(x) = \left(1 + x^{\frac{1}{17}}\right)^{1/3}, -1 \leqslant x \leqslant 0.$

(43) $f(x) = \operatorname{sech} x.$

Answer: $\operatorname{sech}^{-1} x = \ln\left(\dfrac{1 + \sqrt{1 - x^2}}{x}\right)$,

$0 < x \leqslant 1.$

In problems (44)–(47), find the derivatives.

(44) $f(x) = \sinh^{-1} 2x_2.$

Answer: $\dfrac{2}{\sqrt{4x^2 + 1}}.$

(45) $f(x) = \tanh^{-1} \sqrt{x}.$

Answer: $\dfrac{1}{2}\sqrt{x}(1 - x).$

(46) $f(x) = \operatorname{sech}^{-1}\left(\dfrac{1}{x}\right).$

Answer: $\dfrac{1}{\sqrt{x^2 - 1}}.$

(47) $f(x) = \ln(\tanh^{-1} x).$

Answer: $\dfrac{1}{[(1 - x^2)\tanh^{-1} x]}.$

In problems (48)–(50), find the derivative $\dfrac{dy}{dx}$.

(48) $x = \sin^2 t; y = \cos^2 t.$

Answer: $-1 (0 < x < 1).$

(49) $x = a\cos t; y = b\sin t.$

Answer: $-\dfrac{b}{a}\cot t (0 < |t| < \pi).$

(50) $x = a\cos^3 t; y = a\sin^3 t.$

Answer: $-\tan t \ (t + \dfrac{2k+1}{2}\pi, k \text{ an integer}).$

In problems (51)–(53), find the indicated higher order derivatives.

(51) If $f(x) = \ln x$ find, $f^{(5)}(x).$

Answer: $f^{(5)}(x) = -\dfrac{6}{x^4} \ (x > 0).$

(52) If $f(x) = e^x \cos x$, find $f^{(iv)}(x).$

Answer: $f^{(iv)}(x) = -4e^x \cos x.$

(53) If $f(x) = x^2 \sin x$, find $f^{(50)}(x).$

Answer: $f^{(50)}(x) = 2^{50}\left(-x^2 \sin 2x + 50x\cos 2x + \dfrac{1225}{2}\sin 2x\right).$

In problems (54)–(56), find the indicated higher order differentials.

(54) If $f(x) = e^x \ln x$, find $d^4 f.$

Answer: $e^x\left(\ln x + \dfrac{4}{x} - \dfrac{6}{x^2} + \dfrac{8}{x^3} - \dfrac{6}{x^4}\right)dx^4.$

(55) If $f(x) = x\cos 2x$, find $d^{10} f.$

Answer: $-1024(x\cos 2x + 5\sin 2x)dx^{10}.$

(56) If $f(x) = x^n e^x$ find $d^n f.$

Answer: $d^n f = e^x\left[x^n + n^2 x^{n-1} + \dfrac{n^2(n-1)^2}{2!}x^{n-2} + \cdots + n!\right]dx^n.$

In problems (57)–(59), use the linear approximation formula to estimate.

(57) $\sqrt[3]{1.02}$.   *Answer:* 1.007.

(58) $\cos 151^o$.   *Answer:* −0.8747.

(59) $\sin 29^o$.   *Answer:* 0.4849.

(60) Prove that for the function $y = a^x$ $(a > 0)$, the subtangent S_T (see Definition 4.5) is constant.

(61) Suppose L is tangent to the curve $y^2 = 2px$ at (x_0, y_0). Prove that
   (a) the subtangent $S_T = 2x_0$,
   (b) the subnormal S_N (see Definition 4.5) is constant.

(62) Find the tangent and normal lines to the curve given by the parametric equations $x = 2t - t^2$, $y = 3t - t^3$ at the indicated value of t.
   (a) $t = 0$. *Answer:* $3x - 2y = 0$ — tangent line, $2x + 3y = 0$ — normal line.
   (b) $t = 1$. *Answer:* $3x - y - 1 = 0$ — tangent line, $x + 3y - 7 = 0$ — normal line.

(63) Find the equation of the evolvent to the ellipse given in parametric form $x = a \cos t$, $y = b \sin t$.

   Answer: $\left(\dfrac{\xi}{b}\right)^{2/3} + \left(\dfrac{\eta}{a}\right)^{2/3} = \left(\dfrac{a^2 - b^2}{ab}\right)^{2/3}.$

(64) Sketch the graph of the equation $r = \dfrac{16}{5 - 3\cos\theta}$.

(65) Find the equation in polar coordinates of the circle that   *Answer:* $r = 2p\cos(\theta - \alpha)$.
   passes through the origin and is centered at the point (p, α).

Chapter 5

Some Basic Properties of Differentiable Functions

Rolle's theorem and the mean value theorems of Lagrange and Cauchy are proved for differentiable functions. Next, Fermat's theorem concerning relative extreme values is given. Then Taylor's formula is investigated, and in particular, its remainder term is given in the different forms due to Lagrange, Cauchy, and Peano. Furthermore, it is indicated why sometimes the Taylor series of a function does not converge to that fuction. The Maclaurin series for elementary functions is derived and power series in both real and complex variables are studied. Finally, Euler's formulas are proved.

5.1 Relative Extrema and Rolle's Theorem

Definition 5.1. Suppose the function f is defined on an open interval Ω and $x_0 \in \Omega$. If for some $\delta > 0$,

$$f(x) \leqslant f(x_0) \ (f(x) \geqslant f(x_0)), \quad x \in (x_0 - \delta, x_0 + \delta) \subset \Omega,$$

we say that the value $f(x_0)$ is a *local*, or *relative*, *maximum (minimum) value* of the function f.

A *local extremum* is a value of f that is either a local maximum or a local minimum.

Theorem 5.1. *If a function f is differentiable at x_0 and $f'(x_0) > 0$ $(f'(x_0) < 0)$, then the function f is increasing (decreasing) at x_0.*

□ Suppose, for instance, that $f'(x_0) > 0$. By the definition of derivative,

$$f'(x_0) = \lim_{x \to x_0} \frac{f(x) - f(x_0)}{x - x_0}.$$

Then by Cauchy's definition of limits, for $\varepsilon = f'(x_0) > 0$, there exists $\delta > 0$ such that

$$\left| \frac{f(x) - f(x_0)}{x - x_0} - f'(x_0) \right| < f'(x_0), \quad x \in (x_0 - \delta, x_0 + \delta),$$

or

$$0 < \frac{f(x) - f(x_0)}{x - x_0} < 2f'(x_0), \quad x \in (x_0 - \delta, x_0 + \delta).$$

E. Mahmudov, *Single Variable Differential and Integral Calculus*,
DOI: 10.2991/978-94-91216-86-2_5, © Atlantis Press 2013

Hence, the numerator and denominator must be positive or negative at the same time, and so, either

$$f(x) < f(x_0), \quad x < x_0,$$

or

$$f(x) > f(x_0), \quad x > x_0.$$

This means that the function f is increasing at x_0. ∎

Theorem 5.2 (Fermat[1]). *If f is differentiable at $x_0 \in \Omega$ and if $f(x_0)$ is either a local maximum value or a local minimum value of f, then $f'(x_0) = 0$.*

☐ Suppose, for instance, that $f(x_0)$ is a local maximum value. The fact that $f'(x_0)$ exists means that both of the one-sided limits

$$\lim_{x \to x_0+0} \frac{f(x) - f(x_0)}{x - x_0} \quad \text{and} \quad \lim_{x \to x_0-0} \frac{f(x) - f(x_0)}{x - x_0}$$

exist and equal $f'(x_0)$. If $x \in (x_0, x_0 + \delta)$, then

$$\frac{f(x) - f(x_0)}{x - x_0} \leqslant 0, \quad \text{and so} \quad f'(x_0) = \lim_{x \to x_0+0} \frac{f(x) - f(x_0)}{x - x_0} \leqslant 0.$$

Similarly, if $x \in (x_0 - \delta, x_0)$,

$$\frac{f(x) - f(x_0)}{x - x_0} \geqslant 0 \quad \text{and} \quad f'(x_0) = \lim_{x \to x_0-0} \frac{f(x) - f(x_0)}{x - x_0} \geqslant 0.$$

Since both $f'(x_0) \leqslant 0$ and $f'(x_0) \geqslant 0$, we may conclude that $f'(x_0) = 0$. ∎

Remark 5.1. The converse of Theorem 5.2 is false. That is, $f'(x_0) = 0$ does not necessarily imply that $f(x_0)$ is a local extremum. For example, consider the function $f(x) = x^3$. Its derivative $f'(x_0) = 3x_0^2 = 0$ at $x_0 = 0$, but $f(x_0) = 0$ is not a local extremum of x^3.

Theorem 5.3 (Rolle[2]). *Suppose that the function f is continuous on the closed interval $[a, b]$ and is differentiable on the open interval (a, b). If $f(a) = f(b)$, then $f'(c) = 0$ at least for some point c in (a, b).*

☐ Since f is continuous on $[a, b]$, it must, by Theorem 3.12, attain both a maximum, say M, and a minimum, say m, value on $[a, b]$. Now, either $M > m$ or $M = m$. If $M = m$, then the function f is constant and so $f'(x) = 0, x \in (a, b)$. If $M > m$, then since $f(a) = f(b)$, there

---

[1]P. Fermat (1601–1665), French mathematician.
[2]M. Rolle (1652–1719), French mathematician.

is at least one point $c \in (a,b)$ such that $f(c) = M$, i.e., $f(c)$ is a local extremum. Then, by Theorem 5.2, $f'(c) = 0$. ∎

Remark 5.2. The assumption of continuity and differentiability in Theorem 5.3 is necessary, for consider the following examples:

(1) $f(x) = \begin{cases} x, & \text{if } a < x \leqslant b, \\ b, & \text{if } x = a. \end{cases}$

This function is discontinuous at $x = a$ and differentiable on (a,b). In spite of the fact that $f(a) = f(b) = b$, the derivative $f'(x) \equiv 1, x \in (a,b)$.

(2) $f(x) = |x|, x \in [-1,1]$ is continuous at $x \neq 0$ and $f(-1) = f(1) = 1$.

Since $f'(x) = -1, -1 < x < 0$, and $f'(x) = +1, 0 < x < 1$, there is no point x satisfying the equation $f'(x) = 0$.

5.2 The Mean Value Theorem and its Applications

Theorem 5.4 (Lagrange's[3], Mean Value Theorem). *Suppose that the function f is continuous on $[a,b]$ and differentiable on (a,b). Then there exists at least one point $c \in (a,b)$ such that*

$$f(b) - f(a) = f'(c)(b-a). \tag{5.1}$$

☐ Define F on $[a,b]$ as follows:

$$F(x) = f(x) - f(a) - \frac{f(b) - f(a)}{b-a}(x-a).$$

It is not hard to see that this is a continuous function, that $F(a) = F(b) = 0$, and that on the open interval (a,b),

$$F'(x) = f'(x) - \frac{f(b) - f(a)}{b-a}.$$

Thus, F satisfies the hypotheses of Theorem 5.3. Then for some point $c \in (a,b)$,

$$F'(c) = f'(c) = \frac{f(b) - f(a)}{b-a} = 0.$$

This ends the proof of Formula 5.1. ∎

The conclusion of the theorem of Lagrange is that there must be at least one point $(c, f(c))$ on the curve $y = f(x)$ at which the tangent line is parallel to the line joining the end points $(a, f(a))$ and $(b, f(b))$.

---

[3]J.-L. Lagrange (1736–1813), French mathematician.

Remark 5.3. Observe that in the special case where $f(a) = f(b)$, Theorem 5.4 coincides with Rolle's Theorem.

Remark 5.4. Take a point $x_0 \in [a,b]$. If $x_0 + \Delta x \in [a,b]$, then by Formula 5.1, there is a point $c \in (x_0, x_0 + \Delta x)$ such that

$$f(x_0 + \Delta x) - f(x_0) = \Delta x f'(c),$$

or

$$f(x_0 + \Delta x) - f(x_0) = \Delta x f'(x_0 + \theta \Delta x) \quad (0 < \theta < 1), \tag{5.2}$$

$$(c = x_0 + \theta \Delta x).$$

As is seen from (5.2), the name, "finite increment formula," is justified.

Corollary 5.1. *If f is differentiable on the interval Ω and $f'(x) = 0$ on Ω, then $f(x)$ is constant on Ω.*

□ Let x_0 be an arbitrary point of Ω. Then since $f'(x) = 0$ on Ω, for arbitrary Δx it follows from (5.2) that $f(x + \Delta x) = f(x_0) = \text{const.}$ ∎

Corollary 5.2. *Let f be a function that is differentiable on (a,b). Then f is nondecreasing (nonincreasing) if and only if $f'(x) \geqslant 0$ ($f'(x) \leqslant 0$) for all $x \in (a,b)$.*

□ (1) Suppose, for definiteness, that $f'(x) \geqslant 0$ for all x in (a,b). If x_1 and x_2 are points in (a,b), then we must show that $f(x_1) \leqslant f(x_2)$ (the nonincreasing case would be similar). We apply Theorem 5.1 to f, but on the interval $[x_1, x_2]$. This gives

$$f(x_2) - f(x_1) = (x_2 - x_1) f'(c), \quad c \in (x_1, x_2). \tag{5.3}$$

Since $x_2 > x_1$ and $f'(c) \geqslant 0$, it follows that $f(x_2) - f(x_1) \geqslant 0$ or $f(x_2) \geqslant f(x_1)$, as we wanted to show.

(2) Let f be differentiable on (a,b) and nondecreasing (nonincreasing) on this interval. We show that $f'(x) \geqslant 0$ ($f'(x) \leqslant 0$), $x \in (a,b)$. Indeed, since f is nondecreasing (nonincreasing) on (a,b), f is nondecreasing (resp., nonincreasing) at every point in (a,b). Then, by Theorem 5.1, we have $f'(x) \geqslant 0$ ($f'(x) \leqslant 0$), $x \in (a,b)$. ∎

Corollary 5.3. *Suppose f is differentiable on (a,b). Then for f to be increasing (decreasing), it is sufficient that $f'(x) > 0$ ($f'(x) < 0$) on this interval.*

☐ Let x_1, $x_2 \in (a,b)$ be points such that $x_1 < x_2$. Then since $f'(x) > 0$, it follows from (5.3) that $f(x_1) < f(x_2)$ ($f'(c) > 0$). ■

Remark 5.5. The condition $f'(x) > 0$ ($f'(x) < 0$) on (a,b) is not necessary for the function to be increasing (decreasing). For example, the function given by $f(x) = x^3$ is increasing on $(-1,1)$, but its derivative $f'(x) = 3x^2$ at $x = 0$ is zero.

Corollary 5.4. *Let f be a function that is continuous on $[a,b]$ and differentiable on (a,b). Then every point x_0 in $[a,b]$ is either a point of continuity of $f'(x)$ or a discontinuity of the second kind. In other words, x_0 can not be either a removable discontinuity or a discontinuity of the first kind of the derivative function.*

☐ If the limit $\lim_{x \to x_0+0} f'(x)$ ($x_0 \in [a,b]$) exists, then since $x_0 + \theta(x - x_0) \to x_0$ as $(0 < \theta < 1)$ the limit

$$\lim_{x \to x_0+0} f'(x_0 + \theta(x - x_0)) = \lim_{x \to x_0+0} f'(x) \qquad (5.4)$$

exists.

On the other hand, setting $x_0 + \Delta x = x$, we have, from (5.2), that

$$f(x) - f(x_0) = f'(x_0 + \theta(x - x_0))(x - x_0).$$

Then it follows from (5.4) that

$$\lim_{x \to x_0+0} f'(x) = \lim_{x \to x_0+0} f'(x_0 + \theta(x - x_0)) = \lim_{x \to x_0+0} \frac{f(x) - f(x_0)}{x - x_0} = f'(x_0).$$

This means that $f'(x)$ is continuous from the right at x_0.

Similarly, it can be shown that if $\lim_{x \to x_0-0} f'(x)$ exists, then $\lim_{x \to x_0-0} f'(x) = f'(x_0)$ and so $f'(x)$ is continuous from the left at point $x = x_0$. Consequently, $f'(x)$ is continuous at x_0. If at least one of these limits does not exists, then x_0 is a point of discontinuity of the second kind. ■

Example 5.1. The function given by

$$f(x) = \begin{cases} x^2 \sin \dfrac{1}{x}, & \text{if } x \neq 0, \\ 0, & \text{if } x = 0, \end{cases}$$

is differentiable on $(-\infty, +\infty)$. Indeed,

$$f'(x) = 2x \sin \frac{1}{x} - \cos \frac{1}{x}, \quad x \neq 0, \quad \text{and}$$

$$f'(0) = \lim_{\Delta x \to 0} \frac{f(\Delta x) - f(0)}{\Delta x} = \lim_{\Delta x \to 0} \frac{(\Delta x)^2 \sin \dfrac{1}{\Delta x}}{\Delta x} = 0, \quad x = 0.$$

Thus, $f'(x)$ is defined on $(-\infty, +\infty)$ and is continuous at any point $x \neq 0$. But as is easily seen from the expression of $f'(x)$, its limit does not exists a $x \to 0$. Therefore, the point $x = 0$ is a discontinuity of the second kind.

Example 5.2. Prove that $|\sin x_1 - \sin x_2| \leqslant |x_1 - x_2|$. We apply the mean value theorem (Theorem 5.4) to $f(x) = \sin x$ on the interval $[x_1, x_2]$. Then

$$\sin x_1 - \sin x_2 = (x_1 - x_2)f'(c), \quad f'(x) = \cos x, \quad c \in (x_1, x_2).$$

Since $|\cos c| \leqslant 1$ for every c, the desired inequality follows immediately from this.

Example 5.3. Prove that $|\cot^{-1} x_1 - \cot^{-1} x_2| \leqslant |x_1 - x_2|$. As in the previous example, by Theorem 5.4,

$$\cot^{-1} x_1 - \cot^{-1} x_2 = (x_1 - x_2)f'(c),$$
$$f(x) = \cot^{-1} x, \quad c \in (x_1, x_2).$$

It only remains to apply the inequality $f'(c) = \dfrac{1}{1+c^2} \leqslant 1$.

5.3   Cauchy's Mean Value Theorem

Theorem 5.5 (Cauchy). *Suppose that f and g are continuous on the closed interval $[a,b]$ and differentiable on the open interval (a,b). Moreover, suppose that $g'(x) \neq 0$ for all x in (a,b). Then there exists at least one point c in (a,b) such that*

$$\frac{f(b)-f(a)}{g(b)-g(a)} = \frac{f'(c)}{g'(c)}. \tag{5.5}$$

☐   We first show that $g(a) - g(b) \neq 0$. Indeed, if $g(a) - g(b) = 0$ or $g(a) = g(b)$, then by Rolle's theorem (Theorem 5.3), there exists $c \in (a,b)$ such that $g'(c) = 0$. But by hypothesis, $g'(x) \neq 0$, $x \in (a,b)$. This contradiction proves that $g(a) - g(b) \neq 0$. Hence we can define the following function

$$F(x) = f(x) - f(a) - \frac{f(b)-f(a)}{g(b)-g(a)}[g(x)-g(a)]. \tag{5.6}$$

It is clear that F is continuous on $[a,b]$ and differentiable on (a,b). Since, by Rolle's theorem there exists a point c in (a,b) with

$$F'(c) = 0. \tag{5.7}$$

Then we have, from (5.6) and (5.7),

$$f'(c) - \frac{f(b)-f(a)}{g(a)-g(b)}g'(c) = 0.$$

If we divide this relation by $g'(c) \neq 0$, we obtain (5.5). ∎

Remark 5.6. To see that (5.5) is a generalization of Lagrange's mean value theorem, we take $g(x) = x$. Then $g'(x) = 1$ and (5.5) reduces to (5.1).

Theorem 5.6 (Darboux[4]). *Let f be differentiable on the closed interval $[a,b]$ and*

$$f'(a) < \xi < f'(b) \quad (f'(a) > \xi > f'(b)).$$

Then there exists at least one point $c \in (a,b)$ such that

$$f'(c) = \xi.$$

□ Consider the function g defined by $g(x) = f(x) - \xi x$.
Obviously, $g(x)$ is differentiable on $[a,b]$ and so continuous on it.
We show that there exists $\delta > 0$ such that $g(x) < g(a)$ for all x satisfying $a < x < a + \delta$.
Indeed, since

$$g'(a) = \lim_{x \to a} \frac{g(x) - g(a)}{x - a},$$

the function defined as

$$h(x) = \begin{cases} \dfrac{g(x) - g(a)}{x - a}, & \text{if } x \neq a, \\ g'(a), & \text{if } x = a, \end{cases} \tag{5.8}$$

is continuous at the point $x = a$. By Theorem 3.8, there exists a $\delta > 0$ such that $h(x) < 0$ $(g'(a) < 0)$, $a < x < a + \delta$. Then, by (5.8), $g(x) < g(a)$, $a < x < a + \delta$. Similarly, we can show that $g(x) < g(b)$, $b - \delta_1 < x < b$ for some $\delta_1 > 0$. Hence, by Theorem 3.12, there exists $c \in (a,b)$ such that $g(c) \leqslant g(x)$. Thus, according to Theorem 5.2, $g'(c) = 0$, or, $f'(c) = \xi$. ∎

Remark 5.7. It follows from the theorem that the derivative $f'(x)$ attains every intermediate value between any two of its values. In other words, the set of values $f'(x)$ is an interval.

5.4 Taylor's Formula

In Chapter 2 we concentrated on infinite series of constant terms. But much of the practical importance of infinite series derives from the fact that many functions have useful representations as infinite series with variable terms. Many such series representations can be deduced from Taylor's formula.

---
[4]G. Darboux (1842–1917), French mathematician.

Theorem 5.7 (Taylor[5]). *Suppose that n is an arbitrary positive integer and the $(n+1)$-st derivative of the function f exists throughout some neighborhood of a. Suppose further that x is any point of this neighborhood, and that p is an arbitrary positive number. Then there exists a point ξ between a and x such that*

$$f(x) = f(a) + \frac{f'(a)}{1!}(x-a) + \frac{f''(a)}{2!}(x-a)^2 + \cdots + \frac{f^{(n)}(a)}{n!}(x-a)^n + R_{n+1}(x), \quad (5.9)$$

where

$$R_{n+1}(x) = \left(\frac{x-a}{x-\xi}\right)^p \frac{(x-\xi)^{n+1}}{n!p} f^{(n+1)}(\xi). \quad (5.10)$$

Formula (5.9) is called *Taylor's formula* and $R_{n+1}(x)$ is called the *remainder term in general form.*

□ Let $F(x,a)$ be the n-th degree polynomial of the right hand side of (5.9):

$$F(x,a) = f(a) + \frac{f'(a)}{1!}(x-a) + \cdots + \frac{f^{(n)}(a)}{n!}(x-a)^n. \quad (5.11)$$

Denote by $R_{n+1}(x)$ the difference

$$R_{n+1}(x) = f(x) - F(x,a). \quad (5.12)$$

We must show that $R_{n+1}(x)$ defined as in (5.12) can be represented in the form (5.10). Let x be a fixed point in the neighborhood of a and for definiteness, suppose $x > a$. Denote by φ the function defined on $[a,x]$ as follows:

$$\varphi(t) = f(x) - F(x,t) - (x-t)^p Q(x), \quad (5.13)$$

$$Q(x) = \frac{R_{n+1}(x)}{(x-a)^p}, \quad t \in [a,x], \quad (5.14)$$

or, in more detail:

$$\varphi(t) = f(x) - f(t) - \frac{f'(t)}{1!}(x-t) - \cdots - \frac{f^{(n)}(t)}{n!}(x-t)^n - (x-t)^p Q(x). \quad (5.15)$$

It follows from (5.15) that φ is continuous on $[a,x]$ and differentiable on (a,x). Setting $t = a$ in (5.13), and taking into account (5.14), we have

$$\varphi(a) = f(x) - F(x,a) - R_{n+1}(x).$$

Then, by (5.12), $\varphi(a) = 0$. On the other hand, at once from (5.15) we have $\varphi(x) = 0$. Thus φ satisfies all the hypotheses of Rolle's theorem (Theorem 5.3) on $[a,x]$. Consequently, for some $c \in (a,x)$,

$$\varphi'(c) = 0. \quad (5.16)$$

---

[5]B. Taylor (1685–1731), English mathematician.

By differentiation of (5.15), we have

$$\varphi'(t) = -f'(t) + \frac{f'(t)}{1!} - \frac{f''(t)}{1!}(x-t) + \frac{f''(t)}{2} \cdot 2(x-t)$$

$$- \cdots + \frac{f^{(n)}(t)}{n!}n(x-t)^{n-1} - \frac{f^{(n+1)}(t)}{n!}(x-t)^n + p(x-t)^{n-1}Q(x),$$

or, after simplification,

$$\varphi(t) = -\frac{f^{(n+1)}(t)}{n!}(x-t)^n + p(x-t)^{(p-1)}Q(x). \tag{5.17}$$

Taking $t = c$ in (5.17) and using (5.16), we see that

$$Q(x) = \frac{(x-\xi)^{n-p+1}}{n!p}f^{(n+1)}(\xi). \tag{5.18}$$

Comparing (5.14) and (5.18), we have the required formula, (5.10), for $R_{n+1}(x)$. The case $x < a$ is proved analogously. ∎

Example 5.4. To an n-th degree polynomial

$$f(x) = A_0x^n + A_1x^{n-1} + \cdots + A_{n-1}x + A_n,$$

with known coefficients $A_0, A_1, \ldots, A_n$, we can apply Taylor's formula (5.9). Since $f^{(n+1)}(x) \equiv 0$, (5.10) implies $R_{n+1}(x) \equiv 0$. So (5.10) has the form:

$$f(x) = f(a) + \frac{f'(a)}{1!}(x-a) + \frac{f''(a)}{2!}(x-a)^2 + \cdots + \frac{f^{(n)}(a)}{n!}(x-a)^n.$$

5.5 Lagrange's, Cauchy's, and Peano's Forms for the Remainder Term; Maclaurin's formula

First of all we investigate the remainder $R_{n+1}(x)$ given by formula (5.10). Since ξ is a point between a and x, there exists $0 < \theta < 1$ such that $\xi = a + \theta(x-a)$. It should be pointed out that ξ, and hence θ, depend on p. Then it is evident that $x - \xi = (x-a)(1-\theta)$, and (5.10) can be rewritten as:

$$R_{n+1}(x) = \frac{(x-a)^{(n+1)}(1-\theta)^{(n-p+1)}}{n!p}f^{(n+1)}(a+\theta(x-a)). \tag{5.19}$$

Consider the formula (5.19) with $p = n+1$ and $p = 1$. In the case of $p = n+1$, we have formulas of Lagrange and Cauchy, respectively, for the remainder:

$$R_{n+1}(x) = \frac{(x-a)^{n+1}}{(n+1)!}f^{(n+1)}(a+\theta(x-a)), \tag{5.20}$$

$$R_{n+1}(x) = \frac{(x-a)^{n+1}(1-\theta)^n}{n!}f^{(n+1)}(a+\theta(x-a)). \tag{5.21}$$

In general, (5.20) and (5.21) can be used to obtain approximate values for $f(x)$ at some fixed point x $(x \neq a)$. Sometimes in numerical analysis it is very important to determine the order of the error with respect to small $(x - a)$. To this end, the so-called Peano[6] form of the remainder term is used.

Suppose that $f^{(n+1)}(x)$ is continuous at the point $x = a$. Since ξ depends on x, the composite, $f^{(n+1)}(\xi(x))$, is continuous at a. Moreover, $\xi(x) \to a$ as $x \to a$, and so

$$\lim_{x \to a} \left(f^{(n+1)}(\xi(x)) - f^{(n+1)}(a) \right) = 0.$$

Using this last equality in (5.20), we can write

$$R_{n+1}(x) = \frac{f^{(n+1)}(a)}{(n+1)!}(x-a)^{n+1} + \frac{f^{(n+1)}(\xi) - f^{(n+1)}(a)}{(n+1)!}(x-a)^{n+1}$$

$$= \frac{f^{(n+1)}(a)}{(n+1)!}(x-a)^{n+1} + o\left((x-a)^{n+1}\right).$$

Thus, Taylor's formula has the form:

$$f(x) = f(a) + \frac{f'(a)}{1!}(x-a) + \cdots + \frac{f^{(n+1)}(a)}{(n+1)!}(x-a)^{n+1} + o\left((x-a)^{n+1}\right).$$

Here,

$$R_{n+1}(x) = o\left((x-a)^{n+1}\right) \tag{5.22}$$

is Peano's form of the remainder.

If we take, in (5.9), $a = 0$, then we have the Maclaurin[7] formula:

$$f(x) = f(0) + \frac{f'(0)}{1!}x + \cdots + \frac{f^{(n)}(0)}{n!}x^n + R_{n+1}(x). \tag{5.23}$$

The Lagrange, Cauchy, and Peano forms of the remainder are, respectively:

$$R_{n+1}(x) = \frac{x^{n+1}}{(n+1)!}f^{(n+1)}(\theta x) \qquad (0 < \theta < 1),$$

$$R_{n+1}(x) = \frac{x^{n+1}(1-\theta)}{n!}f^{(n+1)}(\theta x) \quad (0 < \theta < 1), \tag{5.24}$$

$$R_{n+1}(x) = o\left(x^{n+1}\right).$$

Remark 5.8. By Taylor's formula for an $(n+1)$-times differentiable function, we may use an easily computable polynomial $F(x, a)$, see (5.11), and different remainder formulas, to approximate $f(x)$ to any desired degree of accuracy simply by choosing n sufficiently large.

[6]G. Peano (1858–1932), Italian mathematician.
[7]C. Maclaurin (1698–1746), Scottish mathematician.

5.6   Taylor Series; Maclaurin Series Expansion of Some Elementary Functions

1. Taylor and Maclaurin series.   Suppose that f is a function with continuous derivatives of all orders in a neighborhood of a point a of an interval Ω. Then for every $n \geqslant 0$, we can write Taylor's formula, (5.9), as follows:

$$f(x) = f(a) + \frac{f'(a)}{1!}(x-a) + \cdots + \frac{f^{(n)}(a)}{n!} + R_{n+1}(x). \qquad (5.25)$$

Definition 5.2. The infinite series

$$f(a) + \frac{f'(a)}{1!}(x-a) + \cdots + \frac{f^{(n)}(a)}{n!}(x-a)^n + \cdots = \sum_{n=0}^{\infty} \frac{f^{(n)}(a)}{n!}(x-a)^n \qquad (5.26)$$

is called the *Taylor series* of the function f at the point $x = a$ of the interval Ω. By (5.25), the Taylor series (5.26) converges to $f(x)$ for a fixed value of $x \in \Omega$ if and only if

$$\lim_{n \to \infty} R_{n+1}(x) = 0. \qquad (5.27)$$

In particular, if $a = 0$, the series (5.26) is called the *Maclaurin series*.

We give necessary and sufficient conditions for the convergence of the series (5.26) to the value $f(x)$.

Theorem 5.8. *Suppose that the function f has derivatives of all orders on an interval Ω and there exists a number $M > 0$ such that $\left| f^{(n)}(x) \right| \leqslant M^n$ for all $n \geqslant 0$ and arbitrary $x \in \Omega$. Then the Taylor series converges to the value $f(x)$ throughout the interval Ω.*

☐   By hypothesis, it is sufficient to show the validity of (5.27). We use Lagrange's form of the remainder for Taylor's formula:

$$R_{n+1}(x) = \frac{(x-a)^{n+1} f^{(n+1)}(\xi)}{(n+1)!}.$$

Since $\left| f^{(n)}(\xi) \right| \leqslant M^n$, it follows that

$$|R_{n+1}(x)| \leqslant \frac{(M|x-a|)^{n+1}}{(n+1)!}. \qquad (5.28)$$

As was shown before (see Example 2.8),

$$\lim_{n \to \infty} \frac{(M|x-a|)^{n+1}}{(n+1)!} = 0$$

and so, by (5.28), $\lim\limits_{n \to \infty} R_{n+1}(x) = 0.$ ∎

Corollary 5.5. *Suppose that f and its derivatives of all orders are bounded on the interval Ω, that is, there exists a number $M > 0$ such that*

$$\left| f^{(n)}(x) \right| \leqslant M$$

for all $n = 0, 1, 2, \ldots$ and arbitrary $x \in \Omega$. Then the series (5.26) converges to the value $f(x)$ throughout Ω.

Remark 5.9. Sometimes it occurs that the Taylor series of a function does not converge to the function. For example, consider

$$f(x) = \begin{cases} e^{-\frac{1}{x^2}}, & \text{if } x \neq 0, \\ 0, & \text{if } x = 0. \end{cases}$$

This function has derivatives of every order. Indeed, for $x \neq 0$, this is clear: the derivatives of f at $x \neq 0$ consist of

$$f^{(n)}(x) = P_n\left(\frac{1}{x}\right) e^{-\frac{1}{x^2}} \quad (n = 0, 1, 2, \ldots) \tag{5.29}$$

where $P_n(t)$ ($t = 1/x$) is a polynomial. Thus the derivatives $f^{(n)}(x)$ are linear combinations of expressions $\frac{1}{x^k} e^{-\frac{1}{x^2}}$ ($k = 0, 1, 2, \ldots$). We will show that

$$\lim_{x \to 0} \frac{1}{x^k} e^{-\frac{1}{x^2}} = 0. \tag{5.30}$$

By substituting $t = 1/x^2$,

$$\lim_{x \to 0} \left| \frac{1}{x^k} e^{-\frac{1}{x^2}} \right| = \lim_{t \to \infty} \frac{t^{\frac{k}{2}}}{e^t} = 0 \quad (k = 0, 1, 2, \ldots),$$

and so we have (5.30). By using the definition, we next compute the derivatives at $x = 0$. We have, by the above,

$$\lim_{x \to 0} P_n\left(\frac{1}{x}\right) e^{-\frac{1}{x^2}} = 0 \quad (n = 0, 1, 2, \ldots).$$

By mathematical induction, it is easy to see that $f^{(n)}(0) = 0$ ($n = 0, 1, 2, \ldots$). Indeed, for $n = 0$ we have $f(0) = 0$. Suppose, then, that. Then

$$f^{(n)}(0) = \lim_{x \to 0} \frac{f^{(n-1)}(x) - f^{(n-1)}(0)}{x} = \lim_{x \to 0} \frac{1}{x} P_{n-1}\left(\frac{1}{x}\right) e^{-\frac{1}{x^2}} = 0.$$

Consequently, the Taylor series of f at the point $a = 0$ is equal to zero. But this implies that the series does not converge to $f(x)$ on any interval.

(1) Consider the exponential function, $f(x) = e^x$. Now, $f^{(n)}(x) = e^x$ ($n = 0, 1, 2, \ldots$). Hence, at $a = 0$, we have that the Maclaurin series for the function e^x is

$$1 + \frac{x}{1!} + \cdots + \frac{x^n}{n!} + \cdots = \sum_{n=0}^{\infty} \frac{x^n}{n!}. \tag{5.31}$$

Clearly, for any r and any, the inequality $\left| f^{(n)}(x) \right| \leqslant e^r$ ($n = 0, 1, 2, \ldots$) is satisfied. Then by Theorem 5.8, the sum of the series (5.31) on the interval $(-r, r)$ is e^x. Since r is arbitrary, the Maclaurin series of the exponential function,

$$e^x = 1 + \frac{x}{1!} + \cdots + \frac{x^n}{n!} + \cdots = \sum_{n=0}^{\infty} \frac{x^n}{n!} \quad (-\infty < x < +\infty),$$

converges everywhere.

(2) The function $f(x) = \sin x$ is defined everywhere and has derivatives of all order:

$$f^{(n)}(x) = \sin\left(x + n\frac{\pi}{2}\right), \quad f^{(2k)}(0) = 0, \quad f^{(2k-1)}(0) = (-1)^{k+1}.$$

Putting $a = 0$, we have the Maclaurin series,

$$x - \frac{x^3}{3!} + \frac{x^5}{5!} + \cdots + (-1)^{k+1}\frac{x^{2k-1}}{(2k-1)!} + \cdots = \sum_{k=1}^{\infty}(-1)^{k+1}\frac{x^{2k-1}}{(2k-1)!}. \tag{5.32}$$

Since $|f^{(n)}(x)| \leqslant 1$ $(n \geqslant 0)$, $x \in (-\infty, +\infty)$, the series (5.32) converges to the function $\sin x$ by Theorem 5.8:

$$\sin x = \sum_{k=1}^{\infty}(-1)^{k+1}\frac{x^{2k-1}}{(2k-1)!} \quad (-\infty < x < +\infty).$$

Similarly, for the function $f(x) = \cos x$, the Maclaurin series converges to the function,

$$\cos x = \sum_{k=0}^{\infty}(-1)^{k+1}\frac{x^{2k}}{2k!}, \quad \text{for } -\infty < x < +\infty.$$

(3) Considering the function $f(x) = \ln(1+x)$ $(x > -1)$, we have:

$$f^{(n)}(x) = (-1)^{n+1}\frac{(n-1)!}{(1+x)^n} \quad (n = 1, 2, \ldots).$$

In particular, $f^{(n)}(0) = (-1)^{n+1}(n-1)!$ at $x = 0$ and so the Maclaurin series is

$$x - \frac{x^2}{2} + \cdots + (-1)^{n+1}\left(\frac{x^n}{n}\right) + \cdots = \sum_{n=1}^{\infty}(-1)^{n+1}\frac{x^n}{n}. \tag{5.33}$$

On the other hand, it follows from Lagrange's form of the remainder that

$$|R_{n+1}(x)| = \frac{|f^{(n+1)}(\xi)x^{n+1}|}{(n+1)!} = \frac{x^{n+1}}{(1+\xi)^{n+1}(n+1)} \leqslant \frac{1}{n+1} \quad (0 < \xi < 1).$$

Therefore $\lim_{n\to\infty} R_{n+1}(x) = 0$ when $0 \leqslant x \leqslant 1$. Now, for any $x \in (-1, 0)$, we choose a number r $(0 < r < 1)$ satisfying $-r < x < 0$. By using Cauchy's form for the remainder, it follows that

$$|R_{n+1}(x)| = \left|\frac{f^{(n+1)}(\xi)(x-\xi)^n x}{n!}\right| = \left|\frac{(x-\xi)^n x}{(1+\xi)^{n+1}}\right| < \frac{r^n}{1-r} \quad (x < \xi < 0),$$

because

$$|x - \xi| = \xi - x < -x < r \text{ and } 1 + \xi > 1 + x > 1 - r.$$

Thus,

$$\lim_{n\to\infty} R_{n+1}(x) = 0 \quad (-1 < x < 0),$$

and, finally, this gives us the convergence:

$$\ln(1+x) = \sum_{n=1}^{\infty}(-1)^{n+1}\frac{x^n}{n} \quad (-1 < x \leqslant 1). \tag{5.34}$$

At $x = 1$, we obtain the remarkable formula,

$$\ln 2 = 1 - \frac{1}{2} + \frac{1}{3} - \frac{1}{4} + \cdots + \frac{(-1)^{n+1}}{n} + \cdots.$$

Remark 5.10. By D'Alembert's test (see Corollary 2.2), we see that the series (5.34) diverges for points satisfying the inequality $|x| > 1$. On the other hand, at $x = -1$ we have the harmonic series, which is divergent. Thus, in spite of the fact that the function $f(x) = \ln(1+x)$ is defined on the interval $(-1, +\infty)$, the representation (5.34) is true only on the interval $(-1, 1]$.

(4) Let $f(x) = (1+x)^\alpha$ ($x > -1$, α a real number).

Since $f^{(n)}(x) = \alpha(\alpha - 1) \cdots (\alpha - n + 1)(1 + x^{\alpha-n})$, it follows that $f^{(n)}(0) = \alpha(\alpha - 1) \cdots (\alpha - n + 1)$. For this reason, the Maclaurin series is

$$(1+x)^\alpha = 1 + \alpha x + \frac{\alpha(\alpha - 1)}{2!} x^2 + \cdots + \frac{\alpha(\alpha - 1) \cdots (\alpha - n + 1)}{n!} x^n + \cdots. \qquad (5.35)$$

If, here, $\alpha = m$ is a positive integer, then the coefficient of x^n in (5.35) vanishes for $n > m$ and this series reduces to Newton's binomial formula:

$$(1+x)^m = 1 + mx + \cdots + \frac{m(m - 1) \cdots (m - n + 1)}{n!} x^n + \cdots + x^m$$

(in order to write the binomial formula of $(a+x)^n$ we use the formula $(a+x)^n = a^n \left(1 + \frac{x}{a}\right)^n$). To determine the interval of convergence of the binomial series, we let

$$u_n = \frac{\alpha(\alpha - 1)(\alpha - 2) \cdots (\alpha - n + 1)}{n!} x^n.$$

Then it is easy to see that

$$\lim_{n \to \infty} \left| \frac{u_{n+1}}{u_n} \right| = \lim_{n \to \infty} \left| \frac{(\alpha - n)x}{n + 1} \right| = |x|.$$

Hence the D'Alembert ratio test shows that the series (5.35) converges absolutely if $|x| < 1$ and diverges if $|x| > 1$. Its convergence at the points $x = \pm 1$ depends upon the value of α; we shall not pursue this.

In particular, if in (5.35) we take $\alpha = -1$, the series simplifies to the geometric series,

$$\frac{1}{1+x} = 1 - x + x^2 - x^3 + \cdots + (-1)^n x^n + \cdots.$$

(5) Let $f(x) = \tan^{-1} x$ It is not hard to establish that (see (4.29))

$$f^{(n)}(x) = \frac{(n - 1)!}{(1 + x^2)^{\frac{n}{2}}} \cdot \sin \left[n \left(\tan^{-1} x + \frac{\pi}{2} \right) \right].$$

Then

$$f^{(n)}(0) = \begin{cases} 0, & \text{if } n \text{ is even,} \\ (-1)^{\frac{n-1}{2}} (n - 1)!, & \text{if } n \text{ is odd.} \end{cases}$$

And so the Maclaurin series is

$$x - \frac{x^3}{3} + \frac{x^5}{5} - \frac{x^7}{7} + \cdots + (-1)^{\frac{n-1}{2}} \frac{x^n}{n} + \cdots.$$

Here n is an odd number.

By Lagrange's form for the remainder,

$$R_{n+2}(x) = \frac{x^{n+2}}{(n+2)} \frac{\sin\left[(n+2)\left(\tan^{-1}\theta x + \frac{\pi}{2}\right)\right]}{\left[1+(\theta x)^2\right]^{\frac{n+2}{2}}} \quad (0 < \theta < 1).$$

Thus for every $x \in [-r, r]$ $(r > 0)$,

$$|R_{n+2}(x)| < \frac{r^{n+2}}{n+2}. \tag{5.36}$$

If, in (5.36), we take $r \leqslant 1$, then $\lim\limits_{n\to\infty} R_{n+2}(x) = 0$, and so

$$\tan^{-1}(x) = x - \frac{x^3}{3} + \frac{x^5}{5} - \frac{x^7}{7} + \cdots + (-1)^{\frac{n-1}{2}} \frac{x^n}{n} + \cdots.$$

If we substitute $x = 1$ into this series, we obtain Leibniz's series

$$\frac{\pi}{4} = 1 - \frac{1}{3} + \frac{1}{5} - \frac{1}{7} + \cdots.$$

(6) In order to find the Maclaurin series for $\sinh x$ and $\cosh x$, we should use the Maclaurin series for the exponential functions e^x and e^{-x}. Thus,

$$\sinh x = \sum_{k=0}^{\infty} \frac{x^{2k+1}}{(2k+1)!} = x + \frac{x^3}{3!} + \frac{x^5}{5!} + \cdots + \frac{x^{2k+1}}{(2k+1)!} + \cdots, \quad -\infty < x < +\infty,$$

and

$$\cosh x = \sum_{k=0}^{\infty} \frac{x^{2k}}{2k!} = 1 + \frac{x^2}{2!} + \frac{x^4}{4!} + \cdots + \frac{x^{2k}}{(2k)!} + \cdots, \quad -\infty < x < +\infty.$$

2. Computations Using Maclaurin Series.

As is shown in Section 2.3, the sequence $\left\{\left(1+\frac{1}{n}\right)^n\right\}$ is monotone and its limit is e. On the other hand, if we take $x = 1$ in (5.31),

$$e = 1 + \frac{1}{1!} + \frac{1}{2!} + \cdots + \frac{1}{n!} + \cdots. \tag{5.37}$$

Since $k! \geqslant 2^{k-1}$ for $k \geqslant 2$, it follows from (5.37) that

$$e < 2 + \frac{1}{2} + \frac{1}{2^2} + \cdots + \frac{1}{2^{n-1}} + \cdots = 2 + 1 = 3.$$

Thus $2 \leqslant e \leqslant 3$.

By the first formula of (5.24),

$$R_{n+1}(x) = \frac{x^{n+1}}{(n+1)!} e^{\theta x} \quad (0 < \theta < 1),$$

and so the inequality

$$|R_{n+1}(1)| \leqslant \frac{e}{(n+1)!} < \frac{3}{(n+1)!} \tag{5.38}$$

is satisfied. On the other hand, by Maclaurin's formula,

$$e = 1 + \frac{1}{1!} + \frac{1}{2!} + \cdots + \frac{1}{n!} + R_{n+1}(1). \tag{5.39}$$

Therefore, it follows from (5.38) and (5.39) that the larger we choose n, the more accurate the approximation. An appropriate value for n may be found by trial and error. For example, using a hand-held calculator or computer, one can evalute $\dfrac{3}{(n+1)!}$ for $n = 0, 1, 2, \ldots$ until a value of n satisfying (5.38) is obtained. Now, using (5.39), we show that the number e is irrational.

From (5.38) and the expression $R_{n+1}(1) = \dfrac{e^{\theta}}{(n+1)!}$ $(0 < \theta < 1)$, we have

$$\frac{1}{(n+1)!} < R_{n+1}(1) < \frac{3}{(n+1)!}. \tag{5.40}$$

Suppose that e is rational and has the form $e = \dfrac{m}{n}$ (where m and n are positive integers). Without loss of generality, we may take $n \geqslant 2$. Let us choose n as the denumerator of the quotient $e = \dfrac{m}{n}$ and multiply both sides (5.39) by $n!$. Then $n!e$ and $n!\left(1 + \dfrac{1}{1!} + \cdots + \dfrac{1}{n!}\right)$ are integers, but at the same time, $n!R_{n+1}(1)$ satisfies the inequality (5.40), so $\dfrac{1}{n+1} < n!R_{n+1}(1) < \dfrac{3}{n+1}$ is not an integer. Thus, by multiplying the Maclaurin series (5.39) by $n!$, we have the inequality

$$n!e - n!\left(1 + \frac{1}{1!} + \cdots + \frac{1}{n!}\right) = n!R_{n+1}(1),$$

where the left hand side is an integer, but the right hand side is not integral. This contradiction proves that the number e is irrational.

Next we show how the Maclaurin series can be used to obtain approximate values for trigonometric functions. Since the values of the trigonometric functions $\sin x$ and $\cos x$ at any point x can be expressed through their values on $\left[0, \dfrac{\pi}{4}\right]$, without loss of generality we may assume $x \in \left[0, \dfrac{\pi}{4}\right]$. For example, for four decimal place accuracy, we choose $n = 5$, $r = \dfrac{\pi}{4}$. Lagrange's form for the remainder term has the form

$$R_{n+1}(x) = \frac{x^{n+2}}{(n+2)!} \sin\left(\theta x + n\frac{\pi}{2} + \pi\right) \quad (0 < \theta < 1),$$

and so, throughout the interval $[-r, r]$ $(r > 0)$,

$$|R_{n+1}(x)| \leqslant \frac{r^{n+2}}{(n+2)!}. \tag{5.41}$$

Thus, for given values of n and r, we have, from (5.41),

$$|R_7(x)| \leqslant \frac{\left(\dfrac{\pi}{4}\right)^7}{7} < 10^{-4}.$$

Consequently, for $|x| \leqslant \dfrac{\pi}{4}$, we have, to within an accuracy of 10^{-4},

$$\sin x \approx x - \frac{x^3}{6} + \frac{x^5}{120}.$$

Similarly, taking $n = 6$ and $r = \dfrac{\pi}{4}$ in Maclaurin's series for $\cos x$, we obtain

$$|R_{n+2}(x)| = |R_8| \leqslant \frac{\left(\dfrac{\pi}{4}\right)^8}{8!} < 10^{-5},$$

and so, for x satisfying the inequality $|x| \leqslant \dfrac{\pi}{4}$, we have, to within an accuracy of 10^{-5},

$$\cos x \approx 1 - \frac{x^2}{2} + \frac{x^4}{24} - \frac{x^6}{720}.$$

5.7   Power Series

1. Series with Real Terms.   **Definition 5.3.** If a_k $(k = 0, 1, 2, \ldots)$ are constants and x is a real variable, then a series of the form

$$a_0 + a_1 x + a_2 x^2 + \cdots + a_n x^n + \cdots = \sum_{k=0}^{\infty} a_k x^k \tag{5.42}$$

is called a *power series* in x.

The following theorem is the fundamental result on the convergence of power series.

Theorem 5.9 (Abel[8]).   *For any power series* (5.42) *in* x,
(1) *if the series converges for some* $x_0 \neq 0$, *then, for all* x *satisfying the inequality* $|x| < |x_0|$, *the series converges absolutely;*
(2) *if the series diverges at* x_1, *then the series diverges for all* x *satisfying* $|x| > |x_1|$.

◻ (1) By hypothesis, the numerical series

$$a_0 + a_1 x_0 + a_2 x_0^2 + \cdots + a_n x_0^n + \cdots \tag{5.43}$$

converges, and hence $\lim\limits_{n \to \infty} a_n x_0^n = 0$. Then there exists a number $C > 0$ such that $|a_n x_0^n| < C$ $(n = 0, 1, 2, \ldots)$.

At first we rewrite (5.42) as follows:

$$a_0 + a_1 x_0 \left(\frac{x}{x_0}\right) + a_2 x_0^2 \left(\frac{x}{x_0}\right)^2 + \cdots + a_n x_0^n \left(\frac{x}{x_0}\right)^n + \cdots \tag{5.44}$$

---

[8]H. Abel (1802–1829), Norwegian mathematician.

and consider the series

$$C + C\left|\frac{x}{x_0}\right| + C\left|\frac{x}{x_0}\right|^2 + \cdots + C\left|\frac{x}{x_0}\right|^n + \cdots. \tag{5.45}$$

At once, we see that for $|x| < |x_0|$, the series (5.45) is a geometric series, and converges. It follows that the series (5.44), and so (5.42), converges absolutely.

(2) Suppose that for some x_1 the series (5.42) diverges. Then (5.42) diverges for all x satisfying $|x| > |x_1|$. Indeed, if for some x such that $|x| > |x_1|$ the series (5.42) converges, then case (1) of the theorem shows that for all $|x_1| < |x|$, the series converges. This contradiction ends the proof of theorem.

Thus, by Abel's theorem, the series (5.42) converges absolutely for all x in the open interval $(-|x_0|, |x_0|)$ if it converges at x_0, and diverges for all x in the set $\mathbb{R} \setminus [-|x_1|, |x_1|]$ if it diverges at some x_1.   ■

Definition 5.4. If there exists a number $R > 0$ such that the power series (5.42) converges absolutely for $|x| < R$ and diverges for $|x| > R$, then R is called the *radius of convergence* of the power series (5.42).

In order to determine R, we use D'Alembert's test (Proposition 2.2) for a series with the terms $|a_n||x|^n$. We find that in order to have

$$\lim_{n\to\infty}\left|\frac{a_{n+1}x^{n+1}}{a_n x^n}\right| = \lim_{n\to\infty}\left|\frac{a_{n+1}}{a_n}\right||x| < 1,$$

we must have

$$|x| < \lim_{n\to\infty}\left|\frac{a_n}{a_{n+1}}\right| = R.$$

Then, since $\sum_{n=0}^{\infty}|a_n||x|^n$ converges for $-R < x < R$, the series $\sum_{n=0}^{\infty}a_n x^n$ converges absolutely. Similarly, by Cauchy's test (Proposition 2.3), the radius of convergence is

$$R = \frac{1}{\lim_{n\to\infty}\sqrt[n]{|a_n|}}.$$

In particular, R may be 0 or $+\infty$. From the last two formulas, we see that

$$\lim_{n\to\infty}\left|\frac{a_{n+1}}{a_n}\right| = \lim_{n\to\infty}\sqrt[n]{|a_n|}.$$

Indeed, the validity of this formula is implied by problems (13) and (15) of Section 2.7, taking $x_n = \ln|a_n|$, $y_n = n$ and $x_n = \left|\frac{a_{n+1}}{a_n}\right|$, respectively.

We formulate this result as the following theorem.

Theorem 5.10. *The interval of convergence of a power series is $(-R, R)$. At the points $x = \pm R$, the series may converge absolutely, converge conditionally, or diverge, depending on the particular series.*

Example 5.5. Find the interval of convergence and radius of convergence of the series $\sum\limits_{n=0}^{\infty} \dfrac{x^n}{n!}$.

By applying the ratio test, we have

$$R = \lim_{n \to \infty} \left| \frac{a_n}{a_{n+1}} \right| = \lim_{n \to \infty} \frac{(n+1)!}{n!} = \lim_{n \to \infty} (n+1) = \infty.$$

Thus, the radius of convergence is $R = +\infty$ and the interval of convergence is $(-\infty, +\infty)$.

Example 5.6. Find the interval of convergence and radius of convergence of $\sum\limits_{n=0}^{\infty} x^n$.

In this case, $R = \lim\limits_{n \to \infty} \left| \dfrac{a_n}{a_{n+1}} \right| = 1$, and so the radius of convergence is $(-1,1)$. Clearly, at the points $x = \pm 1$, the series converges.

Example 5.7. For the series $\sum\limits_{n=1}^{\infty} (-1)^{n+1} \dfrac{(2x)^n}{n}$, the radius of convergence is $R = \lim\limits_{n \to \infty} \dfrac{2^n(n+1)}{2^{n+1}n} = \dfrac{1}{2}$ and the interval of convergence is $\left(-\dfrac{1}{2}, \dfrac{1}{2} \right)$. It is easy to see that at $x = \dfrac{1}{2}$, the series is the alternating harmonic series and so converges conditionally. At $x = -\dfrac{1}{2}$, the series diverges, as it is the harmonic series.

Corollary 5.6. *For $r < R$, the power series (5.42) converges uniformly on $[-r, r]$, and the sum of the series is a continuous function.*

$\square$ In fact, for $|x| < r$, the series $\sum\limits_{n=0}^{\infty} |a_n| r^n$ majorizes the power series (5.42). Then (Theorems 3.18 and 3.20) the series converges uniformly on the indicated interval and its sum is continuous. $\blacksquare$

We formulate the next theorem without proof.

Theorem 5.11. *(1) Let $\{f_n(x)\}$ be a sequence of differentiable functions defined on an interval, $[a, b]$. Suppose that $\{f_n(x_0)\}$ converges for some $x_0 \in [a, b]$, and $\{f'(x)\}$ is uniformly convergent on $[a, b]$. Then $\{f_n(x)\}$ converges uniformly to $f(x)$ on $[a, b]$ and*

$$f'(x) = \lim_{n \to \infty} f_n'(x), \quad x \in [a, b].$$

(2) Let $\sum\limits_{n=1}^{\infty} u_n(x)$ be a series with terms which are differentiable on $[a, b]$ and suppose that the series $\sum\limits_{n=0}^{\infty} u_n(x_0)$ converges for some $x_0 \in [a, b]$, and that $\sum\limits_{n=0}^{\infty} u_n'(x)$ converges uniformly on $[a, b]$. Then the series $\sum\limits_{n=1}^{\infty} u_n(x)$ converges uniformly on $[a, b]$ to a function $s(x)$ and

$$s'(x) = \sum_{n=1}^{\infty} u_n'(x), \quad x \in [a, b].$$

Theorem 5.12. *The power series* (5.42) *is termwise differentiable to all orders on an open interval* $(-R,R)$, *where R is a nonzero radius of convergence.*

☐   We will show that the series (5.42) and the series obtained from it by termwise differentiation have the same radius of convergence. Indeed, if R' is the radius of convergence of the series $\sum\limits_{n=1}^{\infty} na_n x^{n-1}$, then

$$R' = \frac{1}{\lim\limits_{n\to\infty} \sqrt[n]{n|a_n|}} = \frac{1}{\lim\limits_{n\to\infty} \sqrt[n]{n} \cdot \lim\limits_{n\to\infty} \sqrt[n]{|a_n|}} = R.$$

On the other hand, by Corollary 5.6, for any $x \in [-r,r]$ $(r < R)$, the termwise differentiated series converges uniformly on $[-r,r]$ and the series (5.42) is termwise differentiable on $[-r,r]$ (Theorem 5.11). For higher order derivatives, the proof is analogous (by mathematical induction).   ■

Remark 5.11. The termwise integration of a power series will be considered in Chapter 9.

Example 5.8. Find a power series representation for the function $\dfrac{1}{1+x^2}$.

We write the power series for the arctangent function

$$\tan^{-1} x = x - \frac{x^3}{3} + \cdots + (-1)^k \frac{x^{2k+1}}{2k+1} + \cdots \quad (|x| < 1).$$

Since $(\tan^{-1} x)' = \dfrac{1}{1+x^2}$, it follows from Theorem 5.12 that

$$\frac{1}{1+x^2} = 1 - x^2 + \cdots + (-1)^k x^{2k} + \cdots \quad (|x| < 1).$$

Example 5.9. Let $\{f_n\}$ be the sequence on $[0,1]$ given by $f_n(x) = \dfrac{\sin nx}{\sqrt{n}}$ $(n = 1, 2, \ldots)$. Obviously, $f(x) = \lim\limits_{n\to\infty} f_n(x) = 0$. On the other hand, since $f'_n(0) = \sqrt{n} \to +\infty$, $n \to \infty$, the sequence $f'_n(x) = \sqrt{n} \cos nx$ is not uniformly convergent on $[0,1]$. Moreover, the sequence $f'_n(x)$ does not converge pointwise to $f'(x)$ (since $f'(x) = 0$).

Corollary 5.7. *The power series* (5.42) *is the Taylor series for its sum on the interval of convergence.*

☐   Let $s(x) = a_0 + a_1 x + \cdots + a_n x^n + \cdots$. Then, by differentiating n-times,

$$s^{(n)}(x) = n!a_n + (n+1)n \cdots 2a_{n+1}x + \cdots.$$

Now, by substituting $x = 0$, we obtain that $a_n = \dfrac{s^{(n)}(0)}{n!}$.   ■

Remark 5.12. In order to investigate the power series of the form $\sum\limits_{n=0}^{\infty} a_n(x-x_0)^n$, we use the substitution $x - x_0 = X$. Then the power series obtained, $\sum\limits_{n=0}^{\infty} a_n X^n$, has the form (5.42). All the above properties for such power series hold in the interval of convergence $(x_0 - R, x_0 + R)$. For example, consider the series $\sum\limits_{n=1}^{\infty} (x-3)^n$. By substituting $x - 3 = X$, we get a new power series of the form $\sum\limits_{n=1}^{\infty} X^n$. This series converges absolutely on the interval $(-1, 1)$. Consequently, the given power series has the interval of convergence $-1 < x - 3 < 1$, or, $2 < x < 4$.

2. Power Series with Complex Terms. Let $\{z_n\} = \{x_n + iy_n\}$ be a sequence of complex numbers (see Section 3.9). Consider the series

$$z_1 + z_2 + \cdots + z_n + \cdots = \sum_{n=1}^{\infty} z_n \qquad (5.46)$$

with complex terms z_n. The limit (provided that this limit exists) of the n-th partial sums $Z_n = \sum\limits_{k=1}^{n} z_k$ of the series (5.46) is the sum of the series. If the series of absolute values $\sum\limits_{n=1}^{\infty} |z_n|$ converges, then the series (5.46) is said to converge absolutely.

Let $w = f(z)$, $z \in D$, be a complex valued function defined on a domain D.

The basic definitions and conceptions for real-valued functions (for example, the limit of functions, continuity, derivative, differential, etc.) can be generalized to the case of complex valued functions. For example, by mathematical induction it is not hard to see that the derivative of the function $f(z) = z^n$ is $(z^n)' = nz^{n-1}$. In particular, for a sequence of complex valued functions $\{f_n(z)\}$ (or series $\sum\limits_{n=1}^{\infty} u_n(z)$), the usual results (as above) are true.

Definition 5.5. If $\{a_n\}$ is a sequence of complex numbers then the series $\sum\limits_{n=0}^{\infty} a_n(z-z_0)^n$ is said to be a series with complex terms. If its radius of convergence $R = 0$, then this series converges only at the point z_0. Note that for finite $R > 0$ the series converges on the circle $|z - z_0| < R$. And $R = +\infty$ means that the series converges on the entire complex plane.

Definition 5.6. For a complex number z, the exponential functon is defined as the sum of the series $e^z = \sum\limits_{n=0}^{\infty} \dfrac{z^n}{n!}$.

By D'Alembert's ratio test, this series converges for each z, because

$$\lim_{n\to\infty} \left| \frac{z^{n+1}n!}{(n+1)!z^n} \right| = \lim_{n\to\infty} \frac{|z|}{n+1} = 0 < 1.$$

It is easy to see that, for arbitrary z_1 and z_2, we have $e^{z_1} \cdot e^{z_2} = e^{z_1 + z_2}$. By substituting $z = iy$ (for y on the real axis) in the expression of e^z, we can write

$$e^{iy} = 1 + \frac{iy}{1!} - \frac{y^2}{2!} - \frac{iy^3}{3!} + \frac{y^4}{4!} + \cdots + \frac{i^n y^n}{n!} + \cdots$$

$$= \left(1 - \frac{y^2}{2!} + \frac{y^4}{4!} + \cdots + (-1)^k \frac{y^{2k}}{(2k)!} + \cdots\right) + i\left(\frac{y}{1!} - \frac{y^3}{3!} + \cdots + (-1)^{k+1} \frac{y^{2k-1}}{(2k-1)!} + \cdots\right)$$

$$= \cos y + i \sin y.$$

Finally,

$$e^{iy} = \cos y + i \sin y. \tag{5.47}$$

In particular, $e^{\pi i} = -1$, $e^{2\pi i} = 1$, and $e^{z + 2\pi i} = e^z \cdot e^{2\pi i} = e^z$.

From (5.47) follow the famous formulas of Euler,

$$\cos y = \frac{e^{iy} + e^{-iy}}{2} \quad \text{and} \quad \sin y = \frac{e^{iy} - e^{-iy}}{2i}.$$

Let us introduce the hyperbolic functions defined on the complex plane:

$$\sinh z = \frac{e^z - e^{-z}}{2i}, \quad \cosh z = \frac{e^z + e^{-z}}{2}.$$

By comparing with Euler's formula, we derive

$$\cos z = \cosh iz, \quad \cosh z = \cos iz, \quad \sin z = -i \sinh iz, \quad \text{and} \quad \sinh z = -i \sin iz.$$

Let $z = |z|(\cos \theta + i \sin \theta)$ ($\theta = \arg z$) be the polar form of the complex number z. Note that the polar form is expressed in terms of the magnitude $r = |z|$. Then, by (5.47), we have the exponential form $z = |z|e^{i\theta}$ of the complex number z. On the other hand, for $z = x + iy$,

$$e^{x+iy} = e^x e^{iy} = e^x(\cos y + i \sin y),$$

and so this formula can be accepted as an alternative definition of e^z.

In summary, since a power series is differentiable to every order in the circle of convergence, from Definition 5.6 it follows that

$$(e^z)' = \frac{1}{1!} + \frac{2z}{2!} + \cdots + \frac{nz^{n-1}}{n!} + \cdots = 1 + \frac{z}{1!} + \cdots + \frac{z^{n-1}}{(n-1)!} + \cdots = e^z,$$

or

$$(e^z)' = e^z.$$

5.8　Problems

In problems (1)–(5), show that the given function satisfies the hypotheses of Rolle's theorem on the indicated interval $[a, b]$, and find all numbers c in $[a, b]$ that satisfy the conclusion of the theorem.

(1)　$f(x) = (x-1)(x-2)(x-3)$,　$x \in [1, 3]$.

(2)　$f(x) = x^2 - 3x + 2$,　　　　$x \in [1, 2]$.

(3)　$f(x) = \sin^2 x$,　　　　　　$x \in [0, \pi]$.

(4)　$f(x) = x^3 + 5x^2 - 6x$,　　$x \in [0, 1]$.

(5)　$f(x) = \cos^2 x$,　　　　　$x \in \left[-\dfrac{\pi}{4}, \dfrac{\pi}{4} \right]$.

(6) Suppose that the function f, defined on the interval (a, b), has a finite derivative f', and that $\lim\limits_{x \to a+0} f(x) = \lim\limits_{x \to b-0} f(x)$. Prove that $f'(c) = 0$ for some $c \in (a, b)$.

(7) Show that between the roots of the function f defined by $f(x) = \sqrt[3]{x^2 - 5x + 6}$, there exists one root of the equation $f'(x) = 0$.

(8) Show that the function $f(x) = 1 - \sqrt[3]{x^2}$ defined on the interval $[-1, 1]$ does not satisfy the conclusion of Rolle's theorem on the indicated interval.

(9) Suppose a secant line passes through the two points, $(-1, 1)$ and $(2, 8)$, of the curve $f(x) = x^3$. Determine the tangent line to the curve which is parallel to this secant. *Answer*: $A(-1, -1); B(1, 1)$.

(10) At which point is the tangent line to the curve $f(x) = \ln x$ parallel to the secant passing through the points $A(1, 0)$ and $B(e, 1)$? *Answer*: $(e - 1, \ln(e - 1))$.

(11) Suppose the function f is differentiable everywhere, and for all x and Δx $f(x + \Delta x) - f(x) \equiv \Delta x f'(x)$. Show that $f(x) = kx + b$ (with k and b constants).

(12) If all roots of the polynomial $P_n(x) = c_0 x^n + c_1 x^{n-1} + \cdots + c_n$ $(c_0 \neq 0)$ with real coefficients are real, then show that the roots of the polynomials $P_n'(x), P_n''(x), \ldots, P_n^{(n-1)}(x)$ are real.

(13) Suppose the function f is differentiable on (a, b) and its derivative f' is continuous on this interval. Then, for arbitrary $c \in (a, b)$, do there exist points x_1, x_2 such that

$$f(x_2) - f(x_1) = f'(c)(x_2 - x_1)?$$

Answer: Such points do not always exist.

(14) If $0 < b < a$, prove that $\dfrac{a - b}{a} < \ln \dfrac{a}{b} < \dfrac{a - b}{b}$.

(15) If the function f is differentiable on the interval $(a, +\infty)$ and $\lim\limits_{x \to +\infty} f'(x) = 0$, then prove that, as $x \to +\infty$, $f(x) = o(x)$.

(16) Suppose that the function f is differentiable on $[a,b]$ and $ab > 0$. Prove that

$$\frac{1}{a-b}\begin{vmatrix} a & b \\ f(a) & f(b) \end{vmatrix} = f(c) - cf'(c),$$

where $a < c < b$.

(17) Determine whether each of the functions $f(x) = x^2$ and $g(x) = x^3$ defined on $[-1,1]$ satisfies or does not satisfy the conclusion of the mean value theorem.

(18) Suppose that a function f defined on (a,b) has a bounded derivative f'. Prove that this function is uniformly continuous on this interval.

(19) If for all x, $f'(x) = k$ (k a constant), then show that $f(x) = kx + b$.

(20) Suppose for a function f the following conditions are satisfied:

(a) The second derivative, $f''(x)$, exists on $[a,b]$.

(b) $f'(a) = f'(b) = 0$.

Then prove that at least for one point $c \in (a,b)$, we have the inequality $|f''(c)| \geqslant \dfrac{4}{(b-a)^2}|f(b) - f(a)|$.

(21) Suppose that the function f is differentiable on $[a,b]$, $f(a) = 0$, and that there exists a real number A such that $|f'(x)| \leqslant A|f(x)|$ for all x in $[a,b]$. Prove that $f(x) = 0$, $x \in [a,b]$.

Suggestion: Take $M_0 = \sup\limits_{a \leqslant x \leqslant x_0} |f(x)|$ and $M_1 = \sup\limits_{a \leqslant x \leqslant x_0} |f'(x)|$ for some fixed $x_0 \in [a,b]$. Then $|f(x)| \leqslant M_1(x_0 - a) \leqslant A(x_0 - a)M_0$ for all x in $[a,x_0]$. And $A(x_0 - a) \leqslant 1$ implies that $M_0 = 0$.

(22) Express the polynomial $x^4 - 5x^3 + 5x^2 + x + 2$ in a power series in $(x-2)$.

Answer: $-7(x-2) - (x-2)^2 + 3(x-2)^3 + (x-2)^4$.

(23) Express the polynomial $1 + 3x + 5x^2 - 2x^3$ in a power series in $(x+1)$.

Answer: $5 - 13(x+1) + 11(x+1)^2 - 2(x+1)^3$.

(24) Write the Taylor formula for the function $f(x) = \sqrt{x}$ when $a = 1$ and $n = 3$.

Answer: $\sqrt{x} = 1 + \dfrac{x-1}{1} \cdot \dfrac{1}{2} - \dfrac{(x-1)^2}{1 \cdot 2} \cdot \dfrac{1}{4} + \dfrac{(x-1)^3}{1 \cdot 2 \cdot 3} \cdot \dfrac{3}{8}$

$\qquad\qquad - \dfrac{(x-1)^4}{4!} \cdot \dfrac{15}{16}[1 + \theta(x-1)]^{-\frac{7}{2}}, 0 < \theta < 1$.

(25) Write the Maclaurin formula of $f(x) = \sqrt{1+x}$ when $n = 2$.

Answer: $\sqrt{1+x} = 1 + \dfrac{1}{2}x - \dfrac{1}{8}x^2 + \dfrac{x^3}{16(1+\theta x)^{\frac{5}{2}}}, 0 < \theta < 1$.

In problems (26)–(28), estimate the error.

(26) $e^x \approx 1 + x + \dfrac{x^2}{2!} + \cdots + \dfrac{x^n}{n!}$, $0 \leqslant x \leqslant 1$. *Answer*: $< \dfrac{3}{(n+1)!}$.

(27) $\tan x \approx x + \dfrac{x^3}{3}$. *Answer*: $< 2 \cdot 10^{-6}$.

(28) $\sin x \approx x - \dfrac{x^3}{6}$. *Answer*: $\leqslant \dfrac{1}{3840}$.

(29) (a) Determine the number of decimal places of accuracy the given approximation formula yields for $|x| \leqslant 0.1$.

$$\ln(1+x) \approx x - \frac{1}{2}x^2 + \frac{1}{3}x^3 - \frac{1}{4}x^4.$$

Answer: Five-place accuracy.

(b) For what range of values of x can $\sin x$ be aproximated by $x - \dfrac{x^3}{3!}$ with three decimal place accuracy? *Answer*: $|x| < 0.569$.

In problems (30)–(34), determine for which values of x the series converges.

(30) $1 + \dfrac{x}{2} + \dfrac{x^2}{4} + \cdots + \dfrac{x^n}{2^n} + \cdots$. *Answer*: $-2 < x < 2$.

(31) $3x + 3^4 x^4 + 3^9 x^9 + \cdots + 3^{n^2} x^{n^2} + \cdots$. *Answer*: $|x| < \dfrac{1}{3}$.

(32) $x - \dfrac{x^2}{2^2} + \dfrac{x^3}{3^2} - \dfrac{x^4}{4^2} + \cdots + (-1)^{n+1}\dfrac{x^n}{n^2} + \cdots$. *Answer*: $-1 \leqslant x \leqslant 1$.

(33) $\sin x + 2\sin\dfrac{x}{3} + 4\sin\dfrac{x}{9} + \cdots + 2^n \sin\dfrac{x}{3^n} + \cdots$. *Answer*: $-\infty < x < \infty$.

(34) $x + \dfrac{2!}{2^2}x^2 + \dfrac{3!}{3^3}x^3 + \cdots + \dfrac{n!}{n^n}x^n + \cdots$. *Answer*: $-e < x < e$.

(35) $x + 2x^2 + \cdots + nx^n + \cdots$. *Answer*: $|x| < 1$.

(36) $\dfrac{x}{1+\sqrt{1}} + \dfrac{x^2}{2+\sqrt{2}} + \cdots + \dfrac{x^n}{n+\sqrt{n}} + \cdots$. *Answer*: $-1 \leqslant x < 1$.

In each of the problems (37)–(39), determine whether the series is majorized or not.

(37) $1 + \dfrac{x}{1} + \dfrac{x^2}{2} + \cdots + \dfrac{x^n}{n} + \cdots$, $(0 \leqslant x \leqslant 1)$. *Answer*: No.

(38) $\dfrac{\sin x}{1^2} + \dfrac{\sin 2x}{2^2} + \cdots + \dfrac{\sin nx}{n^2} + \cdots$, $[0, 2\pi]$. *Answer*: Yes.

(39) $1 + \dfrac{x}{1^2} + \dfrac{x^2}{2^2} + \cdots + \dfrac{x^n}{n^2} + \cdots$, $(0 \leqslant x \leqslant 1)$. *Answer*: Yes.

(40) If $S(x) = \sum\limits_{n=0}^{\infty} \dfrac{x^n}{n!}$, then prove that $S(x) \cdot S(y) = S(x+y)$.

(41) Using the series $\sin x = \sum\limits_{n=0}^{\infty} (-1)^n \dfrac{x^{2n+1}}{(2n+1)!}$ and $\cos x = \sum\limits_{n=0}^{\infty} (-1)^n \dfrac{x^{2n}}{(2n)!}$, prove that

(a) $\sin x \cos x = \dfrac{1}{2}\sin 2x$ and

(b) $\sin^2 x + \cos^2 x = 1$.

(42) Suppose that R_1, R_2 are the radii of convergences of the series $\sum\limits_{n=0}^{\infty} a_n x^n$ and $\sum\limits_{n=0}^{\infty} b_n x^n$, respectively. Determine the radius of convergence R for the following series.

(a) $\sum\limits_{n=0}^{\infty} (a_n + b_n) x^n$. *Answer:* $R \geqslant \min(R_1, R_2)$.

(b) $\sum\limits_{n=0}^{\infty} a_n b_n x^n$. *Answer:* $R \geqslant R_1 \cdot R_2$.

In problems (43)–(45), using termwise differentiation, find the sum of the series.

(43) $x + \dfrac{x^3}{3} + \dfrac{x^5}{5} + \cdots$. *Answer:* $\dfrac{1}{2} \ln \dfrac{1+x}{1-x}$ $(|x| < 1)$.

(44) $1 + \dfrac{x^2}{2!} + \dfrac{x^4}{4!} + \cdots$. *Answer:* $\cosh x$ $(|x| < \infty)$.

(45) $\dfrac{x}{1 \cdot 2} + \dfrac{x^2}{2 \cdot 3} + \dfrac{x^3}{3 \cdot 4} + \cdots$. *Answer:* $1 + \dfrac{1-x}{x} \ln(1-x)$ $(|x| \leqslant 1)$.

(46) Find the n-th degree Taylor polynomial of $f(x) = \ln x$ at $a = 1$. *Answer:* $P_n(x) = (x-1) - \dfrac{1}{2}(x-1)^2 + \cdots + \dfrac{(-1)^{n-1}}{n}(x-1)^n$.

(47) Find the Taylor series representation for the function $\tan x$ around $a = \dfrac{\pi}{4}$. *Answer:*
$1 + 2\left(x - \dfrac{\pi}{4}\right) + 2\left(x - \dfrac{\pi}{4}\right)^2 + \cdots$.

(48) Find the inverse functions of the following functions.

(a) $w = \sin z$. *Answer:* $\sin^{-1} z = -i\ln\left(iz + \sqrt{1 - z^2}\right)$.

(b) $w = \cos z$. *Answer:* $\cos^{-1} z = -i\ln\left(z + \sqrt{z^2 - 1}\right)$.

(49) Use Euler's formula to prove that $\cos^2 y = \dfrac{1}{2}(1 + \cos 2y)$.

(50) Find the exponential representation of the complex number $-i$. *Answer:* $e^{-\frac{\pi}{2}i}$.

Chapter 6

Polynomials and Interpolations

If selected values of a function are given in a tabular representation, it is often desirable to obtain an easily computable formula which yields both those given values and approximations for in-between points, not included in the table. This procedure is called interpolation. In this chapter, we obtain formulas of Lagrange and Newton for polynomials which accomplish this interpolation. This necessitates the study of the factorization of polynomials. Finally, we study the approximation by polynomials of a function defined on a closed interval.

6.1 Factorization of Polynomials

If n is a positive integer, then a function $P(x)$ of the form

$$P(x) = a_0 x^n + a_1 x^{n-1} + \cdots + a_n \tag{6.1}$$

is called a *polynomial function of degree n*. Here, the coefficients a_i $(i = 0, 1, \ldots, n)$ may be real or complex numbers. The number a_0 is called the *leading coefficient*. The independent variable x may be real or complex. A value of x satisfying the equation $P(x) = 0$ is called a *root of the polynomial*.

Theorem 6.1. *If α is any number, then $P(x) = (x - \alpha)P_1(x) + R$, where $P_1(x)$ is the polynomial of degree $n - 1$ and $R = P(\alpha)$ is the remainder.*

☐ If we divide the polynomial $P(x)$ by $x - \alpha$, we have the polynomial $P_1(x)$ of degree $n - 1$ and the remainder R, which is constant. Thus we can write

$$P(x) = (x - \alpha)P_1(x) + R. \tag{6.2}$$

This equality is true for all x such that $x \neq a$. Then by passing to the limit in (6.2) as $x \to \alpha$, we have $R = P(\alpha)$. ∎

E. Mahmudov, *Single Variable Differential and Integral Calculus*,
DOI: 10.2991/978-94-91216-86-2_6, © Atlantis Press 2013

Corollary 6.1. *If α is a root of the polynomial* (6.1), *that is,* $P(\alpha) = 0$, *then in* (6.2) *the remainder is zero, i.e.,* $R = 0$, *and so* $x - \alpha$ *is a factor of* $P(x)$:

$$P(x) = (x - \alpha)P_1(x),$$

where $P_1(x)$ *is a polynomial of degree* $(n - 1)$.

The Fundamental Theorem of Algebra is easy to state but difficult to prove. That is why the proof of theorem is omitted.

Theorem 6.2 (The Fundamental Theorem of Algebra). *The equation* $P(x) = 0$ *has at least one (real or complex) root.*

Example 6.1. We give an example that shows that if P is not a polynomial, the conclusion might not be true. Consider the nonalgebraic equation $e^x = 0$. This equation has no roots. On the contrary, suppose $x_0 = a + bi$ is a root, that is, $e^{a+bi} = 0$. By Euler's formula (Section 5.7), we have $e^a(\cos b + i \sin b) = 0$. Since for any b the square root $\sqrt{\cos^2 b + \sin^2 b} = 1$, it follows that $\cos b + i \sin b \neq 0$. On the other hand, $e^a \neq 0$ for any real number a. Thus $e^{a+bi} \neq 0$, and $x_0 = a + bi$ is not a root. The obtained contradiction ends the proof.

Using Theorem 6.2, we can prove the following theorem.

Theorem 6.3. *Every polynomial* (6.1) *of degree* $n \geq 1$ *can be represented as a product of* n *linear factors of the form* $x - \alpha_i$ $(i = 1, 2, \ldots, n)$ *and the leading coefficient* a_0,

$$P(x) = a_0(x - \alpha_1)(x - \alpha_2) \cdots (x - \alpha_n).$$

☐ By Theorem 6.2 the polynomial $P(x) = a_0 x^n + a_1 x^{n-1} + \cdots + a_n$ of degree n has at least one root α_1. Then by Corollary 6.1

$$P(x) = (x - \alpha_1)P_1(x).$$

Here, $P_1(x)$ is a polynomial of degree $n - 1$. We denote the root of $P_1(x)$ by α_2. Then

$$P_1(x) = (x - \alpha_2)P_2(x)$$

and $P_2(x)$ is a polynomial of degree $n - 2$. Similarly,

$$P_2(x) = (x - \alpha_3)P_3(x).$$

The process evidently can be continued until we arrive at a polynomial P_n of degree zero. Thus, we have

$$P_{n-1}(x) = (x - \alpha_n)P_n.$$

It is clear that $P_n = a_0$.

Therefore, it follows from these equalities that

$$P(x) = a_0(x - \alpha_1)(x - \alpha_2) \cdots (x - \alpha_n). \tag{6.3}$$

As can be seen from (6.3), the α_i ($i = 1, \ldots, n$) are roots of the polynomial $P(x)$. ∎

Furthermore, no $\alpha \neq \alpha_i$ ($i = 1, \ldots, n$) can be a root of the polynomial $P(x)$, because in such a case the right hand side of (6.3) at the point $x = \alpha$ is not zero.

Corollary 6.2. *A polynomial of degree $n \geqslant 1$ has at most n distinct roots.*

Example 6.2. (a) The polynomial $P(x) = x^3 - 6x^2 + 11x - 6$ of degree three is equal to zero at $x = 1$, that is, $P(1) = 0$. Then $R = 0$ and

$$x^3 - 6x^2 + 11x - 6 = (x - 1)(x^2 - 5x + 6).$$

(b) The roots of the same polynomial are $\alpha_1 = 1$, $\alpha_2 = 2$, $\alpha_3 = 3$. Since in this polynomial $a_0 = 1$, by Theorem 6.2 we have the following factorization:

$$x^3 - 6x^2 + 11x - 6 = (x - 1)(x - 2)(x - 3).$$

Theorem 6.4. *If the values of two polynomials $P_1(x)$ and $P_2(x)$ of degree $n \geqslant 1$ are equal at $n + 1$ distinct points, $\alpha_0, \alpha_1, \ldots, \alpha_n$, then they are identical.*

☐ Denote the difference $P_1(x)$ and $P_2(x)$ by $P(x)$:

$$P(x) = P_1(x) - P_2(x). \tag{6.4}$$

Obviously, the degree of $P(x)$ is not more than n, and $P_1(\alpha_i) = P_2(\alpha_i)$ ($i = 1, \ldots, n$). Then,

$$P(x) = a_0(x - \alpha_1)(x - \alpha_2) \cdots (x - \alpha_n). \tag{6.5}$$

On the other hand, by hypothesis, $P(\alpha_0) = 0$, and at α_0 all the factors in (6.5) are nonzero. Therefore $a_0 = 0$, and so $P(x) \equiv 0$. Then (6.4) gives the desired result, $P_1(x) \equiv P_2(x)$. ∎

Theorem 6.5. *If a polynomial $P(x) = a_0x^n + a_1x^{n-1} + \cdots + a_n$ is identically equal to zero, then all its coefficients are equal to zero, i.e., $a_i = 0$, $i = 0, \ldots, n$.*

☐ Since $P(x) \equiv 0$ identically, then by Theorem 6.3,

$$a_0(x - \alpha_1)(x - \alpha_2) \cdots (x - \alpha_n) = 0. \tag{6.6}$$

On the other hand, $P(x) \equiv 0$ implies that the equality (6.6) is true at any point $x \neq \alpha_i$ $(i = 1, \ldots, n)$. Since the factors in (6.6) are nonzero for such an x, it follows that $a_0 = 0$. Similarly, we can prove that $a_1 = 0$, $a_2 = 0$, etc.   ∎

Theorem 6.6. *If two polynomials are equal identically, then the corresponding coefficients of these polynomials are equal to each other.*

☐   Indeed since the difference of those polynomials is zero identically, it follows from Theorem 6.5 that the differences of the corresponding coefficients are equal to zero.   ∎

Example 6.3. If the polynomial $a_0 x^3 + a_1 x^2 + a_2 x + a_3$ is identically equal to $x^2 - 5$, then $a_0 = 0$, $a_1 = 1$, $a_2 = -5$, $a_3 = 0$.

6.2   Repeated Roots of Polynomials

In the factorization of a polynomial (6.5) of degree n, some of the factors can be repeated. Therefore, a polynomial (6.5) in general has the form

$$P(x) = a_0(x - \alpha_1)^{k_1} (x - \alpha_2)^{k_2} \cdots (x - \alpha_m)^{k_m}, \tag{6.7}$$

where

$$k_1 + k_2 + \cdots + k_m = n.$$

According to (6.7), α_i $(i = 1, \ldots, m)$ is a repeated root of multiplicity k_i $(i = 1, \ldots, m)$. In other words, the existence of a repeated root α of multiplicity k means that the number of equal roots α of the polynomial $P(x)$ is k.

The definition of repeated root and Theorem 6.3 imply the following result.

Theorem 6.7. *The number of roots of a polynomial (6.1) of degree n is equal to exactly n (not necessarily real or distinct).*

Theorem 6.8. *If α_1 is a root of multiplicity $k_1 > 1$ of a polynomial $P(x)$, then for the derivative $P'(x)$, the number α_1 is a root of multiplicity $k_1 - 1$.*

☐   By hypothesis, α_1 is a repeated root of multiplicity k_1. Hence, by (6.7),

$$P(x) = (x - \alpha_1)^{k_1} \varphi(x), \tag{6.8}$$

where $\varphi(x) = (x - \alpha_2)^{k_2} \cdots (x - \alpha_m)^{k_m}$ is nonzero at $x = \alpha_1$, that is, $\varphi(\alpha_1) \neq 0$. By differentiating (6.7), we have

$$P'(x) = k_1 (x - \alpha_1)^{k_1 - 1} \varphi(x) + (x - \alpha_1)^{k_1} \varphi'(x) = (x - \alpha_1)^{k_1 - 1} \left[k_1 \varphi(x) + (x - \alpha_1) \varphi'(x) \right].$$

Writing $\psi(x) = k_1 \varphi(x) + (x - \alpha_1)\varphi'(x)$, we have

$$P'(x) = (x - \alpha_1)^{k_1 - 1} \psi(x).$$

Here, $\psi(\alpha_1) = k_1 \varphi(\alpha_1) + (\alpha_1 - \alpha_1)\varphi'(\alpha_1) = k_1 \varphi(\alpha_1) \neq 0$. This means that the number $x = \alpha_1$ is a repeated root of multiplicity $k_1 - 1$ of the polynomial $P'(x)$. At the same time, from the proof we see that if α_1 is a simple root of $P(x)$, then α_1 can not be a root of $P'(x)$. ∎

Also, it is not hard to see that

$$P(\alpha_1) = 0, \ P'(\alpha_1) = 0, \ P''(\alpha_1) = 0, \ldots, \ P^{(k_1-1)}(\alpha_1) = 0, \ P^{(k_1)}(\alpha_1) \neq 0.$$

The definition of repeated root and Theorem 6.3 imply the following result.

6.3 Complex Roots of Polynomials

As can be seen from Formula (6.5), the roots $\alpha_1, \alpha_2, \ldots, \alpha_n$ may be real or complex.

Theorem 6.9. *If the polynomial $P(x)$ has real coefficients, then if $a + ib$ is a complex root of $P(x)$, the conjugate $a - ib$ is also a root of $P(x)$.*

□ It is not hard to see that if we substitute $a + ib$ in polynomial $P(x)$, then after necessary transformations we see that $P(a + ib)$ is a complex number having the form

$$P(a + ib) = A + iB, \tag{6.9}$$

where the expressions A and B do not involve the imaginary unit i.

By hypothesis, $a + ib$ is a root, that is, $P(a + ib) = 0$. Hence, $A + iB = 0$, which implies that $A = 0$ and $B = 0$.

On the other hand it is easy to see that by substitution $x = a - ib$ in $P(x)$ as a result of arithmetic operations with complex numbers (see Section 1.5) it follows from (6.9) that we obtain simply its conjugate number, that is,

$$P(a - ib) = A - iB. \tag{6.10}$$

But since $A = B = 0$, then (6.9) implies that $P(a - ib) = 0$. Consequently, $a - ib$ is also a root of the real algebraic equation $P(x) = 0$. Thus, if in the factorization

$$P(x) = a_0(x - \alpha_1) \cdots (x - \alpha_n)$$

there exists a complex root $a + ib$, then its conjugate $a - ib$ is a root, too.

Now, take $\alpha = a + ib$, $\overline{\alpha} = a - ib$ and compute $(x - \alpha)(x - \overline{\alpha})$:

$$(x - \alpha)(x - \overline{\alpha}) = [x - (a + ib)] \cdot [x - (a - ib)]$$
$$= [(x - a) - ib] \cdot [(x - a) + ib]$$
$$= (x - a)^2 + b^2 = x^2 + px + q$$

with and $q = a^2 + b^2$.

Here, if $\alpha = a + bi$ is a repeated root of multiplicity k, then $\overline{\alpha} = a - bi$ is a repeated root of the same multiplicity k. Therefore, in the factorization of $P(x)$ there appear the factors $(x - \alpha)^k$ and $(x - \overline{\alpha})^k$. Thus, the factorization of a polynomial $P(x)$ with real coefficients has the form

$$P(x) = a_0(x - \alpha_1)^{k_1}(x - \alpha_2)^{k_2} \cdots (x - \alpha_m)^{k_m}(x^2 + p_1 x + q_1)^{l_1} \cdots (x^2 + p_r x + q_r)^{l_r},$$

where

$$k_1 + k_2 + \cdots + k_m + 2l_1 + \cdots + 2l_r = n.$$

■

Example 6.4. (a) The polynomial $P(x) = x^5 - 9x^4 + 31x^3 - 51x^2 + 40x - 12$ of degree five has the factorization

$$P(x) = (x - 2)(x - 2)(x - 1)(x - 1)(x - 3), ,$$

or

$$P(x) = (x - 2)^2(x - 1)^2(x - 3),$$

where $\alpha_1 = 2$ and $\alpha_2 = 1$ are repeated roots of multiplicity two and $\alpha_3 = 3$ is a simple root.

(b) The polynomial $P(x) = x^2 - 2x + 5$ of degree two has the factorization $P(x) = x^2 - 2x + 5 = (x - 1 + 2i)(x - 1 - 2i)$ and $\alpha = 1 - 2i$ and its conjugate $\overline{\alpha} = 1 + 2i$ are complex roots. On the other hand, since $x^4 - 4x^3 + 14x^2 - 20x + 25 = (x^2 - 2x + 5)^2$ it is obvious that $\alpha = 1 - 2i$ and $\overline{\alpha} = 1 + 2i$ are repeated roots of the polynomial $x^4 - 4x^3 + 14x^2 - 20x + 25$ of multiplicity $k = 2$.

6.4 Polynomials and Interpolations

In Section 3.1 we discussed that a function can be given by values in tabular form. In such cases we will try to construct a sufficiently simple computable function approximating the given function $y = f(x)$. As a rule, we construct a polynomial function. Among the various

types of polynomial approximation that are in use, the one that is most flexible and most
often constructed is the interpolating polynomial. This procedure is called interpolation.

Suppose that a real function f is defined on an interval $[a,b]$, and let $x_0, x_1, \ldots, x_n$ be
$n+1$ distinct points of $[a,b]$. Then we wish to find the interpolating polynomial $L_n(x)$ of
degree not exceeding n, such that

$$L_n(x_i) = f(x_i), \quad i = 0, 1, 2, \ldots, n. \tag{6.11}$$

In order to show the uniqueness of the interpolating polynomial, assume there exist two
interpolating polynomials $L_n^1(x)$ and $L_n^2(x)$. Then their difference, $L_n(x) = L_n^1(x) - L_n^2(x)$
being the difference of two polynomials of degree not exceeding n, is again a polynomial
of degree not exceeding n. Moreover, by (6.10),

$$L_n^1(x_i) = L_n^2(x_i) = f(x_i), \quad i = 0, 1, \ldots, n.$$

Their difference thus has $n+1$ roots and hence, being a polynomial of degree not exceeding
n, must vanish identically. By Theorem 6.4, it follows that $L_n^1(x) \equiv L_n^2(x)$. Next, we construct the polynomial $L_n(x)$. At first, note that the polynomial $L_{n,j}(x)$ of degree n satisfying
the condition

$$L_{n,j}(x_j) = 1, \quad L_{n,j}(x_k) = 0, \quad k \neq j,$$

is equal to zero at all points $x = x_k$ ($k \neq j$), n in number, and so divides the polynomial

$$\prod_{k \neq j} (x - x_k) \equiv (x - x_0)(x - x_1) \cdots (x - x_{j-1})(x - x_{j+1}) \cdots (x - x_n).$$

Therefore,

$$L_{n,j}(x) = c_j \prod_{k \neq j} (x - x_k). \tag{6.12}$$

Here, the constant c_j is defined by the condition that $L_{n,j}(x_j) = 1$.
Then,

$$c_j = \frac{1}{\prod_{k \neq j} (x_j - x_k)}, \quad j = 0, 1, \ldots, n. \tag{6.13}$$

Let us take the sum of products $f(x_j) L_{n,j}(x)$ over j to construct $L_n(x) =$
$\sum_{j=0}^{n} f(x_j) L_{n,j}(x)$. Finally, by substituting (6.12) and (6.13) into this expression we have

$$L_n(x) = \sum_{j=0}^{n} f(x_j) \frac{\prod_{k \neq j} (x - x_k)}{\prod_{k \neq j} (x_j - x_k)}. \tag{6.14}$$

The $L_{n,j}(x)$ are called Lagrange's interpolation coefficients. By construction, the values of the polynomial given by (6.14), at the points $x = x_0, x_1, \ldots, x_n$ are equal to the corresponding values of the given function. This polynomial is called Lagrange's interpolation polynomial.

Let $f(x)$, defined on $[a, b]$, be an $n + 1$-times differentiable function. Consider its approximating polynomial $L_n(x)$, and let us now estimate the error $R_n(x)$:

$$f(x) = L_n(x) + R_n(x).$$

Take the points $\bar{x}, x_k \in [a, b]$, $k = 0, 1, \ldots, n$, where $\bar{x} \neq x_k$, and estimate the error at $x = \bar{x}$. From the values of $f(x)$ at the given points $x_0, x_1, \ldots, x_n, \bar{x}$, we compute the interpolating polynomial $L_{n+1}(x)$ of degree $n + 1$ satisfying the following conditions:

$$L_{n+1}(x_i) = f(x_i), \quad i = 0, 1, \ldots, n; \quad L_{n+1}(\bar{x}) = f(\bar{x}).$$

By (6.14), the polynomial sought, $L_{n+1}(x)$, has the form

$$L_{n+1}(x) = f(\bar{x}) \frac{\prod_{k=0}^{n}(x - x_k)}{\prod_{k=0}^{n}(\bar{x} - x_k)} + \sum_{j=0}^{n} f(x_j) \frac{(x - \bar{x}) \prod_{k \neq j}(x - x_k)}{(x_j - \bar{x}) \prod_{k \neq j}(x_j - x_k)}.$$

The function $\varphi(x) = f(x) - L_{n+1}(x)$ has $n + 2$ distinct roots $x_0, x_1, \ldots, x_n, \bar{x}$ in the interval (a, b). By Rolle's theorem (Theorem 5.3), the derivative φ' must have at least $n + 1$ roots $c_k^{(1)}$ ($k = 0, 1, \ldots, n$), that is, $\varphi'(c_k^{(1)}) = 0$ in the smallest interval containing x and x_k, the second derivative must have no less than n roots, that is $\varphi''(c_k^{(2)}) = 0$ for points $c_k^{(2)} \in (a, b)$ ($k = 0, 1, \ldots, n - 1$), and, finally, the $(n + 1)$st derivative must have at least one root, i.e., there exists a point $c \in (a, b)$ such that $\varphi^{(n+1)}(c) = 0$.

Next, we compute the $n + 1^{\text{st}}$ order derivative, $[L_{n+1}(x)]^{(n+1)}$. It is easy to see that the products $\prod_{k=0}^{n}(x - x_k)$ and $(x - \bar{x}) \prod_{k \neq j}(x - x_k)$ are polynomials of degree $n + 1$, where the coefficient of x^{n+1} is one. Hence the $n + 1^{\text{st}}$ order derivatives of these polynomials are equal to $n + 1!$. Thus,

$$[L_{n+1}(x)]^{(n+1)} = (n+1)! \left[\frac{f(\bar{x})}{\prod_{k=0}^{n}(\bar{x} - x_k)} + \sum_{j=0}^{n} \frac{f(x_j)}{(x_j - \bar{x}) \prod_{k \neq j}(x_j - x_k)} \right]$$

$$= \frac{(n+1)!}{\prod_{k=0}^{n}(\bar{x} - x_k)} \left[f(\bar{x}) - \sum_{j=0}^{n} f(x_j) \frac{\prod_{k \neq j}(\bar{x} - x_k)}{\prod_{k \neq j}(x_j - x_k)} \right],$$

because

$$\frac{\prod\limits_{k=0}^{n}(\bar{x}-x_k)}{x_j-\bar{x}} = \frac{\bar{x}-x_j}{x_j-\bar{x}}\prod_{k\neq j}(\bar{x}-x_k) = -\prod_{k\neq j}(\bar{x}-x_k).$$

Consequently,

$$[L_{n+1}(x)]^{(n+1)} = \frac{(n+1)!}{\prod\limits_{k=0}^{n}(\bar{x}-x_k)}[f(\bar{x})-L_n(\bar{x})] = \frac{(n+1)!R_n(\bar{x})}{\prod\limits_{k=0}^{n}(\bar{x}-x_k)}.$$

Note that $\varphi^{(n+1)}(c) = 0$ implies the relationship

$$R_n(\bar{x}) = \frac{f^{(n+1)}(c)}{(n+1)!}\prod_{k=0}^{n}(\bar{x}-x_k) \quad (c\in(a,b)). \tag{6.15}$$

Thus, if we take Lagrange's interpolating polynomial $L_n(x)$ calculated by using the points x_k, $k=0,1,\dots,n$ in place of $f(x)$, the error which arises has the form (6.15).

Example 6.5. Use Lagrange's interpolation formula for the function $f(x) = \sin x$, with given points $x=0$, $x=\pi/6$, and $x=\pi/2$, to approximate $\sin(\pi/4)$.

For $n=2$, we have

$$L_2(x) = f(0)\frac{(x-\pi/6)(x-\pi2)}{(-\pi/6)(-\pi/2)} + f\left(\frac{\pi}{6}\right)\frac{(x-0)(x-\pi/2)}{(\pi/6-0)(\pi/6-\pi/2)}$$

$$+f\left(\frac{\pi}{2}\right)\frac{(x-0)(x-\pi/6)}{(\pi/2-0)(\pi/2-\pi/6)} = \frac{1}{2}\frac{(\pi/2-x)x}{\pi^2/18} + 1\cdot\frac{x(x-\pi/6)}{\pi^2/18}.$$

Then $L_2\left(\frac{\pi}{4}\right) = \frac{11}{16} \approx \frac{\sqrt{2}}{2} = \sin\left(\frac{\pi}{4}\right)$. On the other hand, $(\sin x)^{(3)} = -\cos x$, and so

$$R_2\left(\frac{\pi}{4}\right) = -\left(\frac{\cos c}{3!}\right)\left(\frac{\pi}{4}-0\right)\left(\frac{\pi}{4}-\frac{\pi}{6}\right)\left(\frac{\pi}{4}-\frac{\pi}{2}\right), \quad c\in\left(0,\frac{\pi}{2}\right).$$

This means that

$$|R_2| \leqslant \frac{\pi^3}{1152} < 0.03.$$

Example 6.6. Use the values $f(1)=3$, $f(2)=-5$, and $f(-4)=4$, of $y=f(x)$ to determine Lagrange's interpolation formula of degree two.

If $n=2$, it follows from (6.14) that

$$L_2(x) = 3\frac{(x-2)(x+4)}{(1-2)(1+4)} + (-5)\frac{(x-1)(x+4)}{(2-1)(2+4)} + 4\frac{(x-1)(x-2)}{(-4-1)(-4-2)}$$

$$= -\frac{39}{30}x^2 - \frac{123}{30}x + \frac{252}{30}.$$

Next, suppose we have $n+1$ points x_i $(i = 0, 1, \ldots, n)$, and a function $y = f(x)$ defined at these points with values $y_i = f(x_i)$. Suppose further that $x_{i+1} - x_i = h$ $(i = 0, 1, \ldots, n-1)$ is constant. Then

$$y_i = f(x_0 + ih) \quad (i = 0, 1, \ldots, n).$$

In order to determine an interpolation polynomial of degree not more than n, we denote the higher order differences as follows:

$$\Delta y_0 = y_1 - y_0, \quad \Delta y_1 = y_2 - y_1, \quad \Delta y_2 = y_3 - y_2, \ldots,$$

$$\Delta^2 y_0 = y_2 - 2y_1 + y_0 = \Delta y_1 - \Delta y_0, \quad \Delta^2 y_1 = \Delta y_2 - \Delta y_1, \ldots,$$

$$\Delta^3 y_0 = y_3 - 3y_2 + 3y_1 - y_0 = \Delta^2 y_1 - \Delta^2 y_0, \ldots,$$

$$\cdots\cdots\cdots$$

$$\Delta^n y_0 = \Delta^{n-1} y_1 - \Delta^{n-1} y_0.$$

For the polynomial of first degree, we put

$$L_1(x) = y_0 + \Delta y_0 \frac{x - x_0}{h}. \tag{6.16}$$

It is obvious that

$$L_1(x_0) = y_0, \quad L_1(x_1) = y_0 + \Delta y_0 \frac{h}{h} = y_0 + (y_1 - y_0) = y_1.$$

For the polynomial $L_2(x)$ of degree two,

$$L_2(x) = y_0 + \Delta y_0 \frac{x - x_0}{h} + \frac{\Delta^2 y_0}{2!} \frac{x - x_0}{h} \left(\frac{x - x_0}{h} - 1 \right) \tag{6.17}$$

$$L_2(x_0) = y_0, \quad L_2(x_1) = y_1,$$

$$L_2(x_2) = y_0 + \Delta y_0 2 + \frac{\Delta^2 y_0}{2!} \frac{2h}{h} \left(\frac{2h}{h} - 1 \right) = y_2.$$

In turn, for the polynomial of degree three,

$$L_3(x) = y_0 + \Delta y_0 \frac{x - x_0}{h} + \frac{\Delta^2 y_0}{2!} \frac{x - x_0}{h} \left(\frac{x - x_0}{h} - 1 \right)$$

$$+ \frac{\Delta^3 y_0}{3!} \frac{x - x_0}{h} \left(\frac{x - x_0}{h} - 1 \right) \left[\frac{x - x_0}{h} - 2 \right]. \tag{6.18}$$

Finally, we have Newton's polynomial of degree n satisfying $L_n(x_i) = y_i$ $(i = 0, 1, \ldots, n)$:

$$L_n(x_i) = y_0 + \Delta y_0 \frac{x - x_0}{h} + \frac{\Delta^2 y_0}{2!} \frac{x - x_0}{h} \left(\frac{x - x_0}{h} - 1 \right)$$

$$+ \cdots + \frac{\Delta^n y_0}{n!} \frac{x - x_0}{h} \left(\frac{x - x_0}{h} - 1 \right) \cdots \left[\frac{x - x_0}{h} - (n-1) \right]. \tag{6.19}$$

By Theorem 6.4, Lagrange's and Newton's polynomials are equal. In general, from a practical point of view, Newton's interpolation formula (6.19) is convenient, because in passing from the polynomial of degree k to the polynomial of degree $(k+1)$, the first $(k+1)$ terms remain unchanged. The new term added is zero at all previous values of x.

As before, suppose given a function $y = f(x)$ by means of table of values,

$$y_i = f(x_0 + ih), \quad i = 0, 1, \ldots, n, \quad x_{i+1} = x_i + h.$$

We can compute numerical estimates for the values of the derivative of f. We can use either Lagrange's or Newton's polynomials. For example, let us use Newton's polynomial. Suppose we have values given at three points: $y_i = f(x_i)$, $i = 0, 1, 2$.

By differentiating (6.17) in the interval $x_0 \leqslant x \leqslant x_2$, we obtain the approximation,

$$f'(x) \approx L_2'(x) = \frac{\Delta y_0}{h} + \frac{\Delta^2 y_0}{2h} \left(2\frac{x - x_0}{h} - 1 \right). \tag{6.20}$$

At $x = x_0$,

$$f'(x_0) \approx L_2'(x_0) = \frac{\Delta y_0}{h} - \frac{\Delta^2 y_0}{2h}. \tag{6.21}$$

If we use the polynomial of degree three, then

$$f'(x) \approx L_3'(x) = \frac{\Delta y_0}{h} + \frac{\Delta^2 y_0}{2h} \left(2\frac{x - x_0}{h} - 1 \right) + \frac{\Delta^3 y_0}{2 \cdot 3h} \left[3 \left(\frac{x - x_0}{h} \right)^2 - 6\frac{x - x_0}{h} + 2 \right]. \tag{6.22}$$

In particular, at $x = x_0$,

$$f'(x_0) \approx L_3'(x_0) = \frac{\Delta y_0}{h} - \frac{\Delta^2 y_0}{2h} + \frac{\Delta^3 y_0}{3h}. \tag{6.23}$$

In general for such numerical differentiation, if we take (6.19), then at $x = x_0$,

$$f'(x_0) \approx L_n'(x_0) = \frac{\Delta y_0}{h} - \frac{\Delta^2 y_0}{2h} + \frac{\Delta^3 y_0}{3h} - \frac{\Delta^4 y_0}{4h} + \cdots.$$

Remark 6.1. For a differentiable function, the difference $\Delta^i y_0$, $i = 1, 2, 3$ is infinitely small to i-th order with respect to h.

This study of Lagrange's and Newton's interpolation polynomials suggests the natural question: is there a polynomial $L(x)$ approximating a function f defined on the interval $[a, b]$ with any desired degree of accuracy? In other words, given an arbitrary $\varepsilon > 0$, can we choose a polynomial $L(x)$ such that $|f(x) - L(x)| < \varepsilon$?

The following theorem, formulated without proof, is the answer to this question.

Theorem 6.10 (Weierstrass). *If the function f defined on $[a,b]$ is continuous on this interval, then for every $\varepsilon > 0$ there exists a polynomial $L(x)$ such that*

$$|f(x) - L(x)| < \varepsilon.$$

Bernstein[1] constructed the following polynomials to explicitly solve this problem. For simplicity we take in place of $[a,b]$ the interval $[0,1]$. He defined

$$B_n(x) = \sum_{m=0}^{n} f\left(\frac{m}{n}\right) C_n^m x^m (1-x)^{n-m},$$

which is now called the Bernstein polynomial of degree n. Here, C_n^m is the binomial coefficient. For any $\varepsilon > 0$ there exists an n such that

$$|B_n(x) - f(x)| < \varepsilon.$$

Earlier, Chebyshev[2] had found in the set of polynomials with a fixed degree, that polynomial, the absolute value of which is the best approximation to zero. Such polynomials are called Chebyshev polynomials.

6.5   Problems

(1) Find $P(x) = x^3 - 4x^2 + 8x - 1$ divided by $x + 4$.

   Answer: $P(x) = (x+4)(x^2 - 8x + 40) - 161$.

(2) Find $P(x) = x^4 + 12x^3 + 54x^2 + 108x + 81$ divided by $x + 3$.

   Answer: $P(x) = (x+3)(x^3 + 9x^2 + 27x + 27)$.

(3) Find $P(x) = x^7 - 1$ divided by $x - 1$.

   Answer: $P(x) = (x-1)(x^6 + x^5 + x^4 + x^3 + x^2 + x + 1)$.

In each of problems (4)–(7), use the factor theorem to find the factorization.

(4) $P(x) = 2x^8 - 10x^7 + 8x^6 + 8x^5 + 6x^4 + 18x^3$.

   Answer: $P(x) = 2x^3(x+1)(x-3)^2(x^2+1)$.

(5) $P(x) = x^4 - 1$.

   Answer: $P(x) = (x-1)(x+1)(x^2+1)$.

(6) $P(x) = x^2 - x - 2$.

   Answer: $P(x) = (x-2)(x+1)$.

---

[1]S. Bernstein, (1880–1968), Russian mathematician.
[2]P. Chebyshev (1821–1894), Russian mathematician.

(7) $P(x) = x^3 + 1$.

      Answer: $P(x) = (x+1)(x^2 - x + 1)$.

(8) Suppose $y = f(x)$ has the following tabular representation:

$$y_1 = 4 \text{ at } x_1 = 0,$$
$$y_2 = 6 \text{ at } x_2 = 1,$$
$$y_3 = 10 \text{ at } x_3 = 2.$$

    Find Lagrange's polynomial of degree two for its interpolation.

      Answer: $L_2(x) = x^2 + x + 4$.

(9) At the points $x = 1, 2, 3, 4$, and 5, the values of a function are 2, 1, -1, 5, and 0, respectively. Find Lagrange's polynomial $L_4(x)$.

      Answer: $-\dfrac{7}{6}x^4 + \dfrac{79}{6}x^3 - \dfrac{151}{3}x^2 + \dfrac{226}{3}x - 35$.

(10) At the points $x = 2, 4, 5$, and 10, the values of a function are 3, 7, 9, and 10, respectively. Find Lagrange's polynomial.

      Answer: $2x - 1$.

(11) Find the Bernstein polynomials $B_n(x)$ for the function $y = \sin \pi x$ defined on $[0, 1]$, where $n = 1, 2, 3, 4$.

      Answer: $B_1(x) = 0$; $B_2(x) = 2x(1-x)$; $B_3(x) = \dfrac{3\sqrt{3}}{2}x(1-x)$;

      $B_4(x) = 2x(1-x)\left[(2\sqrt{2}-3)x^2 - (2\sqrt{2}-3)x + \sqrt{2}\right]$.

Chapter 7

Applications of Differential Calculus to Limit Calculations and Extremum Problems

Certain kinds of limit computations, those involving the so-called indeterminate forms, can be accomplished using two rules explained in this chapter, L'Hopital's rule and a rule based on the Taylor–Maclaurin series. Again, studying the Taylor series around critical points gives sufficient conditions for the presence of a local (relative) extremum. Furthermore, whether the curve is bending (upward or downward) depends on the sign of higher order derivatives. Necessary and sufficient conditions for existence of an asymptote to a curve are proved. Finally, such investigations are used in curve sketching.

7.1 L'Hopital's Rule for the Indeterminate Form $\frac{0}{0}$

Definition 7.1. If, for given functions f and g, $\lim_{x \to a} f(x) = \lim_{x \to a} g(x) = 0$, then we say that the quotient $\frac{f(x)}{g(x)}$ has the indeterminate form $\frac{0}{0}$ as $x \to a$. Similarly, for the indeterminate form $\frac{0}{0}$ as $x \to a \pm 0$ or $x \to \pm\infty$.

Theorem 7.1 (L'Hopital's[1] First Rule). *Suppose that the functions f and g are differentiable in a deleted δ-neighborhood $(a - \delta, a + \delta) \smallsetminus a$ $(\delta > 0)$ of a point a. Moreover, suppose that*

$$\lim_{x \to a} f(x) = \lim_{x \to a} g(x) = 0 \tag{7.1}$$

and $g'(x) \neq 0$ in the deleted δ-neighborhood of the point a. Then if the finite or infinite limit

$$\lim_{x \to a} \frac{f'(x)}{g'(x)} \tag{7.2}$$

exists, then so does

$$\lim_{x \to a} \frac{f(x)}{g(x)}, \tag{7.3}$$

---

[1]Guillaume F.A.L'Hopital(1661-1704), French mathematician.

E. Mahmudov, *Single Variable Differential and Integral Calculus*,
DOI: 10.2991/978-94-91216-86-2_7, © Atlantis Press 2013

and

$$\lim_{x \to a} \frac{f(x)}{g(x)} = \lim_{x \to a} \frac{f'(x)}{g'(x)}. \tag{7.4}$$

☐ Extend the definition of the functions f and g to include the point a by setting $f(a) = g(a) = 0$. Then f and g are continuous on the interval $(a - \delta, a + \delta)$. Let $\{x_n\}$ ($x_n \neq a$, $n = 1, 2, \ldots$) be an arbitrary sequence that converges to a. Clearly, f and g are continuous on the closed interval $[a, x_n]$. More precisely, f and g are continuous on the open interval (a, x_n) and $g'(x) \neq 0$ on (a, x_n). Thus, by Theorem 5.5, there is a point $\xi_n \in (a, x_n)$ such that

$$\frac{f(x_n) - f(a)}{g(x_n) - g(a)} = \frac{f'(\xi_n)}{g'(\xi_n)}. \tag{7.5}$$

Since $f(a) = g(a) = 0$, it follows from (7.5) that

$$\frac{f(x_n)}{g(x_n)} = \frac{f'(\xi_n)}{g'(\xi_n)}. \tag{7.6}$$

On the other hand, $x_n \to a$ ($n \to \infty$) implies $\xi_n \to a$. Thus, using Definition 3.2 and passing to the limit in (7.6), we conclude that

$$\lim_{x \to a} \frac{f(x)}{g(x)} = \lim_{x_n \to a} \frac{f(x_n)}{g(x_n)} = \lim_{\xi_n \to a} \frac{f'(\xi_n)}{g'(\xi_n)} = \lim_{x \to a} \frac{f'(x)}{g'(x)}.$$

∎

Remark 7.1. Even if the limit $\lim\limits_{x \to a} \dfrac{f'(x)}{g'(x)}$ does not exists, the limit $\lim\limits_{x \to a} \dfrac{f(x)}{g(x)}$ may exists. Indeed, if $f(x) = x^2 \cos \dfrac{1}{x}$ and $g(x) = \sin x$, then

$$\lim_{x \to 0} \frac{f(x)}{g(x)} = \lim_{x \to 0} \frac{x}{\sin x} \lim_{x \to 0} x \cos \frac{1}{x} = 0$$

while the limit

$$\lim_{x \to 0} \frac{f'(x)}{g'(x)} = \lim_{x \to 0} \frac{2x \cos \dfrac{1}{x} + \sin \dfrac{1}{x}}{\cos x}$$

does not exist.

Remark 7.2. If the derivatives f' and g' are continuous at a, and $g'(a) \neq 0$, then (see Formula 7.4),

$$\lim_{x \to a} \frac{f(x)}{g(x)} = \frac{f'(a)}{g'(a)}$$

Remark 7.3. If it turns out that the quotient $\dfrac{f'(x)}{g'(x)}$ is again indeterminate, then L'Hopital's rule can be applied a second, a third, etc., time.

$$\lim_{x \to a} \frac{f(x)}{g(x)} = \lim_{x \to a} \frac{f'(x)}{g'(x)} = \lim_{x \to a} \frac{f''(x)}{g''(x)}.$$

Example 7.1. Compute the limit $\lim\limits_{x \to 0} \dfrac{\cos x - 1}{x^2}$. The quotient has the indeterminate form $\dfrac{0}{0}$, and so $\lim\limits_{x \to 0} \dfrac{\cos x - 1}{x^2} = \lim\limits_{x \to 0} \dfrac{-\sin x}{2x} = -\dfrac{1}{2}$.

Example 7.2. By applying L'Hopital's rule twice, we have $\lim\limits_{x \to 0} \dfrac{x - \sin x}{x^3} = \lim\limits_{x \to 0} \dfrac{1 - \cos x}{3x^2} = \lim\limits_{x \to 0} \dfrac{\sin x}{6x} = \dfrac{1}{6}$.

Example 7.3. In this case we apply L'Hopital's rule three times, that is,

$$\lim_{x \to 0} \frac{x^4}{x^2 + 2\cos x - 2} = \lim_{x \to 0} \frac{4x^3}{2x - 2\sin x} = \lim_{x \to 0} \frac{12x^2}{2 - 2\cos x} = \lim_{x \to 0} \frac{24x}{2\sin x} = 12.$$

Example 7.4. As in Example 7.3,

$$\lim_{x \to 0} \frac{e^x - e^{-x} - 2x}{x - \sin x} = \lim_{x \to 0} \frac{e^x + e^{-x} - 2}{1 - \cos x} = \lim_{x \to 0} \frac{e^x - e^{-x}}{\sin x} = \lim_{x \to 0} \frac{e^x + e^{-x}}{\cos x} = \frac{2}{1} = 2.$$

Remark 7.4. Suppose that f and g are defined on the set $R \setminus [-\delta, \delta]$ and $g'(x) \neq 0$, $x \in R \setminus [-\delta, \delta]$ ($\delta > 0$). Assume that the limit

$$\lim_{x \to \infty} \frac{f(x)}{g(x)} \tag{7.7}$$

exists. Then the functions $g_1(t) = g\left(\dfrac{1}{t}\right) = g(x)$ and $f_1(t) = f\left(\dfrac{1}{t}\right) = f(x)$ $(t = \dfrac{1}{x})$ are defined and differentiable on $-\dfrac{1}{\delta} < t < \dfrac{1}{\delta}$ $(t \neq 0)$. Besides, the derivative $g_1'(t) \neq 0$ in the deleted $\dfrac{1}{\delta}$ neighborhood of the point $t = 0$ takes the form

$$g_1'(t) = g'\left(\frac{1}{t}\right)\left(-\frac{1}{t^2}\right) = g'(x)(-x^2) \neq 0.$$

On the other hand, the existence of the limit (7.7) implies that the following limit exists:

$$\lim_{t \to 0} \frac{f_1'(t)}{g_1'(t)} = \lim_{t \to 0} \frac{f'\left(\dfrac{1}{t}\right)}{g'\left(\dfrac{1}{t}\right)} = \lim_{x \to \infty} \frac{f'(x)}{g'(x)}. \tag{7.8}$$

Therefore, by Theorem 7.1, the limit

$$\lim_{t \to 0} \frac{f_1(t)}{g_1(t)} = \lim_{x \to \infty} \frac{f(x)}{g(x)} \tag{7.9}$$

exists, and

$$\lim_{x \to \infty} \frac{f(x)}{g(x)} = \lim_{t \to 0} \frac{f_1(t)}{g_1(t)} = \lim_{t \to 0} \frac{f_1'(t)}{g_1'(t)} = \lim_{x \to \infty} \frac{f'(t)}{g'(t)}$$

(see (7.3), (7.8), and (7.9)).

The other limit statements can be investigated similarly.

Example 7.5. Find $\lim\limits_{x \to 0+0} \dfrac{x^2}{\ln(1+x)}$. (Indeterminate form $\dfrac{0}{0}$.)

By Theorem 7.11, we have

$$\lim_{x \to 0+0} \frac{x^2}{\ln(1+x)} = \lim_{x \to 0+0} \frac{2x}{\dfrac{1}{1+x}} = \lim_{x \to 0+0} 2x(1+x) = 0.$$

Example 7.6. Find $\lim\limits_{x \to \infty} x \ln\left(1 + \dfrac{1}{x}\right)$. Obviously,

$$\lim_{x \to \infty} x \ln\left(1 + \frac{1}{x}\right) = \lim_{x \to \infty} \frac{\ln\left(1 + \dfrac{1}{x}\right)}{\dfrac{1}{x}} = \lim_{x \to \infty} \frac{-\dfrac{1}{x(x+1)}}{-\dfrac{1}{x^2}} = \lim_{x \to \infty} \frac{x}{x+1} = 1.$$

Example 7.7.

$$\lim_{x \to \infty} \frac{\dfrac{\pi}{4} - \tan^{-1}\left(1 - \dfrac{1}{x}\right)}{\sin\dfrac{1}{x}} = \lim_{x \to \infty} \frac{-\dfrac{1}{x^2\left[1 + \left(1 - \dfrac{1}{x}\right)^2\right]}}{-\dfrac{1}{x^2}\cos\dfrac{1}{x}}$$

$$= \lim_{x \to \infty} \frac{\dfrac{1}{1 + \left(1 - \dfrac{1}{x}\right)^2}}{\cos\dfrac{1}{x}} = \frac{1}{2}.$$

7.2   L'Hopital's Rule for the Indeterminate Form $\dfrac{\infty}{\infty}$

Definition 7.2. Suppose two functions, f and g, are defined on a neighborhood of a point $x = a$, such that

$$\lim_{x \to a} f(x) = \infty, \quad \lim_{x \to a} g(x) = \infty \tag{7.10}$$

Then we say that the quotient $\dfrac{f(x)}{g(x)}$ has the indeterminate form $\dfrac{\infty}{\infty}$ at $x = a$.

Theorem 7.2 (L'Hopital's Second Rule). *Suppose that the functions f and g are differentiable in a deleted neighborhood of a, $(a - \delta, a + \delta) \setminus a$ ($\delta > 0$), and that condition (7.10) is satisfied. Suppose also that $g'(x) \neq 0$ in that neighborhood and that*

$$\lim_{x \to a} \frac{f'(x)}{g'(x)} \tag{7.11}$$

exists. Then

$$\lim_{x \to a} \frac{f(x)}{g(x)} \tag{7.12}$$

also exists, and

$$\lim_{x \to a} \frac{f(x)}{g(x)} = \lim_{x \to a} \frac{f'(x)}{g'(x)}. \tag{7.13}$$

☐ First assume that the limit (7.11) exists and is equal to c. We show that the limit (7.12) exists and is equal to c. Let $\{x_n\}$ be an arbitrary sequence such that $x_n \to a$ and either all $x_n < a$ or all $x_n < a$. Since $x_n \in (a - \delta, a + \delta)$ ($n = 1, 2, \ldots$), then for any m and n, by applying Theorem 5.5 to the closed interval $[x_m, x_n]$, we have

$$\frac{f(x_n) - f(x_m)}{g(x_n) - g(x_m)} = \frac{f(x_n)}{g(x_n)} \frac{1 - \dfrac{f(x_m)}{f(x_n)}}{1 - \dfrac{g(x_m)}{g(x_n)}} = \frac{f'(\eta_m^n)}{g'(\eta_m^n)}, \quad \eta_m^n \in (x_m, x_n),$$

Thus,

$$\frac{f(x_n)}{g(x_n)} = \frac{f'(\eta_m^n)}{g'(\eta_m^n)} \frac{1 - \dfrac{g(x_m)}{g(x_n)}}{1 - \dfrac{f(x_m)}{f(x_n)}}. \tag{7.14}$$

Now, for any fixed $\varepsilon > 0$ there exists an integer m such that for all $n \geqslant m$

$$\frac{f'(\eta_m^n)}{g'(\eta_m^n)} = c + \alpha_m^n, \quad \text{where } |\alpha_m^n| < \frac{\varepsilon}{2}. \tag{7.15}$$

On the other hand, (see Formula 7.10), for fixed m

$$\lim_{n \to \infty} \frac{1 - \dfrac{g(x_m)}{g(x_n)}}{1 - \dfrac{f(x_m)}{f(x_n)}} = 1.$$

Hence, for positive $\dfrac{\varepsilon/2}{|c| + \varepsilon/2}$ and fixed m, there is an $n_0 > m$ such that

$$\frac{1 - \dfrac{g(x_m)}{g(x_n)}}{1 - \dfrac{f(x_m)}{f(x_n)}} = 1 + \beta_m^n, \quad \text{where } |\beta_m^n| < \frac{\varepsilon/2}{|c| + \varepsilon/2} \tag{7.16}$$

for all $n \geqslant n_0$.

Then we derive from (7.14)–(7.16) that

$$\frac{f(x_n)}{g(x_n)} = (c + \alpha_m^n)(1 + \beta_m^n) = c + (c + \alpha_m^n)\beta_m^n + \alpha_m^n.$$

Therefore,

$$\left| \frac{f(x_n)}{g(x_n)} - c \right| \leqslant (|c| + |\alpha_m^n|) |\beta_m^n| + |\alpha_m^n|.$$

Hence, for all $n \geqslant n_0$,

$$(|c| + |\alpha_m^n|)|\beta_m^n| + |\alpha_m^n| < \left(|c| + \frac{\varepsilon}{2} \right) \frac{\varepsilon/2}{|c| + \varepsilon/2} + \frac{\varepsilon}{2} = \varepsilon.$$

Consequently, for any positive $\varepsilon > 0$ there is an n_0 such that

$$\left| \frac{f(x_n)}{g(x_n)} - c \right| < \varepsilon$$

for all $n \geqslant n_0$. It follows that $\lim\limits_{x \to a} \dfrac{f(x)}{g(x)} = c$, and (7.13) is valid.   ∎

Remark 7.5. If the limit (7.11) becomes ∞, then, since $\lim\limits_{x \to a} \dfrac{g'(x)}{f'(x)} = 0$, it follows that $\lim\limits_{x \to a} \dfrac{g(x)}{f(x)} = 0$. Hence, taking into account (7.10), we have $\lim\limits_{x \to a} \dfrac{f(x)}{g(x)} = \infty$.

Remark 7.6. If in Theorem 7.2, instead of $(a - \delta, a + \delta)$, we take $(a, a + \delta)$, or $(a - \delta, a)$, then in the relations (7.11)–(7.13), $x \to a$ must be replaced by $x \to a + 0$ (or $x \to a - 0$). Moreover, Theorem 7.2 is true for the case $x \to \infty$.

Example 7.8. Find $\lim\limits_{x \to 0+0} \sqrt{x}\ln x$.

Applying Theorem 7.1, we can write

$$\lim_{x \to 0+0} \sqrt{x}\ln x = \lim_{x \to 0+0} \frac{\ln x}{x^{-\frac{1}{2}}} = \lim_{x \to 0+0} \frac{\frac{1}{x}}{\left(-\frac{1}{2} \right) x^{-\frac{3}{2}}} = -2 \lim_{x \to 0+0} \sqrt{x} = 0.$$

Example 7.9. Find $\lim\limits_{x \to +\infty} \dfrac{\ln x}{x}$.

$$\lim_{x \to +\infty} \frac{\ln x}{x} = \lim_{x \to +\infty} \frac{\frac{1}{x}}{1} = \lim_{x \to +\infty} \frac{1}{x} = 0.$$

Example 7.10. Find $\lim\limits_{x \to \infty} \dfrac{x^n}{e^x}$.

$$\lim_{x \to \infty} \frac{x^n}{e^x} = \lim_{x \to \infty} \frac{nx^{n-1}}{e^x} = \cdots = \lim_{x \to \infty} \frac{n!}{e^x} = 0.$$

Besides the indeterminate forms $\dfrac{0}{0}$ and $\dfrac{\infty}{\infty}$ we often see the indeterminate forms $0 \cdot \infty$, $\infty - \infty$, 1^{∞}, 0^{0}, and ∞^{0}. All these indeterminate forms can be converted by algebraic substitutions into a form of type $\dfrac{0}{0}$ or $\dfrac{\infty}{\infty}$. For example, let us investigate the last three cases. Suppose that we try to find the limit of $y = f(x)^{g(x)}$ ($f(x) > 0$) where the limits of f and g as $x \to a$ are such that one of the indeterminate forms 1^{∞}, 0^{0}, or ∞^{0} is produced. First, we calculate the logarithm:

$$\ln y = g(x) \ln f(x).$$

For each of these three cases, $\ln y = g(x) \ln f(x)$ has the form $0 \cdot \infty$ as $x \to a$. Thus, it is easy to see that if

$$z = p(x) \cdot q(x), \tag{7.17}$$

where

$$\lim_{x \to a} p(x) = 0, \quad \lim_{x \to a} q(x) = \infty,$$

the indeterminate form $0 \cdot \infty$ can be converted by algebraic substitutions into either the form $\dfrac{0}{0}$ or $\dfrac{\infty}{\infty}$. Indeed, expressing (7.17) in the form

$$z = \frac{p(x)}{\dfrac{1}{q(x)}}, \tag{7.18}$$

or

$$z = \frac{q(x)}{\dfrac{1}{p(x)}}, \tag{7.19}$$

we see that (7.18) and (7.19) have the indeterminate forms $\dfrac{0}{0}$ and $\dfrac{\infty}{\infty}$, respectively.

Example 7.11. Find $\lim\limits_{x \to 0} x^{n} \ln x$ (indeterminate form $0 \cdot \infty$).

$$\lim_{x \to 0} x^{n} \ln x = \lim_{x \to 0} \frac{\ln x}{\dfrac{1}{x^{n}}} = \lim_{x \to 0} \frac{\dfrac{1}{x}}{-\dfrac{n}{x^{n+1}}} = -\lim_{x \to 0} \frac{x^{n}}{n} = 0.$$

Example 7.12. Find $\lim\limits_{x \to 0} x^{x}$ (indeterminate form 0^{0}).

Let $y = x^{x}$. Then $\ln y = x \ln x = \dfrac{\ln x}{\dfrac{1}{x}}$, and by passing to the limit as $x \to 0$, we have

$$\lim_{x \to 0} (x \ln x) = \lim_{x \to 0} \frac{\ln x}{\dfrac{1}{x}} = \lim_{x \to 0} \frac{\dfrac{1}{x}}{-\dfrac{1}{x^{2}}} = -\lim_{x \to 0} x = 0.$$

Therefore $\ln \lim\limits_{x \to 0} y = 0$, and so $\lim\limits_{x \to 0} x^{x} = 1$.

Example 7.13. Find $\lim\limits_{x\to 0}(1+x^2)^{\frac{1}{e^x-1-x}}$ (indeterminate form 1^∞).

Let $y = (1+x^2)^{\frac{1}{e^x-1-x}}$. Then

$$\ln y = \frac{1}{e^x-1-x}\ln(1+x^2)$$

$$\lim_{x\to 0}(\ln y) = \lim_{x\to 0}\frac{\ln(1+x^2)}{e^x-1-x} = \lim_{x\to 0}\frac{\dfrac{2x}{1+x^2}}{e^x-1} = \lim_{x\to 0}\frac{2x}{(e^x-1)(1+x^2)}$$

$$= \lim_{x\to 0}\frac{2}{e^x(1+x^2)+(e^x-1)2x} = 2.$$

Thus $\lim\limits_{x\to 0} y = e^2$.

7.3 Taylor–Maclaurin Series and Limit Calculations

In many limit calculations, Taylor-Maclaurin series can be applied successfully. For the functions e^x, $\ln(1+x)$, $(1+x)^\alpha$, $\sin x$, $\cos x$, and $\tan^{-1} x$, we give the Maclaurin formula with Peano's form of the remainder term:

$$e^x = 1+\frac{x}{1!}+\cdots+\frac{x^n}{n!}+o(x^n),$$

$$\sin x = x-\frac{x^3}{3!}+\frac{x^5}{5!}-\frac{x^7}{7!}+\cdots+(-1)^n\frac{x^{2n+1}}{(2n+1)!}+o(x^{2n+2}),$$

$$\cos x = 1-\frac{x^2}{2!}+\frac{x^4}{4!}-\frac{x^6}{6!}+\cdots+(-1)^n\frac{x^{2n}}{(2n)!}+o(x^{2n+1}),$$

$$(1+x)^\alpha = 1+\frac{\alpha}{1!}x+\frac{\alpha(\alpha-1)}{2!}x^2+\cdots+\frac{\alpha(\alpha-1)\cdots(\alpha-n+1)}{n!}x^n+o(x^n),$$

$$\ln(1+x) = x-\frac{x^2}{2}+\frac{x^3}{3}-\frac{x^5}{5}+\cdots+(-1)^n\frac{x^n}{n}+o(x^n),$$

$$\tan^{-1}x = x-\frac{x^3}{3}+\frac{x^5}{5}-\frac{x^7}{7}+\cdots+(-1)^n\frac{x^{2n+1}}{2n+1}+o(x^{2n+2}).$$

(7.20)

We will show in the following examples how we can apply these formulae.

Example 7.14. Use the second formula of (7.20), with $n = 1$, to compute $\lim\limits_{x\to 0}\dfrac{x-\sin x}{x^3}$. Since

$$\sin x = x-\frac{x^3}{6}+o(x^4),$$

(7.21)

we can write

$$\lim_{x\to 0}\frac{x-\sin x}{x^3} = \lim_{x\to 0}\frac{x-x+\dfrac{x^3}{6}-o(x^4)}{x^3} = \lim_{x\to 0}\left[\frac{1}{6}+o(x)\right] = \frac{1}{6}.$$

Example 7.15. Find $\lim\limits_{x\to 0}\dfrac{e^{-\frac{x^2}{2}}-\cos x}{x^2\sin x}$.

By formulae (7.20) we see that

$$\cos x = 1 - \frac{x^2}{2!} + \frac{x^4}{4!} + o(x^5),$$

$$\sin x = x + o(x), \quad \text{and} \qquad (7.22)$$

$$e^{-\frac{x^2}{2}} = 1 - \frac{x^2}{2} + \frac{x^4}{8} + o(x^4).$$

Then,

$$\lim_{x\to 0}\frac{1-\dfrac{x^2}{2}+\dfrac{x^4}{8}+o(x^4)-1+\dfrac{x^2}{2}-\dfrac{x^4}{24}-o(x^5)}{x^4+o(x^4)} = \lim_{x\to 0}\frac{\dfrac{1}{12}x^4+o(x^4)-o(x^5)}{x^4+o(x^4)} = \frac{1}{12}.$$

Example 7.16. Find $\lim\limits_{x\to 0}\left(\cos x+\dfrac{x^2}{2}\right)^{\frac{1}{x(\sin x-x)}}$.

Let denote $y=\left(\cos x+\dfrac{x^2}{2}\right)^{\frac{1}{x(\sin x-x)}}$.

Because y is positive for small x we can first compute the limit of the natural logarithm, $\ln y$, as $x\to 0$, where

$$\ln y = \frac{1}{x(\sin x-x)}\ln\left(\cos x+\frac{x^2}{2}\right).$$

By using (7.21) and (7.22), we have

$$\lim_{x\to 0}\ln y = \lim_{x\to 0}\frac{\ln\left(1+\dfrac{x^4}{24}+o(x^5)\right)}{-\dfrac{x^4}{6}+o(x^5)}.$$

On the other hand, since $\ln(1+z)=z+o(z)$, it follows that $\ln\left(1+\dfrac{x^4}{24}+o(x^5)\right)=\dfrac{x^4}{24}+o(x^4)$. Hence, e

$$\lim_{x\to 0}\ln y = \lim_{x\to 0}\frac{\dfrac{x^4}{24}+o(x^4)}{-\dfrac{x^4}{6}+o(x^5)} = \lim_{x\to 0}\frac{\dfrac{1}{24}+\dfrac{o(x^4)}{x^4}}{-\dfrac{1}{6}+o(x)} = -\frac{1}{4}.$$

Consequently,

$$\lim_{x\to 0} y = e^{-\frac{1}{4}}.$$

7.4 Sufficient Conditions for a Relative Extremum

Problems concerned with finding the "best" solution are called optimization problems. A wide class of optimization problems can be reduced to finding the extremum (either maximum or minimum) value of a function and determining the points where this values occurs. Often, we shall be concerned with the extremum values of f on some interval, rather than on the entire domain of f.

As we showed in Chapter 5, the monotonicity of a differentiable function depends on the sign of its first derivative. In particular, in order that f be nondecreasing (nonincreasing) on the interval (a, b), it is necessary and sufficient that $f'(x) \geqslant 0$ ($f'(x) \leqslant 0$). For increasing (decreasing) functions $f'(x) > 0$ ($f'(x) < 0$). Besides, by Theorem 5.2, if f is differentiable at x_0 and $f(x_0)$ is a relative extremum of f, then $f'(x_0) = 0$ (Theorem 5.2). Note that the condition $f'(x_0) = 0$ is necessary but not sufficient for the existence of a relative (or local) extremum (see Remark 5.1). We shall give sufficient conditions for local maxima and minima in this chapter. Nevertheless, in the majority of optimization problems, our interest is the global (or absolute) extremum values attained by a given continuous function.

Note that the largest of the relative values of f is a global maximum value, and the smallest is a global minimum value. Also, observe that every global extremum is, of course, relative as well; on the other hand the converse assertion is false.

Definition 7.3. A point x_0 in the domain of the function f is called a *critical point* of f if either $f'(x_0) = 0$ or if $f'(x_0)$ does not exist.

Consequently, the extreme values (either relative or global) of a continuous function f on a closed interval occur either at a critical point of f or at one of the end points of the interval in question.

Theorem 7.3 (The First Sufficient Condition for a Relative Extremum). *Let a function f be differentiable in some neighborhood of a critical point x_0, so that $f'(x_0) = 0$. If for the set of points $x < x_0$ and $x > x_0$ contained in this neighborhood, $f'(x) > 0$ and $f'(x) < 0$ ($f'(x) < 0$ and $f'(x) > 0$), respectively, then $f(x_0)$ is a relative maximum (minimum) value of f.*

Moreover, if in a neighborhood of x_0, the derivative $f'(x)$ has the same sign (except possibly at x_0), then $f(x_0)$ is neither a relative maximum nor a relative minimum of f.

☐ (1) Suppose, for definiteness, $f'(x) > 0$ for $x < x_0$ and $f'(x) < 0$ for $x > x_0$ (provided, of course, that x is contained in the indicated neighborhood). Let's prove that $f(x_0)$ is a relative maximum. (The alternative case is reduced to this one by considering $-f$.) It suffices to prove that for $x_0 \neq x$, $f(x_0) - f(x) > 0$.

By Lagrange's theorem (Theorem 5.4) we have

$$f(x_0) - f(x) = f'(\xi)(x_0 - x), \tag{7.23}$$

where ξ is some point between x_0 and x. Then, because $f'(\xi) > 0$ for $x < x_0$ and $f'(\xi) < 0$ for $x > x_0$, the right hand side of (7.23) is positive.

(2) If, in a neighborhood of the point x_0, the sign of the derivative $f'(x)$ does not change, then the right hand side of (7.23) has opposite signs when $x_0 < x$ and $x_0 > x$. ∎

Corollary 7.1. *Theorem 7.3 says, briefly, that:*

(1) *If the derivative $f'(x)$ is positive (negative) on the left of x_0 and negative (positive) on the right of x_0, then $f(x_0)$ is a relative maximum (minimum).*

(2) *If, in some neighborhood of x_0, the derivative $f'(x)$ does not change sign, then $f(x_0)$ is not a relative extremum.*

Example 7.17. Find the relative extrema of $f(x) = x^3 - 3x^2 - 4$. Observe that the derivative $f'(x) = 3x(x-2)$ is positive on $(-\infty, 0) \cup (2, +\infty)$ and negative on the interval $0 < x < 2$.

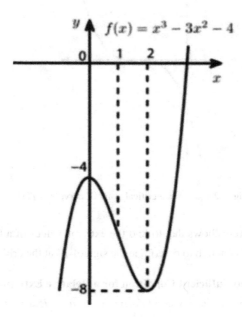

Fig. 7.1   The relative extremum of function of Example 7.17

It follows that the function f is increasing on the intervals $(-\infty,0)$ and $(2,+\infty)$, and decreasing on $(0,2)$. On the other hand the function f has two critical points $(x = 0,$ $x = 2)$. On the left of $x = 0$, the derivative is positive, and on the right, the derivative is negative. Moreover, on the left of $x = 2$, the derivative is negative, and on the right, positive. Therefore, the value $f(0) = -4$ is a relative maximum, and $f(2) = -8$ is a relative minimum (Figure 7.1).

Example 7.18. Find the extrema of $f(x) = (x-1)^3$.
Obviously, for all $x \neq 1$ the derivative $f'(x) = 3(x-1)^2$ is positive and $x = 1$ is the unique point that the derivative is zero. Thus in some neighborhood of the critical point $x = 1$, the derivative does not change sign.
It follows that the value $f(1) = 0$ is neither a relative maximum nor a relative minimum of $f(x) = (x-1)^3$ (the curve is shown in Figure 7.2).

Sometimes, determining the sign of $f'(x)$ in some neighborhood of x_0 is not easy.

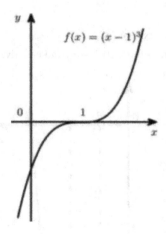

Fig. 7.2 $x = 1$ is a critical point of curve $y = f(x)$.

The following theorem shows that to find the extreme values of a function f which is twice differentiable, it is enough to determine the sign of f'' at the critical points.

Theorem 7.4 (A Second Sufficient Condition for a Relative Extremum).
Suppose a function f has a finite second derivative at a critical point x_0. Then $f(x_0)$ is a relative maximum if $f''(x_0) < 0$ and x_0 is a relative minimum if $f''(x_0) > 0$.

☐ By Theorem 5.1, it follows from the hypothesis $f''(x_0) < 0$ $(f''(x_0) > 0)$ that the first derivative $f'(x)$ is decreasing (increasing) at point x_0. On the other hand, since $f'(x_0) = 0$, there exists a neighborhood of x_0 so that for points smaller than x_0, the first derivative is positive, i.e., $f'(x) > 0$ $(f'(x) < 0)$ and for points bigger than x_0, the derivative is negative, i.e., $f'(x) < 0$ $(f'(x) > 0)$. Then by Theorem 7.3, x_0 is a relative maximum (minimum). ∎

Remark 7.7. If either $f''(x_0) = 0$ or $f''(x)$ does not exist, then Theorem 7.4 is not applicable. In the first case, higher order derivatives should be investigated.

Theorem 7.5 (A Third Sufficient Condition for a Relative Extremum). *Let $n \geqslant 1$ be an odd number and suppose a function $y = f(x)$ has an n-th derivative in a neighborhood of x_0, and an $(n+1)$-th derivative at x_0. Moreover, suppose that*

$$f'(x_0) = f''(x_0) = \cdots = f^{(n)}(x_0) = 0, \quad f^{(n+1)}(x_0) \neq 0. \qquad (7.24)$$

Then the value $f(x_0)$ is a relative maximum of f if $f^{(n+1)}(x_0) < 0$; and $f(x_0)$ is a relative minimum if $f^{(n+1)}(x_0) > 0$.

☐ The case $n = 1$ of Theorem 7.5 is just Theorem 7.4. So let $n \geqslant 3$ and for definiteness, assume $f^{(n+1)}(x_0) > 0$. We show that $f(x_0)$ is a relative minimum of f.

Since $f^{(n+1)}(x_0) > 0$, by Theorem 5.1 the n-th derivative $f^{(n)}(x)$ is increasing at point $x = x_0$. By hypothesis, $f^{(n)}(x_0) = 0$, and so there exists a neighborhood of x_0 such that $f^{(n)}(x) < 0$ for $x < x_0$ and $f^{(n)}(x) > 0$ for $x > x_0$ from that neighborhood.

For f', Taylor's formula with Lagrange's form for the remainder gives us:

$$f'(x) = f'(x_0) + \frac{f''(x_0)}{1!}(x - x_0) + \cdots + \frac{f^{(n-1)}(x_0)}{(n-2)!}(x - x_0)^{n-2} + \frac{f^{(n)}(\xi)}{(n-1)!}(x - x_0)^{n-1},$$

where ξ is some point between x and x_0. Substituting (7.24) in this formula we have

$$f'(x) = \frac{f^{(n)}(\xi)}{(n-1)!}(x - x_0)^{n-1}. \qquad (7.25)$$

As seen before, $f^{(n)}(x) < 0$ $(x < x_0)$ and $f^{(n)}(x) > 0$ $(x > x_0)$ for x sufficiently close to x_0. Because ξ is a point between x and x_0 it follows that $f^{(n)}(\xi) < 0$, if $\xi < x_0$, and $f^{(n)}(\xi) > 0$, if $\xi > x_0$. Therefore, since the number n is odd, by (7.25), $f'(x) < 0$ for points close to x_0 on the left and $f'(x) > 0$ for points close to x_0 on the right. Thus, by Theorem 7.3, x_0 is a relative minimum. Similarly, if $f^{(n+1)}(x_0) < 0$, then x_0 is a relative minimum. ∎

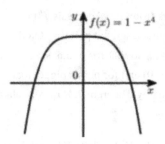

Fig. 7.3 $f(0) = 1$ is the maximum value of the function of Example 7.19.

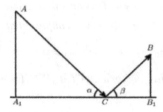

Fig. 7.4   Reflection at C of a light ray by a mirror $A_1 B_1$.

Example 7.19. Find the relative extrema of $f(x) = 1 - x^4$.

First, find the critical points. Since

$$f'(x) = -4x^3 = 0$$

the critical point is $x = 0$. Moreover, $f'(x) > 0$ if $x < 0$ and $f'(x) < 0$ if $x > 0$. By Theorem 7.3, it follows that $f(0) = 1$ is a relative maximum (see Figure 7.3). On the other hand, because $f''(x) = -12x^2$, $f'''(x) = -24x$, $f^{(4)}(x) = -24 < 0$ and $f'(0) = f''(0) = f'''(0) = 0$ by the use of Theorem 7.5 $(n = 3)$ we have the same result.

Example 7.20. Consider the reflection of a ray of light by a mirror $A_1 B_1$, as shown in Figure 7.4, which depicts a ray traveling from point A to point B via reflection in $A_1 B_1$ at the point C.

Assume that the location of the point of reflection is such as to minimize the total distance $d_1 + d_2$ traveled by the light ray; this is an application of Fermat's principle of least time for the propagation of light.

The problem is to find C.

Let a, b, m, x, d_1 and d_2 denote the lengths of the segments AA_1, BB_1, A_1B_1, A_1C, AC, and CB, respectively.

By the Pythagorean theorem, the distance to be minimized is

$$d_1 + d_2 = f(x) = \sqrt{x^2 + a^2} + \sqrt{(m - x)^2 + b^2}, \quad x \in [0, m].$$

Then

$$f'(x) = \frac{x}{\sqrt{x^2 + a^2}} + \frac{x - m}{\sqrt{(m - x)^2 + b^2}} = \frac{x}{d_1} + \frac{x - m}{d_2}, \quad f''(x) = \frac{1}{d_1} + \frac{1}{d_2} > 0$$

and so we find that any horizontal tangent to the graph of f must occur over the point x determined by

$$\frac{x}{d_1} = \frac{m - x}{d_2}.$$

At such a point, $\cos \alpha = \cos \beta$, where α is the angle of the incident light ray and β is the angle of the reflected ray. Thus we find, that $\alpha = \beta$. Hence, the angle of incidence is equal to the angle of reflection (Figure 7.3).

Theorem 7.6. *Suppose that a function f is continuous at x_0 and differentiable in some neighborhood of x_0, except possibly at x_0.*

Then $f(x_0)$ is a relative maximum (minimum) if $f'(x) > 0$, $x < x_0$ and $f'(x) < 0$, $x > x_0$ ($f'(x) < 0$, $x < x_0$ and $f'(x) > 0$, $x > x_0$) in this neighborhood.

The theorem is proved in a similar way as Theorem 7.3.

Example 7.21. The function $f(x) = |x|$ is continuous at $x = 0$ and is differentiable everywhere with the exception of $x = 0$:

$$f'(x) = \begin{cases} 1, & \text{if } x > 0, \\ -1, & \text{if } x < 0, \end{cases}$$

By Theorem 7.6, the value $f(0) = 0$ is a relative minimum (Figure 7.5).

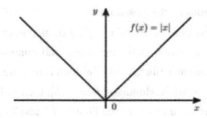

Fig. 7.5   Graph of the function of Example 7.21.

Example 7.22. The function $f(x) = \left(1 - x^{\frac{2}{3}}\right)^{\frac{3}{2}}$ is not differentiable at $x = 0$, because $f'(x) = -\left(1 - x^{\frac{2}{3}}\right)^{\frac{1}{2}} x^{-\frac{1}{3}}$ becomes infinitely large as $x \to 0$. But $f(0) = 1$, $f(x) < 1$, $x \neq 0$, and so $f(0)$ is a maximum (Figure 7.6).

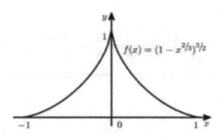

Fig. 7.6   Graph of the function of Example 7.22.

Corollary 7.2. *Let a function f defined on the interval (a,b) (possibly $a = -\infty$ or $b = +\infty$) be continuous and let its derivative f' be continuous everywhere with the exception of a finite number of points. On the other hand, suppose that the number of points x in (a,b) such that $f'(x) = 0$ is finite. This means that the number of critical points of f is finite, say, n and may be enumerated as follows:*

$$a < x_1 < x_2 < \cdots < x_n < b.$$

By hypothesis, the sign of the derivative $f'(x)$ does not change on the intervals $(a,x_1),(x_1,x_2),\ldots,(x_n,b)$.   Therefore, using Theorem 7.6, we can verify whether $x_1, x_2, \ldots, x_n$ is a relative extremum of f.

7.5   Global Extrema of a Function on a Closed Interval

Let a function f be continuous on the closed interval $[a,b]$. We will investigate

the global (absolute) extreme values of $y = f(x)$ on this interval. By Theorem 3.12 (Weierstrass), such functions have a global maximum and minimum. For simplicity, we consider the global maximum of f (the global minimum is investigated similarly).

Observe that if x_0 is a global maximum of f on $[a,b]$, then x_0 is either a critical point of f or one of the end points $x = a$ or $x = b$ (see Figures 7.7 and 7.8).

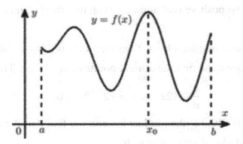

Fig. 7.7 The global maximum $f(x_0)$ occurs at the interior point of $[a,b]$.

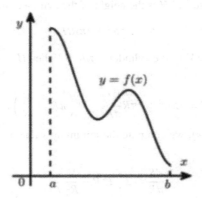

Fig. 7.8 The global maximum $f(b)$ occurs at the end point b.

In Figure 7.7, the global maximum of f on $[a,b]$ occurs at an interior point of $[a,b]$, while in Figure 7.8, it occurs at the end point $x = b$.

As a consequence, we can find the global maximum (minimum) of f on $[a,b]$ by

1. locating all the critical points of f on $[a,b]$,
2. finding the values $f(a)$ and $f(b)$ of f at the two end points of $[a,b]$, and
3. selecting the largest of these values, which must then be the global maximum of f, and the smallest, which will be the global minimum.

Analogously, one may investigate the global extrema of a function f defined on the intervals $(-\infty, b]$, $[a, +\infty)$, or $(-\infty, +\infty)$.

Example 7.23. Find two positive real numbers such that their sum is 12 and their product is as large as possible.

If we denote one of these numbers by x, then the other is $12 - x$. Obviously, in order to maximize the product we must find the critical points of $x(12 - x)$. Thus,

$$\frac{d}{dx}(12x - x^2) = 12 - 2x = 0.$$

Since at the end points of $[0, 12]$ the product is zero, we find that the unique critical point is $x = 6$ and the desired global maximum is 36.

Example 7.24. Find the dimensions of that cylinder of the smallest possible surface area, subject to the constraint that it possesses a given volume, V.

If R is the radius of the base and H is the height of the cylinder, then the surface area is

$$S = 2\pi R^2 + 2\pi RH. \tag{7.26}$$

Note that the given volume V of the cylinder is $\pi R^2 H$. Then $H = \dfrac{V}{\pi R^2}$. By substituting H into (7.26), we obtain

$$S = 2\pi R^2 + 2\pi R \frac{V}{\pi R^2} = 2\left(\pi R^2 + \frac{V}{R}\right).$$

In the open interval $(0, +\infty)$, we compute the minimum value of the surface area S. First we find the derivative:

$$\frac{dS}{dR} = 2\left(2\pi R - \frac{V}{R^2}\right).$$

Equating the derivative to zero, we find that $R_1 = \sqrt[3]{\dfrac{V}{2\pi}}$. Now, at the critical point $R = R_1$, we determine the sign of the second derivative,

$$\left(\frac{d^2 S}{dR^2}\right)_{R=R_1} = 2\left(2\pi + \frac{2V}{R^3}\right)_{R=R_1} > 0.$$

This means that $R = R_1$ is the minimum value of $S = 2\pi R^2 + 2\pi RH$. On the other hand, since $\lim\limits_{R\to 0} S = \infty$ and $\lim\limits_{R\to\infty} S = \infty$, the value $R = R_1$ is the global minimum.

By substituting the value $R = R_1$ into the formula for H, we have

$$H = \frac{V}{\pi R^2} = 2\sqrt[3]{\frac{V}{2\pi}} = 2R_1.$$

Dimensions of the cylinder are $H = 2R_1$, $R_1 = \sqrt[3]{\dfrac{V}{2\pi}}$.

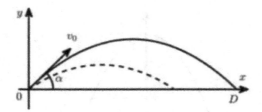

Fig. 7.9 $S\left(\frac{\pi}{4}\right)$ is the global maximum value of Example 7.25.

Example 7.25. Suppose that at the origin O a cannon is fired with initial velocity of the shot v_0 (the air resistance is neglected) (Figure 7.9).

Let $S = OD$ be the distance between the origin and the landing place of the shot fired. As we saw earlier,

$$S = \frac{v_0^2 \sin 2\alpha}{g}.$$

Here $0 \leqslant \alpha \leqslant \dfrac{\pi}{2}$ is the angle between the cannon barrel and the x-axis, g is, as before, the acceleration due to gravity. Find the angle α that makes the distance S maximal. Since

$$\frac{dS}{d\alpha} = \frac{2v_0^2 \cos 2\alpha}{g}, \qquad \frac{2v_0^2 \cos 2\alpha}{g} = 0,$$

it follows that $\alpha = \dfrac{\pi}{4}$. On the other hand,

$$\frac{d^2S}{d\alpha^2} = -\frac{4v_0^2 \sin 2\alpha}{g}, \qquad \left(\frac{d^2S}{d\alpha^2}\right)_{\alpha=\frac{\pi}{4}} = -\frac{4v_0^2}{g} < 0,$$

i.e., the second derivative is negative at $\alpha = \dfrac{\pi}{4}$, and so the value

$$(S)_{\alpha=\frac{\pi}{4}} = \frac{v_0^2}{g} \tag{7.27}$$

is a relative maximum of S (Theorem 7.4). Obviously, at the end points of $\left[0, \dfrac{\pi}{2}\right]$

$$(S)_{\alpha=0} = 0 \ \text{ and } \ (S)_{\alpha=\frac{\pi}{4}} = 0.$$

Thus, the value (7.27) is the global maximum of S.

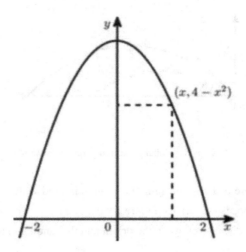

Fig. 7.10   The maximum area of the inscribed rectangle of Example 7.26.

Example 7.26. A rectangle (with vertical and horizontal sides) is inscribed in the region of the first quadrant bounded by the parabola $y = 4 - x^2$. Find the dimensions that yield the rectangle of largest area (Figure 7.10).

Let $(x, 4 - x^2)$ be the vertex that lies on the parabola. Denoting by A the area of the rectangle, we have
$$A = x(4 - x^2) = 4x - x^3,$$
where $x \in [0, 2]$. Then $\dfrac{dA}{dx} = 4 - 3x^2$, $\dfrac{d^2A}{dx^2} = -6x$.

Then the ordinate of the critical point is $y = 4 - \left(\dfrac{2}{\sqrt{3}}\right)^2 = \dfrac{8}{3}$. Note that $\dfrac{d^2A}{dx^2} < 0$ and so,

by Theorem 7.4, the maximum area is $A = \dfrac{2}{\sqrt{3}} \cdot \dfrac{8}{3} = \dfrac{16}{3\sqrt{3}}$.

Example 7.27. Suppose that an automobile factory sells 4,500 automobiles monthly, with a selling price of \$95 000. But last month the price was reduced to \$90 000, and sales went up to 6 000. What should the selling price be made, in order to maximize profit?

Let x denote the selling price and y, the number of automobiles sold. Then let
$$x_1 = 95\,000 - 5\,000 = 90\,000, \quad x_2 = 95\,000, \quad y_1 = 6\,000, \quad \text{and} \quad y_2 = 4\,500.$$

We write an equation of the line that passes through the two points (x_1, y_1) and (x_2, y_2) (assuming that the price function is linear in y). Then
$$\frac{x - 90\,000}{95\,000 - 90\,000} = \frac{y - 6\,000}{4\,500 - 6\,000} \quad \text{and so} \quad y = 6\,000 - \frac{3}{10}(x - 90\,000) = 33\,000 - \frac{3}{10}x.$$

The profit function is $P(x) = x\left(33\,000 - \dfrac{3}{10}x\right)$ and so, in order to find its critical point, we compute its derivative $P'(x) = 33\,000 - \dfrac{6}{10}x$. Thus, from $33\,000 - \dfrac{6}{10}x = 0$, we find that the selling price $x = \$55\,000$ maximizes the month's profit.

Example 7.28. Find the dimensions of that rectangle with perimeter 200 ft and area as large as possible.

Let x and y be the length and width of the rectangle, respectively. Then $A = xy$ is the area of the rectangle. Since the perimeter of the rectangle is 200 ft, we can write $2x + 2y = 200$ or, $x + y = 100$. Substituting $y = 100 - x$ in $A = xy$ yields

$$A = x(100 - x) = 100x - x^2, \quad x \in [0, 100].$$

Then we obtain $\dfrac{dA}{dx} = 100 - 2x$. Setting $100 - 2x = 0$, we find a critical point $x = 50$. Thus the maximum occurs at one of the points $x = 0$, $x = 50$, $x = 100$. It follows that the rectangle of perimeter 200 ft with the greatest area is a square with sides of length 50 ft.

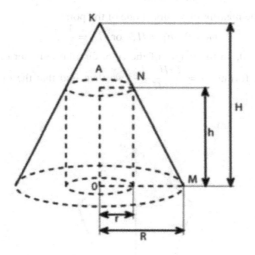

Fig. 7.11   The maximum possible volume of a cylinder inscribed in a right circular cone.

Example 7.29. Find the radius and height of the right-circular cylinder of largest volume that can be inscribed in a right-circular cone with radius R and height H. Denote by h and r the height and base radius of the cylinder, respectively (Figure 7.11).

Using the triangles OMK and AKN we have

$$\frac{R}{r} = \frac{H}{H - h},$$

or
$$r = \frac{R(H-h)}{H}.$$

Substituting r in the formula for the volume of the inscribed cylinder $V = \pi r^2 h$, we can write
$$V = \frac{\pi R^2}{H^2}\left(H^2 h - 2Hh^2 + h^3\right).$$

Now we must find the critical points. Thus, setting
$$V' = \frac{\pi R^2}{H^2}\left(H^2 - 4Hh + 3h^2\right) = 0,$$

we find that
$$3h^2 - 4Hh + H^2 = 0$$

and
$$h_1 = H, \quad h_2 = \frac{1}{3}H.$$

Furthermore, the second derivative is $V'' = \frac{\pi R^2}{H^2}(6h - 4H)$, and so
$$V''(h) > 0, \quad h = H; \quad V''(h) < 0, \quad h = \frac{H}{3}.$$

Because $0 \leqslant h \leqslant H$ the maximum occurs at one of the points
$$h_0 = 0, \quad h_1 = H, \quad \text{or} \quad h_2 = \frac{1}{3}H.$$

But $V(h_0) = V(h_1) = 0$, so the height of the right-circular cylinder of largest volume is $h_2 = \frac{1}{3}H$. Using the formula $r = \frac{R(H-h)}{H}$, we then find that the optimal cylinder has radius $\frac{2R}{3}$.

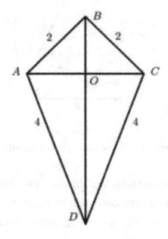

Fig. 7.12   The kite frame.

Example 7.30. A kite frame is to be made of six pieces of wood, as shown in Fig. 7.12. The four outer pieces with the indicated lengths have already been cut. What should be the lengths of the two diagonal pieces in order to maximize the area of the kite?

Let x and y, as indicated in Fig.7.12, denote the lengths of the two diagonal pieces of a kite frame. Obviously, the area of the triangle BCD (or DAB) is equal to $\frac{1}{2} y \cdot \frac{x}{2} = \frac{xy}{4}$. Apply the Pythagorean theorem to the right triangles OCD and BCO in the figure. This yields the equation

$$a = \sqrt{4^2 - \left(\frac{x}{2}\right)^2} = \frac{1}{2}\sqrt{64 - x^2}, \quad b = \sqrt{2^2 - \left(\frac{x}{2}\right)^2} = \frac{1}{2}\sqrt{16 - x^2},$$

where a and b denote the lengths of the segments OD and BO, respectively. Therefore, the area of the kite is given by

$$A = 2 \cdot \frac{xy}{4} = \frac{1}{2}xy = \frac{1}{2}x(a+b) = \frac{1}{4}\left(\sqrt{64 - x^2} + \sqrt{16 - x^2}\right).$$

Here, the derivative is

$$\frac{dA}{dx} = \frac{1}{4}\left[\sqrt{64 - x^2} + \sqrt{16 - x^2} - \frac{x^2}{\sqrt{64 - x^2}} - \frac{x^2}{\sqrt{16 - x^2}}\right]$$

$$= \frac{1}{2}\left[\frac{(32 - x^2)\sqrt{16 - x^2} + (8 - x^2)\sqrt{64 - x^2}}{\sqrt{64 - x^2} \cdot \sqrt{16 - x^2}}\right].$$

This fraction can vanish only when its numerator is zero, that is, when

$$(32 - x^2)\sqrt{16 - x^2} + (8 - x^2)\sqrt{64 - x^2} = 0.$$

After a simple transformation, we have $15x^2 = 192$, and so,

$$x = \pm\sqrt{\frac{192}{15}} = \pm\frac{8}{\sqrt{5}}.$$

Hence, the only critical point in $[0,4]$ is $x = \frac{8}{\sqrt{5}}$. It follows that

$$y = a + b = \frac{1}{2}\left(\sqrt{64 - x^2} + \sqrt{16 - x^2}\right) = \frac{1}{2}\left(\sqrt{64 - \frac{64}{5}} + \sqrt{16 - \frac{64}{5}}\right) = \frac{10}{\sqrt{5}}.$$

Now, we evaluate A at $x = \frac{8}{\sqrt{5}}$ and at the two end points to find

$$A(0) = 0, \quad A\left(\frac{8}{\sqrt{5}}\right) = \frac{1}{2}\frac{8}{\sqrt{5}}\frac{10}{\sqrt{5}} = 8, \quad A(4) = \frac{1}{4} \cdot 4\left(\sqrt{64 - 4^2} + \sqrt{16 - 4^2}\right) = 4\sqrt{3}.$$

Thus, the lengths of the two diagonal pieces that maximize the area of the kite, are

$$x = \frac{8}{\sqrt{5}} \text{ and } y = \frac{10}{\sqrt{5}}.$$

7.6   Higher Derivatives and Concavity

Let f be a differentiable function on an open interval, (a, b). As was seen in Chapter 4, at any point $P(x, f(x))$ of the curve $y = f(x)$, there exists a tangent line with slope $f'(x)$, not parallel to the y-axis since $f'(x)$ is finite at x.

Definition 7.4. If, at every point x in (a, b), the curve $y = f(x)$ lies above (below) the tangent line at $(x, f(x))$, then we say that the function f (or its graph) is *concave upward* (*downward*).

In Figs. 7.13 and 7.14 are shown concave upward and downward curves, respectively.

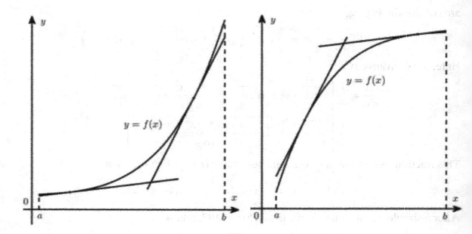

Fig. 7.13   The graph is bending upward.   Fig. 7.14   The graph is bending downward.

Theorem 7.7. *Suppose that the function f has a second derivative on the interval (a, b), and $f''(x) \geqslant 0$ ($f''(x) \leqslant 0$). Then the curve $y = f(x)$ is concave upward (downward).*

☐ Let us investigate the case $f''(x) \geqslant 0$, $x \in (a, b)$ (the case $f''(x) \leqslant 0$ is proved similarly). Take any point $x_0 \in (a, b)$ and write the equation of the tangent line at $(x_0, f(x_0))$

$$Y - f(x_0) = f'(x_0)(x - x_0) \tag{7.28}$$

(Y being the ordinate of a variable point on the tangent line).

For the case $n = 1$, we write the second degree Taylor's formula with remainder for f at $x = x_0$,

$$y = f(x) = f(x_0) + \frac{f'(x_0)}{1!}(x - x_0) + \frac{f''(\xi)}{2!}(x - x_0)^2, \tag{7.29}$$

where ξ is some point between x and x_0. Equations (7.28) and (7.29) imply

$$y - Y = \frac{f''(\xi)}{2}(x - x_0)^2. \tag{7.30}$$

Because $f''(x) \geqslant 0$, $x \in (a, b)$, from (7.30) we obtain $y \geqslant Y$. Consequently, the curve $y = f(x)$ lies above the tangent line at $(x, f(x))$. ∎

Remark 7.8. If $f''(x) = 0$, $x \in (a, b)$, it is easy to see that f is a linear function, i.e., $f(x) = kx + b$ (k, b are constants). This function is concave upward and downward, simultaneously.

Remark 7.9. In general, the definition of concavity for an arbitrary function f can be given as follows; if, for all $0 \leqslant \lambda \leqslant 1$ and x_1, x_2 in (a, b), the inequality

$$f(\lambda x_1 + (1 - \lambda)x_2) \leqslant \lambda f(x_1) + (1 - \lambda)f(x_2)$$

$$\left[f(\lambda x_1 + (1 - \lambda)x_2) \geqslant \lambda f(x_1) + (1 - \lambda)f(x_2)\right]$$

holds, then f is concave upward (downward).

Theorem 7.8. *Suppose that the second derivative f'' of a function f is continuous and positive (negative) at point x_0. Then there is a neighborhood of x_0 such that the function is concave upward (downward).*

☐ It is easy to see that the continuity of f'' at x_0 implies the existence of some neighborhood of x_0 such that for points in this neighborhood, $f''(x) > 0$ ($f''(x) < 0$). Consequently, by Theorem 7.7, for such points x the curve $y = f(x)$ is concave upward (downward). ∎

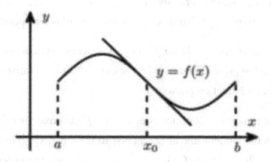

Fig. 7.15 $(x_0, f(x_0))$ is the inflection point of curve $y = f(x)$.

Example 7.31. Determine the intervals of concavity of $f(x) = x^3 - 3x^2 - 4$.
Clearly, $f'(x) = 3x^2 - 6x$, $f''(x) = 6x - 6 = 6(x - 1)$. Then we have that $f''(x) < 0$ for all points $x < 1$, and so the curve is concave downward; and that $f''(x) > 0$ for all points $x > 1$, and so the curve is concave upward (see Figure 7.1).

Definition 7.5. Let f be differentiable on the interval (a,b) and x_0 be a fixed point, $a < x_0 < b$. Suppose that in the intervals (a,x_0) and (x_0,b) the concavity of the curve $y = f(x)$ changes. Then we say that the point x_0 is a *point of inflection* of the function f. More precisely, x_0 is an inflection point of the function f provided that f is concave upward on one side of x_0, concave downward on the other side, and continuous at x_0. We may also refer to $(x_0, f(x_0))$ as an inflection point on the graph of f.

In Figure 7.15 the point x_0 is an inflection point of the curve $y = f(x)$, because f is concave downward on (a,x_0) but concave upward on (x_0,b).

Theorem 7.9 (A Sufficient Condition for an Inflection Point). *Suppose that the function f has a second order derivative in a neighborhood of x_0, such that $f''(x_0) = 0$ and that $f''(x)$ has opposite signs in that neighborhood, when $x < x_0$ and $x > x_0$. Then the point x_0 is an inflection point.*

□   Indeed, by hypothesis, $f'(x_0)$ exists and is finite, so the curve $y = f(x)$ has a tangent line at $(x_0, f(x_0))$. On the other hand, since the sign of $f''(x)$ changes from the left to the right of x_0, by Theorem 7.7 the concavity of the curve $y = f(x)$ changes from $x < x_0$ to $x > x_0$.   ■

Example 7.32. Determine the inflection points of the curve $f(x) = x^3 - 3x^2 - 4$. There is a unique point where the second derivative $f''(x) = 6(x-1)$ is zero. Moreover, $f''(x) < 0$ for $x < 1$ and $f''(x) > 0$ for $x > 1$. Thus, $x = 1$ is the only inflection point.

Remark 7.10. In general, the condition $f''(x) = 0$ is necessary, but not sufficient for the existence of an inflection point. For example, the curve $f(x) = x^4$ satisfies $f''(0) = 0$. But the point $x = 0$ is not an inflection point of f.

Remark 7.11. For a function f continuous at $x = x_0$, $f''(x_0)$ may or may not be finite. If $f''(x_0)$ is finite, then the tangent line at $(x_0, f(x_0))$ to the curve $y = f(x)$ is parallel to the y-axis. For example, the function $f(x) = x^{1/3}$ has a finite second order derivative at all points except $x = 0$. This function is continuous at $x = 0$ and has a tangent line at $(0,0)$. On the other hand, $f''(x) = -\dfrac{2}{9}\dfrac{1}{x^{5/3}}$, and because $f''(x) > 0$ for $x < 0$ and $f''(x) < 0$ for $x > 0$, the point $(0,0)$ is an inflection point of the graph of f.

Theorem 7.10. *Suppose that $n \geqslant 2$ is an even integer, that a function f in some neighborhood of x_0 has an n-th derivative, and that f has an $(n+1)$-th derivative at x_0, such* *that*

$$f''(x_0) = f'''(x_0) = \cdots = f^{(n)}(x_0) = 0, \ f^{(n+1)}(x_0) \neq 0. \tag{7.31}$$

Then $(x_0, f(x_0))$ *is an inflection point of the curve* $y = f(x)$.

□ By Theorem 5.1, $f^{(n)}(x)$ decreases at point $x = x_0$ if $f^{(n+1)}(x_0) < 0$, and increases if $f^{(n+1)}(x_0) > 0$.

Besides, it follows from $f^{(n)}(x_0) = 0$ that in any case there exists a neighborhood of x_0 such that $f^{(n)}(x)$ changes sign from the left $x < x_0$ to the right $x > x_0$ of x_0.

By using Lagrange's form for the remainder term, we write Taylor's formula for $f''(x)$ in a neighborhood of x_0:

$$f''(x) = f''(x_0) + \frac{f'''(x_0)}{1!}(x - x_0) + \cdots + \frac{f^{(n-1)}(x_0)}{(n-3)!}(x - x_0)^{n-3} + \frac{f^{(n)}(\xi)}{(n-2)!}(x - x_0)^{n-2},$$

where ξ is a point between x and x_0. Then, by (7.31), we have

$$f''(x) = \frac{f^{(n)}(\xi)}{(n-2)!}(x - x_0)^{n-2}. \tag{7.32}$$

Since n is even, it follows from (7.32) that the sign of $f''(x)$ depends on the sign of $f^{(n)}(\xi)$. But for points sufficiently close to x_0, the sign of $f^{(n)}(\xi)$ changes as we pass from the region $x < x_0$ to the region $x > x_0$. Then, by Theorem 7.9, the point $(x_0, f(x_0))$ is an inflection point of the curve $y = f(x)$. ∎

Remark 7.12. In Theorem 7.10, the evenness of n is essential for the validity of the conclusion. Indeed, consider, for instance, the function defined by $f(x) = (x-1)^{n+1}$. Here $f(1) = 0$ is a minimum value of f, if n is an odd, and $(1,0)$ is an inflection point of f, if n is even.

Example 7.33. Find an inflection point of the curve $f(x) = \sin x$, $x \in \left(-\frac{\pi}{2}, \frac{\pi}{2}\right)$. Here, the derivatives are $f'(x) = \cos x$, $f''(x) = -\sin x$, and $f'''(x) = -\cos x$. Clearly, $f''(0) = 0$ and $f'''(0) = -1 \neq 0$. Thus, by Theorem 7.10, the point $(0, f(0)) = (0,0)$ is an inflection point of the graph of f.

7.7 Asymptotes of Graphs; Sketching Graphs of Functions

In Section 3.2 we mentioned the possibility of possessing finite limits at infinity, in connection with the behavior of a function as $x \to \pm\infty$. There is also such a thing as the possession of an infinite limit at a finite location.

Definition 7.6. We say that the line $x = a$ is a vertical asymptote for the curve $y = f(x)$ provided that at least one of the limits $\lim\limits_{x \to a+0} f(x)$, $\lim\limits_{x \to a-0} f(x)$ or $\lim\limits_{x \to a} f(x)$, is either $+\infty$ or $-\infty$. Also, we say that the line $y = b$ is a horizontal asymptote for the curve $y = f(x)$ if either

$$\lim_{x \to +\infty} f(x) = b \text{ or } \lim_{x \to -\infty} f(x) = b.$$

Definition 7.7. If the function f has a representation

$$f(x) = kx + b + \alpha(x), \quad \lim_{x \to +\infty} \alpha(x) = 0, \tag{7.33}$$

then we say that the line

$$Y = kx + b \tag{7.34}$$

is an *oblique asymptote* for the curve $y = f(x)$ as $x \to +\infty$ $(x \to -\infty)$.

Theorem 7.11. *For the existence of the inclined asymptote* (7.34) *of the curve* $y = f(x)$ *as* $x \to +\infty$ *it is necessary and sufficient that, there exist the following limits*

(1) $\lim\limits_{x \to +\infty} \dfrac{f(x)}{x} = k$;

(2) $\lim\limits_{x \to +\infty} [f(x) - kx] = b.$

☐ **Necessity.** Suppose that the line (7.34) is an oblique asymptote to the curve $y = f(x)$ as $x \to +\infty$. Then we have a representation (7.33) for $y = f(x)$. Then

$$\lim_{x \to +\infty} \frac{f(x)}{x} = \lim_{x \to +\infty} \frac{kx + b + \alpha(x)}{x} = \lim_{x \to +\infty} \left[k + \frac{b}{x} + \frac{\alpha(x)}{x} \right] = k,$$

and

$$\lim_{x \to +\infty} [f(x) - kx] = \lim_{x \to +\infty} [b - \alpha(x)] = b.$$

Sufficiency. Suppose we have the conditions (1), (2). It follows from (2) that $f(x) - kx - b = \alpha(x)$, and so $\lim\limits_{x \to +\infty} \alpha(x) = 0$. Consequently, f has the form (7.33).

Similarly, Theorem 7.11 can be proved when $x \to -\infty$. ∎

In the following definition we give the concept of a nonlinear asymptote.

Definition 7.8. If the function f can be represented in the form

$$f(x) = a_0 x^n + a_1 x^{n-1} + \cdots + a_{n-1} x + a_n + \alpha(x), \quad \text{with} \quad \lim_{x \to +\infty} \alpha(x) = 0,$$

then we say that the curve

$$Y = a_0 x^n + a_1 x^{n-1} + \cdots + a_{n-1} x + a_n$$

of degree n is *asymptotic* to the curve $y = f(x)$ as $x \to +\infty$.

It is not hard to see that the existence of all of the following limits, $n + 1$ in number, is necessary and sufficient for the existence of such an asymptote as $x \to +\infty$:

$$\lim_{x \to +\infty} \frac{f(x)}{x^n} = a_0, \qquad \lim_{x \to +\infty} \frac{f(x) - a_0 x^n}{x^{n-1}} = a_1,$$

$$\cdots \cdots ,$$

$$\lim_{x \to +\infty} \frac{f(x) - (a_0 x^n + \cdots + a_{n-2} x^2)}{x} = a_{n-1},$$

$$\lim_{x \to +\infty} [f(x) - (a_0 x^n + \cdots + a_{n-1} x)] = a_n.$$

In particular, the case $n = 1$ implies Theorem 7.11.

Example 7.34. Let $f(x) = \dfrac{1}{x-3}$. Note that for this curve, the line $x = 3$ is a vertical asymptote because $\lim\limits_{x \to 3+0} f(x) = +\infty$ and $\lim\limits_{x \to 3-0} f(x) = -\infty$.

Example 7.35. For the curve $f(x) = \tan x$, there are infinitely many asymptotes, one each at $x = \pm\dfrac{\pi}{2}$, $x = \pm\dfrac{3\pi}{2}$, $x = \pm\dfrac{5\pi}{2}, \ldots$, because $\tan x \to \pm\infty$ as x approaches such points (Figure 7.16).

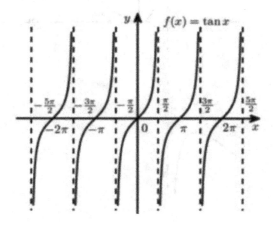

Fig. 7.16   The vertical asymptotes of $y = \tan x$.

Example 7.36. Find the asymptotes of the curve $f(x) = \dfrac{x^2}{x+1}$.

To begin with, we find the limits

$$k = \lim_{x \to \pm\infty} \frac{f(x)}{x} = \lim_{x \to \pm\infty} \frac{x}{x+1} = 1$$

$$b = \lim_{x \to \pm\infty} [f(x) - kx] = \lim_{x \to \pm\infty} \left(\frac{x^2}{x+1} - x \right) = \lim_{x \to \pm\infty} \frac{-x}{x+1} = -1.$$

Therefore, by Theorem 7.11, the line $Y = kx + b = x - 1$ is asymptotic to the curve $y = f(x)$.

Furthermore, $\lim_{x \to -1+0} \dfrac{x^2}{x+1} = +\infty$ and $\lim_{x \to -1-0} \dfrac{x^2}{x+1} = -\infty$. Thus, the line $x = -1$ is a vertical asymptote (Figure 7.17).

Remark 7.13. If $f(x) = \dfrac{P(x)}{Q(x)}$ is a rational function, with the degree of $P(x)$ greater by 1 than that of $Q(x)$, then by long division of $P(x)$ into $Q(x)$ we find that $y = f(x)$ has the form $f(x) = kx + b + \dfrac{P_1(x)}{Q(x)}$, where $\lim_{x \to \infty} \dfrac{P_1(x)}{Q(x)} = 0$, and $Y = kx + b$ is an asymptote for the curve $y = f(x)$(the degree of $P_1(x)$ is less than the degree of $Q(x)$). For example, let $P(x) = x^2$ and $Q(x) = x + 1$ (see Example 7.34). Then, $f(x) = \dfrac{x^2}{x+1} = x - 1 + \dfrac{1}{x+1}$ and $Y = x - 1$ is an oblique asymptote.

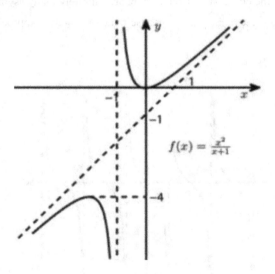

$$f(x) = \tfrac{x^2}{x+1}$$

Fig. 7.17   A function with asymptote the line $Y = x - 1$.

Example 7.37. Find the asymptotes of $f(x) = \tanh x$.

By definition, $\tanh x = \dfrac{e^x - e^{-x}}{e^x + e^{-x}}$, and it is easy to see that there do not exist any vertical or oblique asymptotes. We investigate whether horizontal asymptotes exist or not. Obviously,

$$\lim_{x \to +\infty} \frac{e^x - e^{-x}}{e^x + e^{-x}} = \lim_{x \to +\infty} \frac{e^{2x} - 1}{e^{2x} + 1} = \lim_{x \to +\infty} \frac{e^{2x}}{e^{2x}} = 1.$$

Similarly, $\lim\limits_{x \to -\infty} \dfrac{e^x - e^{-x}}{e^x + e^{-x}} = -1$. Thus, the lines $y = \pm 1$ are two horizontal asymptotes (see Figure 3.4).

For curve-sketching of a function f, we follow these steps, loosely rather than rigidly, and should obtain a rather accurate sketch of the graph of a given function:

1) Determine the domain of f.
2) Determine the points of discontinuity of f.
3) Determine the intervals on which f is increasing and those on which it is decreasing. Find the critical points and extreme values of f.
4) Determine the intervals on which f is concave upward, those on which it is concave downward, and its inflection points.
5) Find the x-intercepts and the y-intercepts of the graph and determine the asymptotes.
6) Determine the range of f.

Example 7.38. Investigate the function defined by $f(x) = \dfrac{x^2}{x+1}$ and sketch its graph. Note that the point $x = -1$ is a point of discontinuity. On the other hand, the curve $f(x) = \dfrac{x^2}{x+1}$ has a vertical asymptote at $x = -1$ and an oblique asymptote $Y = x - 1$ as $x \to \pm\infty$ (see Figure 7.17). Moreover, $f'(x) = \dfrac{x^2 + 2x}{(x+1)^2}$ and $f'(0) = f'(-2) = 0$. Because $f'(x) > 0$ on $(-\infty, -2)$ and $f'(x) < 0$ on $(-2, -1)$, it follows that $f(-2) = -4$ is a relative maximum of f. Similarly, since $f'(x) < 0$ on $(-1, 0)$ and $f'(x) > 0$ on $(0, +\infty)$, we see that $f(0) = 0$ is a minimum of f.

x	$(-\infty, -2)$	$(-2, -1)$	$(-1, 0)$	$(0, +\infty)$
y'	$+$	$-$	$-$	$+$
$f(x)$	increasing	decreasing	decreasing	increasing

An examination of the second derivative shows that $f''(x) = \dfrac{2}{(x+1)^3} > 0$ if $x > -1$, and $f''(x) = \dfrac{2}{(x+1)^3} < 0$ if $x < -1$. Therefore, the curve $y = f(x)$ is concave down on $(-\infty, -1)$ and concave up on $(-1, +\infty)$.

Remark 7.14. Suppose that the function f is even ($f(x) = f(-x)$). Then to sketch the graph of f, it is enough to take positive values of x, because the graph of an even function is symmetric about the y-axis. Similarly, if f is an odd function ($f(-x) = -f(x)$), then, since the graph of an odd function is symmetric with respect to the origin, we may again restrict ourselves to investigating the positive values of x.

7.8 Problems

In problems (1)–(25), use L'Hopital's rule to find the limits.

(1) $\displaystyle\lim_{x\to 0}\frac{\tan x - x}{x - \sin x}$.   *Answer*: 2.

(2) $\displaystyle\lim_{x\to 0}\frac{\sin ax}{\sin bx}$.   *Answer*: $\dfrac{a}{b}$.

(3) $\displaystyle\lim_{x\to \frac{\pi}{4}}\frac{\sqrt[3]{\tan x} - 1}{2\sin^2 x - 1}$.   *Answer*: $\dfrac{1}{3}$.

(4) $\displaystyle\lim_{x\to +\infty}\frac{\ln x}{x^\varepsilon}$ $(\varepsilon > 0)$.   *Answer*: 0.

(5) $\displaystyle\lim_{x\to \frac{\pi}{4}}(\tan x)^{\tan 2x}$.   *Answer*: $\dfrac{1}{e}$.

(6) $\displaystyle\lim_{x\to 0}\frac{1 - \cos^2 x}{x^2 \sin x^2}$.   *Answer*: $\dfrac{1}{2}$.

(7) $\displaystyle\lim_{x\to 0}\frac{\ln(\cos ax)}{\ln(\cos bx)}$.   *Answer*: $\left(\dfrac{a}{b}\right)^2$.

(8) $\displaystyle\lim_{x\to a}\frac{a^x - x^a}{x - a}$ $(a > 0)$.   *Answer*: $a^a(\ln a - 1)$.

(9) $\displaystyle\lim_{x\to 0+0}x^{kx}$ (k is a constant).   *Answer*: 1.

(10) $\displaystyle\lim_{x\to 0+0}\left(\ln\frac{1}{x}\right)^x$.   *Answer*: 1.

(11) $\displaystyle\lim_{x\to 0}\frac{(a+x)^x - a^x}{x^2}$ $(a > 0)$.   *Answer*: $\dfrac{1}{a}$.

(12) $\displaystyle\lim_{x\to 0}\frac{(1+x)^{\frac{1}{x}} - e}{x}$.   *Answer*: $-\dfrac{e}{2}$.

(13) $\displaystyle\lim_{x\to 0}\left(\frac{\sin x}{x}\right)^{\frac{1}{x^2}}$.   *Answer*: $e^{-\frac{1}{6}}$.

(14) $\displaystyle\lim_{x\to 0}\left(\frac{\tan x}{x}\right)^{\frac{1}{x^2}}$.   *Answer*: $e^{\frac{1}{3}}$.

(15) $\displaystyle\lim_{x\to +\infty}\frac{x^{\ln x}}{(\ln x)^x}$.   *Answer*: 0.

(16) $\displaystyle\lim_{x\to 1}x^{\frac{1}{1-x}}$.   *Answer*: $\dfrac{1}{e}$.

(17) $\displaystyle\lim_{x\to \infty}\left(1+\frac{a}{x}\right)^x$.   *Answer*: e^a.

(18) $\displaystyle\lim_{t\to 0}(\cot x)^{\frac{1}{\ln x}}$.   *Answer*: $\dfrac{1}{e}$.

(19) $\lim\limits_{x\to\frac{\pi}{2}} (\sec x - \tan x)$.   *Answer:* 0.

(20) $\lim\limits_{x\to\frac{\pi}{2}} \dfrac{\ln\sin x}{(\pi - 2x)^2}$.   *Answer:* $\dfrac{1}{8}$.

(21) $\lim\limits_{x\to a} \dfrac{\sin x - \sin a}{x - a}$.   *Answer:* $\cos a$.

(22) $\lim\limits_{x\to 0} \dfrac{a^x - b^x}{x}$.   *Answer:* $\ln\dfrac{a}{b}$.

(23) $\lim\limits_{x\to 1} \left(\tan\dfrac{\pi x}{4}\right)^{\tan\frac{\pi x}{2}}$.   *Answer:* $\dfrac{1}{e}$.

(24) $\lim\limits_{x\to 0} \left(\dfrac{1}{x}\right)^{\tan x}$.   *Answer:* 1.

(25) $\lim\limits_{x\to\frac{\pi}{2}} (\cos x)^{\frac{\pi}{2}-x}$.   *Answer:* 1.

(26) Suppose that f posseses a second derivative f'', and then prove that

$$f''(x) = \lim_{h\to 0} \frac{f(x+h) + f(x-h) - 2f(x)}{h^2}.$$

In problems (27)–(28), determine whether L'Hopital's rule is applicable or not.

(27) $\lim\limits_{x\to 0} \dfrac{x^2 \sin\dfrac{1}{x}}{\sin x}$.   *Answer:* 0 (not applicable).

(28) $\lim\limits_{x\to\infty} \dfrac{x - \sin x}{x + \sin x}$.   *Answer:* 1 (not applicable).

In problems (29)–(34) use Taylor's formula to find the limits.

(29) $\lim\limits_{x\to 0} \dfrac{x - \sin x}{e^x - 1 - x - \dfrac{x^2}{2}}$.   *Answer:* 1.

(30) $\lim\limits_{x\to 0} \left[x - x^2 \ln\left(1 + \dfrac{1}{x}\right)\right]$.   *Answer:* 0.

(31) $\lim\limits_{x\to 0} \left(\dfrac{1}{x^2} - \dfrac{\cot x}{x}\right)$.   *Answer:* $\dfrac{1}{3}$.

(32) $\lim\limits_{x\to 0} \dfrac{2(\tan x - \sin x) - x^3}{x^5}$.   *Answer:* $\dfrac{1}{4}$.

(33) $\lim\limits_{x\to 0} \dfrac{a^x + a^{-x} - 2}{x^2}$ $(a > 0)$.   *Answer:* $\ln^2 a$.

(34) $\lim\limits_{x\to 0} \dfrac{1}{x}\left(\dfrac{1}{x} - \cot x\right)$.   *Answer:* $\dfrac{1}{3}$.

In each of the problems (35)–(46), find the relative maximum and minimum values.

(35) $f(x) = x^4 - 8x^2 + 2.$

Answer: $f(0) = 2$ maximum;
$f(\pm 2) = -14$ minimum.

(36) $f(x) = \dfrac{x^3}{3} - 2x^2 + 3x + 1.$

Answer: $f(1) = \dfrac{7}{2}$ maximum;
$f(3) = 1$ minimum.

(37) $f(x) = 2 - (x - 1)^{2/3}.$

Answer: $f(1) = 2$ maximum.

(38) $f(x) = -x^4 + 2x^2.$

Answer: $f(\pm 1) = 1$ maximum;
$f(0) = 0$ minimum.

(39) $f(x) = \dfrac{x^2 - 3x + 2}{x^2 + 3x + 2}.$

Answer: $f(\sqrt{2})$ minimum;
$f(\sqrt{2})$ maximum.

(40) $f(x) = \cos x + \sin x,\ x \in \left[-\dfrac{\pi}{2}, \dfrac{\pi}{2} \right].$

Answer: $f\left(\dfrac{\pi}{4} \right) = \sqrt{2}$ maximum.

(41) $f(x) = e^x \sin x.$

Answer: $f\left(2k\pi - \dfrac{\pi}{4} \right)$ minimum;
$f\left(2k\pi + \dfrac{3}{4}\pi \right)$ maximum.

(42) $f(x) = x + \dfrac{1}{x}.$

Answer: $f(1)$ minimum;
$f(-1)$ maximum.

(43) $f(x) = \dfrac{a^2}{x} + \dfrac{b^2}{a - x},$

Answer: $f\left(\dfrac{a^2}{a - b} \right)$ maximum;
$f\left(\dfrac{a^2}{a + b} \right)$ minimum.

(44) $f(x) = x \ln x.$

Answer: $f\left(\dfrac{1}{e} \right)$ minimum.

(45) $f(x) = 2x + \tan^{-1} x.$

Answer: There are no extrema.

(46) $f(x) = \sin x \cos^2 x.$

Answer: $f\left(\dfrac{\pi}{2} \right)$ minimum;
$f\left(\cos^{-1}\left(\pm\sqrt{\dfrac{2}{3}} \right) \right)$ maximum.

In problems (47)–(50), determine the global extreme values.

(47) $f(x) = -3x^4 + 6x^2 - 1,\ x \in [-2, 2].$ *Answer:* $f(\pm 1) = 2$ global maximum;
$f(\pm 2) = -25$ global minimum.

(48) $f(x) = \dfrac{x-1}{x+1}, x \in [0,4].$ *Answer:* $f(4) = \dfrac{3}{5}$ global maximum;

$f(0) = -1$ global minimum.

(49) $f(x) = \sin 2x - x, x \in \left[-\dfrac{\pi}{2}, \dfrac{\pi}{2}\right].$ *Answer:* $f\left(-\dfrac{\pi}{2}\right) = \dfrac{\pi}{2}$ global maximum;

$f\left(\dfrac{\pi}{2}\right) = -\dfrac{\pi}{2}$ global minimum.

(50) $f(x) = \dfrac{x^3}{3} - 2x^2 + 3x + 1, x \in [-1,5].$ *Answer:* $f(5) = \dfrac{23}{3}$ global maximum;

$f(-1) = -\dfrac{13}{3}$ global minimum.

(51) If $f \geqslant 0$, then show that the functions f and $F = cf^2$ ($c > 0$) have the same extremum values.

(52) A rectangle has a fixed perimeter S. What is the maximum possible area of such a rectangle?

Answer: The square with side $\sqrt{S}$.

(53) Find the dimensions of the rectangle (with vertical and horizontal sides) of maximal area that can be inscribed in the ellipse with equation $\dfrac{x^2}{a^2} + \dfrac{y^2}{b^2} = 1$.

Answer: The rectangle's dimensions are $a\sqrt{2}$ and $b\sqrt{2}$.

(54) Find the minimum distance from the point $M(p,p)$ to the parabola $y^2 = 2px$.

Answer: $p\left(\sqrt[3]{2} - 1\right)\sqrt{\dfrac{2 + \sqrt[3]{2}}{2}}.$

(55) Find the minimum and maximum distances from the point $M(2,0)$ to the circle $x^2 + y^2 = 1$.

Answer: 1; 3.

(56) A cone is made from a circular sheet of radius R by cutting out a sector and gluing the cut edges of the remaining piece together. What is the maximum volume attainable for the cone?

Answer: $\varphi = 2\pi\sqrt{\dfrac{2}{3}}$ (φ being the central angle of the remaining piece).

(57) If $f''(x) \geqslant 0, x \in [a,b]$, prove that for every $x_1, x_2 \in [a,b]$,

$$f\left(\dfrac{x_1 + x_2}{2}\right) \leqslant \dfrac{1}{2}[f(x_1) + f(x_2)].$$

In problems (58)–(63), determine whether the curve is bending upward or bending downward and find the inflection points of the curve.

(58) $f(x) = \dfrac{a^2}{a^2 + x^2}$ $(a > 0)$.   *Answer:* $|x| < \dfrac{a}{\sqrt{3}}$ concave downward;

$\qquad\qquad\qquad\qquad\qquad\qquad$ $|x| > \dfrac{a}{\sqrt{3}}$ concave upward;

$\qquad\qquad\qquad\qquad\qquad\qquad$ $x = \pm\dfrac{a}{\sqrt{3}}$ are the inflection points.

(59) $f(x) = x^x$ $(x > 0)$.   *Answer:* $0 < x < +\infty$ concave upward.

(60) $f(x) = e^{-x^2}$.   *Answer:* $|x| < \dfrac{1}{\sqrt{2}}$ concave downward;

$\qquad\qquad\qquad\qquad\qquad\qquad$ $|x| > \dfrac{1}{\sqrt{2}}$ concave upward;

$\qquad\qquad\qquad\qquad\qquad\qquad$ $x = \pm\dfrac{1}{\sqrt{2}}$ are the inflection points.

(61) $f(x) = 3x^2 - x^3$.   *Answer:* $-\infty < x < 1$ concave upward;

$\qquad\qquad\qquad\qquad\qquad\qquad$ $1 < x < +\infty$ concave downward;

$\qquad\qquad\qquad\qquad\qquad\qquad$ $x = 1$ is the inflection point of f.

(62) $f(x) = \sqrt{1 + x^2}$.   *Answer:* concave upward.

(63) $f(x) = x + \sin x$.   *Answer:* $2k\pi < x < (2k+1)\pi$ concave downward;

$\qquad\qquad\qquad\qquad\qquad\qquad$ $(2k+1)\pi < x < (2k+2)\pi$ concave upward;

$\qquad\qquad\qquad\qquad\qquad\qquad$ $x = k\pi$ $(k = 0, \pm 1, \pm 2, \ldots)$

$\qquad\qquad\qquad\qquad\qquad\qquad$ are the inflection points.

In problems (64)–(68), find the asymptotes of the curve:

(64) $f(x) = \dfrac{1}{(x+2)^3}$.   *Answer:* $x = -2, y = 0$.

(65) $f(x) = c + \dfrac{a^3}{(x-b)^2}$.   *Answer:* $x = b, y = c$.

(66) $f(x) = \ln x$.   *Answer:* $x = 0$.

(67) $f(x) = e^{-2x}\sin x$.   *Answer:* $y = 0$.

(68) $x = \dfrac{2t}{1-t^2}, y = \dfrac{t^2}{1-t^2}$.   *Answer:* $y = \pm\dfrac{1}{2}x - \dfrac{1}{2}$.

Investigate the functions in problems (69)–(74) and sketch their graphs:

(69) $f(x) = \dfrac{4+x}{x^2}$.

(70) $f(x) = \dfrac{x+2}{x^3}$.

(71) $f(x) = e^{-x}\sin x$.

(72) $f(x) = xe^{-x}$.

(73) $f(x) = x\sin x$.

(74) $f(x) = \ln \sin x$.

Chapter 8

The Indefinite Integral

The theory of indefinite integrals is one of the basic topics of mathematical analysis. In this chapter, we study the main properties of the indefinite integral and give tables of the integrals of the main functions of analysis. We will find that sometimes the integrals of the elementary functions cannot be expressed in terms of finite combinations of the familiar algebraic and elementary transcendental functions. We shall discuss various basic techniques of integration: integration by substitution and by parts, which can often be used to transform complicated integration problems into simpler ones. Moreover, we shall discuss methods for integrating arbitrary rational functions; in particular, we will explain the method of Ostrogradsky. Finally, the evaluation of integrals involving irrational algebraic functions, quadratic polynomials, trigonometric functions, etc., will also be considered.

8.1   The Antiderivative and the Indefinite Integral

As we saw in Chapter 4, to a given function f is associated its derivative, $f(x) = F'(x)$. Inversely, the process of finding a function from its derivative is the opposite of differentiation and so is called antidifferentiation. If there exists a function f having f as its derivative, we call f an antiderivative of f.

As in previous chapters, we let Ω denote any one of the intervals (a,b), $(a,b]$, $[a,b)$, $[a,b]$, $(-\infty,a]$, $[a,+\infty)$, or $(-\infty,+\infty)$.

Definition 8.1. Let the function f be defined on Ω. An *antiderivative* of the function f is a function f defined on Ω, such that

$$F'(x) = f(x)$$

for all $x \in \Omega$. Sometimes instead of antiderivative, we call f a primitive of f.

Example 8.1. For $f(x) = x^2$, $x \in (-\infty,+\infty)$, its antiderivative is $F(x) = \dfrac{x^3}{3}$, because $F'(x) = \left(\dfrac{x^3}{3}\right)' = x^2$.

E. Mahmudov, *Single Variable Differential and Integral Calculus*,
DOI: 10.2991/978-94-91216-86-2_8, © Atlantis Press 2013

Example 8.2. Let $f(x) = -\dfrac{x}{\sqrt{1-x^2}}$, $x \in (-1,+1)$. Its antiderivative is $F(x) = \sqrt{1-x^2}$, because $\left(\sqrt{1-x^2}\right)' = -\dfrac{x}{\sqrt{1-x^2}}$.

Example 8.3. Let $f(x) = \sin x$, $x \in (-\infty, +\infty)$. An antiderivative of this function is. Indeed, $(-\cos x)' = \sin x$.

Example 8.4. Similarly, the function $F(x) = \ln x$ is an antiderivative of the function $f(x) = \dfrac{1}{x}$ on $(0,+\infty)$, that is, $(\ln x)' = \dfrac{1}{x}$.

Observe that if $F(x)$ is an antiderivative of f, then so is $F(x)+C$ for any constant. Conversely, if $F(x)$ is one antiderivative of $y = f(x)$ on the interval Ω, then every antiderivative of $f(x)$ on Ω is of the form $F(x)+C$.

Theorem 8.1. *If the functions $F_1(x)$ and $f_2(x)$ are two antiderivatives of the function f on the interval Ω, then $F_1(x) - F_2(x) = C$ for some constant C.*

☐ Note that the function $F(x) = F_1(x) - F_2(x)$ is differentiable on Y. But, $F'(x) = F_1'(x) - F_2'(x) = f(x) - f(x) = 0$. Therefore, by Theorem 5.4, the function $F(x)$ is constant on Ω:

$$F(x) = F_1(x) - F_2(x) = C.$$

∎

Definition 8.2. The process of finding antiderivatives of the form $F(x)+C$ on the interval Ω is called *antidifferentiation* or *integration*. If $F'(x) = f(x)$, we denote this by writing

$$\int f(x)dx = F(x)+C \tag{8.1}$$

Here the symbol $\int$ is called an *indefinite integral sign*. It is easy to see that $f(x)dx = F'(x)dx = dF$. The adjective "indefinite" is used because the right hand side of (8.1) is not a definite function. The function $f(x)$ and constant C are called the *integrand* and *constant of integration*, respectively.

Using this definition, we can write

$$\int \frac{-x}{\sqrt{1-x^2}}dx = \sqrt{1-x^2}+C \quad \text{or} \quad \int \sin x\, dx = -\cos x + C$$

(see Examples 8.2 and 8.3).

8.2   Basic Properties of the Indefinite Integral

First, we discuss the properties of indefinite integrals that follow from the definition.

(1) $d \int f(x)dx = f(x)dx.$

(2) $\int dF(x) = F(x) + C.$

In order to prove (1), it is sufficient take into account $dF(x) = F'(x)dx = f(x)dx$ and differentiate (8.1). Since $dF(x) = f(x)dx$, the proof of (2) immediately follows from (8.1).

The following properties are called the linearity of indefinite integrals.

(3) $\int [f(x) + g(x)]dx = \int f(x)dx + \int g(x)dx.$

(4) $\int [cf(x)]dx = c \int f(x)dx$ (c a constant).

To prove (3), we must show that the left hand side of (3) is an antiderivative of $f(x) + g(x)$; and to prove (4), we must show that the left hand side of (4) is an antiderivative of $cf(x)$. Indeed, if $F(x)$ and $G(x)$ are antiderivatives of the functions $f(x)$ and $g(x)$, respectively, then

$$[F(x) + G(x)]' = F'(x) + G'(x) = f(x) + g(x)$$

$$[cF(x)]' = cF'(x) = cf(x).$$

On the other hand, by using the table of derivatives given in Section 4.6, we may begin to construct a short table of integrals: each line is equivalent to one of the basic derivative formulas.

1. $\int dx = x + C.$

2. $\int x^n dx = \dfrac{x^{n+1}}{n+1} + C$ ($n \neq -1$).

3. $\int \dfrac{dx}{x} = \ln|x| + C$ ($x \neq 0$).

4. $\int a^x dx = \dfrac{a^x}{\ln a} + C$ ($0 < a \neq 1$), $\int e^x dx = e^x + C.$

5. $\int \sin x\, dx = -\cos x + C.$

6. $\int \cos x\, dx = \sin x + C.$

7. $\int \dfrac{dx}{\cos^2 x} = \int \sec^2 x\, dx = \tan x + C$ ($x \neq \dfrac{\pi}{2} + \pi n,\ n = 0, \pm 1, \pm 2, \dots$).

8. $\int \dfrac{dx}{\sin^2 x} = -\cot x + C$ ($x \neq \pi n,\ n = 0, \pm 1, \pm 2, \dots$).

9. $\displaystyle\int \frac{dx}{\sqrt{1-x^2}} = \begin{cases} \sin^{-1}x + C, \\ -\cos^{-1}x + C \end{cases} \quad (-1 < x < 1).$

10. $\displaystyle\int \frac{dx}{1+x^2} = \begin{cases} -\cot^{-1}x + C, \\ \tan^{-1}x + C. \end{cases}$

11. $\displaystyle\int \frac{dx}{\sqrt{x^2 \pm 1}} = \ln\left|x + \sqrt{x^2 \pm 1}\right| + C$ (the negative sign only if $|x| > 1$).

12. $\displaystyle\int \frac{dx}{1-x^2} = \frac{1}{2}\ln\left|\frac{1+x}{1-x}\right| + C\ (|x| \neq 1).$

13. $\displaystyle\int \sinh x\, dx = \cosh x + C.$

14. $\displaystyle\int \cosh x\, dx = \sinh x + C.$

15. $\displaystyle\int \frac{dx}{\cosh^2 x} = \tanh x + C.$

16. $\displaystyle\int \frac{dx}{\sinh^2 x} = -\coth x + C\ (x \neq 0).$

Sometimes a knowledge of elementary derivative formulas will allow us to integrate a given function. This approach can, however, be inefficacious, especially in view of the following surprising fact: There exist indefinite integrals that cannot be evaluated in terms of finite combinations of the familiar elementary functions. We begin with the list of such integrals.

1. $\displaystyle\int e^{-x^2}\, dx.$

2. $\displaystyle\int \cos(x^2)\, dx.$

3. $\displaystyle\int \sin(x^2)\, dx.$

4. $\displaystyle\int \frac{dx}{\ln x}\ (0 < x \neq 1).$

5. $\displaystyle\int \frac{\cos x}{x}\, dx\ (x \neq 0).$

6. $\displaystyle\int \frac{\sin x}{x}\, dx.$

7. $\displaystyle\int \frac{dx}{\sqrt{1 - k^2 \sin^2 x}}\ (0 < k < 1).$

8. $\displaystyle\int \sqrt{1 - k^2 \sin^2 x}\, dx\ (0 < k < 1).$

The existence of these examples indicates that we cannot in all cases reduce integration to a routine process like differentiation. Nevertheless, these integrals play an important role in different physics problems. For example, the so-called Poisson[1] integrals, 1 and 2, are

[1] D. Poisson, (1781–1840), French mathematician.

useful in heat and diffusion problems. The Fresnel[2] integrals, 5 and 6, are used in optics. The Legendre[3] integrals, 7 and 8, are also referred to as the elliptic integrals of the first and second types, respectively. Lacking simple formulas for these antiderivatives, we give them a proper name, such as

$$\varphi(x) = \frac{2}{\sqrt{\pi}} \int e^{-x^2} dx + C_0, \quad \varphi(0) = 0 \quad \text{(Laplace's function)},$$

$$F(k,x) = \int \frac{dx}{\sqrt{1 - k^2 \sin^2 x}} + C_1, \quad F(k,0) = 0,$$

and

$$E(k,x) = \int \sqrt{1 - k^2 \sin^2 x}\, dx + C_2, \quad E(k,0) = 0,$$

and calculate tables of their values for different x.

8.3   Integration by Substitution

The basis for this is the differentiation of composite functions. Thus, to evaluate

$$\int f(x)dx,$$

we use the substitution $x = \varphi(t)$ ($\varphi(t)$ is assumed differentiable), a purely mechanical substitution gives the tentative formula

$$\int f(x)dx = \int f[\varphi(t)]\varphi'(t)dt. \tag{8.2}$$

This substitution transforms the given integral into the simpler integral (8.2). The key to making this simplification lies in observing a composition $f[\varphi(t)]$ inside the given integrand. In order to convert this integrand into a function of x alone, we replace $f[\varphi(t)]$ by the simpler $f(x)$ and also $\varphi'(t)dt$ by dx. In order to prove (8.2), let f be an antiderivative of f, so that $F'(x) = f(x)$ or, equivalently

$$\frac{d}{dx}\left[\int f(x)dx\right] = f(x).$$

Then

$$\frac{d}{dx}[F(\varphi(t)] = F'(\varphi(t)) \cdot \varphi'(t) = f(\varphi(t)) \cdot \varphi'(t)$$

---

[2] A. Fresnel (1788–1827), French physicist and mathematician.
[3] A.-M. Legendre (1752–1833), French mathematician.

by the chain rule (Theorem 4.3), so that

$$\int f(x)dx = F(x) + C = F(\varphi(t)) + C = \int \left[\frac{d}{dt}F(\varphi(t))\right]dt$$

$$= \int F'(\varphi(t)) \cdot \varphi'(t)dt = \int f(\varphi(t)) \cdot \varphi'(t)dt.$$

This ends the proof of the formula (8.2).

Remark 8.1. With a "good" choice of $x = \varphi(t)$, the integral $\int f[\varphi(t)]\varphi'(t)dt$ will be easier to evaluate than the original $\int f(x)dx$. Sometimes in integration by substitution, a substitution of the form $t = \psi(x)$ is more useful than the substitution $x = \varphi(t)$. For example, in order to transform the integral

$$\int \frac{\psi'(x)dx}{\psi(x)},$$

the substitution $t = \psi(x)$ is the key. Indeed, since $dt = \psi'(x)dx$,

$$\int \frac{\psi'(x)dx}{\psi(x)} = \int \frac{dt}{t} = \ln|t| + C = \ln|\psi(x)| + C.$$

Example 8.5. Find $\int \sin 3x\, dx$.

If we choose $t = 3x$, then $dt = 3dx$ and so

$$\int \sin 3x\, dx = \int \frac{1}{3}\sin t\, dt = -\frac{1}{3}\cos t + C = -\frac{1}{3}\cos 3x + C.$$

Example 8.6. Find $\int e^{\cos x} \sin x\, dx$.

We take $t = \cos x$. Since $dt = -\sin x\, dx$,

$$\int e^{\cos x} \sin x\, dx = -\int e^t dt = -e^t + C = -e^{\cos x} + C.$$

Example 8.7. Find $\int \sqrt{\sin x}\cos x\, dx$.

If $t = \sin x$, then $dt = \cos x\, dx$. Hence,

$$\int \sqrt{\sin x}\cos x\, dx = \int \sqrt{t}\, dt = \int t^{1/2} dt = \frac{2t^{3/2}}{3} + C = \frac{2}{3}\sin^{3/2}x + C.$$

Example 8.8. Find $\int \dfrac{dx}{\sqrt{a^2 - x^2}}$.

By substituting $t = \dfrac{x}{a}$, we derive

$$\int \frac{dx}{\sqrt{a^2 - x^2}} = \frac{1}{a}\int \frac{dx}{\sqrt{1 - \left(\frac{x}{a}\right)^2}} = \frac{1}{a}\int \frac{a\,dt}{\sqrt{1 - t^2}} = \arcsin t + C = \arcsin \frac{x}{a} + C.$$

Example 8.9. Find $\int \sqrt{\dfrac{a+x}{a-x}}\,dx$.

If we choose $x = a\cos 2t$, then $dx = -2a\sin 2t\,dt$. Therefore,

$$\int \sqrt{\frac{a+x}{a-x}}\,dx = -4a\int \cos^2 t\,dt = -4a\int \left(\frac{1}{2}+\frac{1}{2}\cos 2t\right)dt = -2at - 2a\int \cos 2t\,dt$$

$$= -2at - a\sin 2t + C = -a\left[\cos^{-1}\frac{x}{a} + \sqrt{1-\left(\frac{x}{a}\right)^2}\right] + C.$$

Example 8.10. Evaluate the integral $\int \cos^2 x\,dx$.

$$\int \cos^2 x\,dx = \int \frac{1+\cos 2x}{2}\,dx = \frac{1}{2}\int (1+\cos 2x)\,dx = \frac{1}{2}\int dx + \frac{1}{2}\int \cos 2x\,dx$$

$$= \frac{1}{2}x + \frac{1}{4}\sin 2x + C.$$

Example 8.11. Use the substitution $t = \sin x$ to evaluate the integral $\int \dfrac{dx}{\cos x}$.

$$\int \frac{dx}{\cos x} = \int \frac{\cos x\,dx}{\cos^2 x} = \int \frac{\cos x\,dx}{1-\sin^2 x} = \int \frac{dt}{1-t^2} = \frac{1}{2}\ln\left|\frac{1+t}{1-t}\right| + C = \ln\left|\tan\left(\frac{x}{2}+\frac{\pi}{4}\right)\right| + C.$$

We next use the antiderivative to investigate Einstein's equation of motion along a line.

By Einstein's relativity theory the linear motion of a body with mass m_0 in vacuum and acted on by a constant force $F > 0$, is governed by the equation, see (4.23),

$$\frac{d}{dt}\frac{v}{\sqrt{1-\left(\dfrac{v^2}{c^2}\right)}} = \frac{F}{m_0}.$$

Here, c is the velocity of light in vacuum ($c = 3\cdot 10^{10}$ cm $\cdot$ s^{-1}). Suppose the velocity v of the body at $t = 0$ is zero. This equation tells us that the function $\dfrac{v}{\sqrt{1-(v^2/c^2)}}$ is an antiderivative of a constant function, $\dfrac{F}{m_0}$. It is zero at $t = 0$, and so

$$\frac{v}{\sqrt{1-\left(\dfrac{v^2}{c^2}\right)}} = \frac{Ft}{m_0}.$$

Therefore,

$$\frac{v^2}{1-\left(\dfrac{v^2}{c^2}\right)} = \frac{F^2 t^2}{m_0^2},$$

or

$$v^2 = c^2\frac{F^2 t^2}{m_0^2 c^2 + F^2 t^2}.$$

The velocity $v = v(t)$ is positive, and

$$v(t) = c \frac{Ft}{\sqrt{m_0^2 c^2 + F^2 t^2}}.$$

Here, the denominator of the quotient is greater than its numerator, and so $v(t) < c$ for arbitrary t. In other words, the velocity of a body with non-zero rest-mass may draw arbitrarily close to the velocity of light, but not become equal to it.

8.4 Integration by Parts

The formula for integration by parts is a simple consequence of the product rule for differentials. Thus, for differentiable functions $u = u(x)$ and $v = v(x)$, we can write

$$d(u \cdot v) = u \, dv + v \, du.$$

Then antidifferentiation gives

$$\int u \, dv = uv - \int v \, du. \tag{8.3}$$

This is the formula for integration by parts. We try to choose the part $\int u \, dv$ in (8.3) such a way that the new integral $\int v \, du$ is easier to compute than the original integral $\int u \, dv$.

Example 8.12. Find $\int \sqrt{x^2 - a^2} \, dx$.

Let

$$\sqrt{x^2 - a^2} = u, \quad dx = dv.$$

Then

$$du = \frac{x \, dx}{\sqrt{x^2 - a^2}}, \quad v = x,$$

and

$$\int \sqrt{x^2 - a^2} \, dx = x\sqrt{x^2 - a^2} - \int \frac{x^2 \, dx}{\sqrt{x^2 - a^2}}$$

$$= x\sqrt{x^2 - a^2} - \int \frac{(x^2 - a^2) \, dx}{\sqrt{x^2 - a^2}} - a^2 \int \frac{dx}{\sqrt{x^2 - a^2}}.$$

Thus,

$$\int \sqrt{x^2 - a^2} \, dx = x\sqrt{x^2 - a^2} - \int \sqrt{x^2 - a^2} \, dx - a^2 \ln \left| x + \sqrt{x^2 - a^2} \right|.$$

Finally, by solving this equation for $\int \sqrt{x^2 - a^2}$, we obtain

$$\int \sqrt{x^2 - a^2} \, dx = \frac{1}{2} x\sqrt{x^2 - a^2} - \frac{a^2}{2} \ln \left| x + \sqrt{x^2 - a^2} \right| + C.$$

Example 8.13. Find $I = \int x^n \ln x \, dx \ (n \neq -1)$.

If we choose $u = \ln x$, $dv = x^n \, dx$, then $du = \dfrac{dx}{x}$, $v = \dfrac{x^{n+1}}{n+1}$.

Hence,

$$I = \frac{x^{n+1}}{n+1} \ln x - \frac{1}{n+1} \int x^n \, dx = \frac{x^{n+1}}{n+1} \left(\ln x - \frac{1}{n+1} \right) + C.$$

Example 8.14. Evaluate $I = \int \tan^{-1} x \, dx$.

Let $u = \tan^{-1} x$, $dv = dx$. It follows that $du = \dfrac{dx}{1+x^2}$, $v = x$, and so

$$I = x \tan^{-1} x - \int \frac{x \, dx}{1+x^2} = x \tan^{-1} x - \frac{1}{2} \ln(1+x^2) + C.$$

Example 8.15. Find $I = \int e^{ax} \cos bx \, dx$ (a, b constants).

If we put $u = e^{ax}$, $dv = \cos bx \, dx$ in (8.3), then we obtain that

$$du = ae^{ax} dx \text{ and } v = \frac{\sin bx}{b}.$$

Thus,

$$I = \frac{e^{ax} \sin bx}{b} - \frac{a}{b} \int e^{ax} \sin bx \, dx.$$

Now, in order to find the integral on the right hand side of this equation, we take $u = e^{ax}$, $dv = \sin bx \, dx$. Then since $du = ae^{ax} dx$ and, by (8.3),

$$I = \frac{e^{ax} \sin bx}{b} + \frac{a}{b^2} e^{ax} \cos bx - \frac{a^2}{b^2} I. \tag{8.4}$$

By solving (8.4) for I, we obtain

$$I = \frac{a \cos bx + b \sin bx}{a^2 + b^2} e^{ax}.$$

Remark 8.2. As was seen above, in general, integration by parts is successful when applied to integrals involving integrands which fall into one of these patterns:

1. The integrand is a product of functions, one of which has a form $\ln x$, $\sin^{-1} x$, $\cos^{-1} x$, $\tan^{-1} x$, $(\tan^{-1} x)^2$, $(\cos^{-1} x)^2$, $\ln \varphi(x)$, etc., and the other is the derivative of another function;

2. The integrand is of the form $(ax+b)^n \cos(cx)$, $(ax+b)^n \sin(cx)$, $(ax+b)^n e^{cx}$, where a, b, and c are constants.

   In this case, integration by parts is applied n times sequentially.

3. The integrand is of the form $e^{ax} \sin bx$, $e^{ax} \cos bx$, $\sin(\ln x)$, $\cos(\ln x)$, etc.

Here, integration by parts is applied twice (see Example 8.15).

Example 8.16. Evaluate $I_l^0 = \int \dfrac{dt}{(t^2 + a^2)^l}$, a a constant, $l = 1, 2, \ldots$.

In order to compute this integral we establish a relation between I_l^0 and I_{l-1}^0. For $l \neq 1$, we can write

$$
\begin{aligned}
I_l^0 &= \frac{1}{a^2} \int \frac{a^2\,dt}{(t^2 + a^2)^l} = \frac{1}{a^2} \int \frac{[(t^2 + a^2) - t^2]dt}{(t^2 + a^2)^l} \\
&= \frac{1}{a^2} \int \frac{dt}{(t^2 + a^2)^{l-1}} - \frac{1}{2a^2} \int t \frac{2t\,dt}{(t^2 + a^2)^l} \\
&= \frac{1}{a^2} I_{l-1}^0 - \frac{1}{2a^2} \int t \frac{d(t^2 + a^2)}{(t^2 + a^2)^l}.
\end{aligned}
$$

We evaluate the integral on the right hand side in turn by taking

$$
u = t, \quad dv = \frac{d(t^2 + a^2)}{(t^2 + a^2)^l}.
$$

Then we have

$$
du = dt, \quad v = -\frac{1}{(l - 1)(t^2 + a^2)^{l-1}}.
$$

Thus,

$$
I_l^0 = \frac{1}{a^2} I_{l-1}^0 + \frac{t}{2a^2(l-1)(t^2 + a^2)^{l-1}} - \frac{1}{2a^2(l-1)} I_{l-1}^0
$$

and

$$
I_l^0 = \frac{t}{2a^2(l-1)(t^2 + a^2)^{l-1}} + \frac{2l - 3}{a^2(2l - 2)} I_{l-1}^0. \tag{8.5}
$$

Observe that the I_l^0 can be evaluated by using the recurrent sequence (8.5) for every $l = 2, 3, \ldots$. Indeed, for I_1^0, we derive

$$
I_1^0 = \int \frac{dt}{t^2 + a^2} = \frac{1}{a} \int \frac{d(t/a)}{(t/a)^2 + 1} = \frac{1}{a} \tan^{-1} \frac{t}{a} + C,
$$

and then for $l = 2$ we define I_2^0 by (8.5) and this process can be continued.

8.5　Partial Fractions; Integrating Rational Functions

Recall that a rational function $R(x)$ is one that can be expressed as a quotient of two polynomials $Q(x)$ and $P(x)$. That is, $R(x) = \dfrac{Q(x)}{P(x)}$. Without loss of generality, we consider the case where $Q(x)$ and $P(x)$ have no common roots.

Definition 8.3. If the degree of the numerator $Q(x)$ is less than that of the denominator $P(x)$, we say that the rational fraction $R(x)$ is a *proper fraction*.

It is clear that if $R(x)$ is not proper, then by "long division" of $P(x)$ into $Q(x)$ we find that

$$R(x) = \frac{Q(x)}{P(x)} = F(x) + \frac{Q_1(x)}{P(x)},$$

where $F(x)$ is a polynomial and $\dfrac{Q_1(x)}{P(x)}$ is a proper fraction.

Lemma 8.1. *Let $R(x)$ be a proper fraction with real coefficients Suppose that the real number α is a repeated root of $P(x)$ of multiplicity k,*

$$P(x) = (x - \alpha)^k \varphi(x), \quad \varphi(x) \neq 0. \tag{8.6}$$

Then $R(x)$ has the representation

$$R(x) = \frac{Q(x)}{P(x)} = \frac{A}{(x-\alpha)^k} + \frac{\psi(x)}{(x-\alpha)^{k-\lambda}\varphi(x)}, \tag{8.7}$$

where $A = \dfrac{Q(\alpha)}{\varphi(\alpha)}$ is a real constant, λ $(\lambda \geqslant 1)$ is an integer, and $\psi(x)$ is the polynomial such that the last fraction in (8.7) is proper.

☐ Since, $A = \dfrac{Q(\alpha)}{\varphi(\alpha)}$ is a finite real number. On the other hand, the number is a repeated root of the polynomial $\varphi(x) = Q(x) - A\varphi(x)$ with multiplicity $\lambda \geqslant 1$, because

$$\varphi(\alpha) = Q(\alpha) - A\varphi(\alpha) = Q(\alpha) - \frac{Q(\alpha)}{\varphi(\alpha)}\varphi(\alpha) = 0.$$

Hence, there exists a representation

$$\varphi(x) = (x - \alpha)^\lambda \psi(x), \quad \psi(\alpha) \neq 0,$$

where $\psi(x)$ is a polynomial.

Then we can write

$$\frac{Q(x)}{P(x)} - \frac{A}{(x-\alpha)^k} = \frac{Q(x) - A\varphi(x)}{(x-\alpha)^k\varphi(x)} = \frac{\varphi(x)}{(x-\alpha)^k\varphi(x)} = \frac{\psi(x)}{(x-\alpha)^{k-\lambda}\varphi(x)}. \tag{8.8}$$

This ends the proof of (8.7). Moreover, as the difference of two proper fractions, the right hand side of (8.8) is a proper fraction. ∎

Lemma 8.2. *Suppose $R(x)$ is a proper fraction and $P(x)$ has a pair of complex conjugate roots $\alpha = a + ib$ and $\overline{\alpha} = a - ib$, of multiplicity l:*

$$P(x) = (x^2 + px + q)^l \varphi(x), \quad \varphi(\alpha) \neq 0, \quad \varphi(\overline{\alpha}) \neq 0,$$

$$p = -2a, \quad q = a^2 + b^2.$$

Then there exists a representation:

$$\frac{Q(x)}{P(x)} = \frac{Mx + N}{(x^2 + px + q)^l} + \frac{\psi(x)}{(x^2 + px + q)^{l-\lambda}\varphi(x)}, \tag{8.9}$$

where M and N are real numbers, λ $(\lambda \geqslant 1)$ is integral, and $\psi(x)$ is the polynomial such that the last fraction in (8.9) is a proper fraction.

☐ Let Re A and Im A be the real and imaginary parts of the complex number A.

Now, let us denote

$$M = \frac{1}{b} \operatorname{Im} \left[\frac{Q(\alpha)}{\varphi(\alpha)} \right], \quad N = \operatorname{Re} \left[\frac{Q(\alpha)}{\varphi(\alpha)} \right] - \frac{a}{b} \operatorname{Im} \left[\frac{Q(\alpha)}{\varphi(\alpha)} \right].$$

It is not hard to see that M and N are the solutions of the equation

$$Q(\alpha) - (M\alpha + N)\varphi(\alpha) = 0. \tag{8.10}$$

Indeed, it follows from (8.10) that

$$Ma + N = \operatorname{Re} \left[\frac{Q(\alpha)}{\varphi(\alpha)} \right], \quad Mb = \operatorname{Im} \left[\frac{Q(\alpha)}{\varphi(\alpha)} \right],$$

where M and N are the numbers defined above.

On the other hand,

$$\frac{Q(x)}{P(x)} - \frac{Mx + N}{(x^2 + px + q)^l} = \frac{Q(x) - (Mx + N)\varphi(x)}{(x^2 + px + q)^l \varphi(x)} = \frac{\phi(x)}{(x^2 + px + q)\varphi(x)}, \tag{8.11}$$

where $\phi(x) = Q(x) - (Mx + N)\varphi(x)$. Then it follows from (8.10) that α and $\overline{\alpha}$ are a pair of complex conjugate roots of $\phi(x)$ of multiplicity $\lambda \geqslant 1$ (see Section 6.3). Consequently, $\phi(x)$ has the representation:

$$\phi(x) = (x^2 + px + q)^l \psi(x). \tag{8.12}$$

Here, α and $\overline{\alpha}$ are not roots of $\psi(x)$, that is, $\psi(\alpha) \neq 0$ and $\psi(\overline{\alpha}) \neq 0$. Thus, by substituting (8.12) into (8.11), we have the representation (8.9). Observe that the last fraction in (8.9), as a difference of two proper fractions, is proper. ∎

Fractions of the forms $\dfrac{A}{(x - \alpha)^k}$ and $\dfrac{Mx + N}{(x^2 + px + q)^{l'}}$, as in (8.7) and (8.9), are called partial fractions of the first and second types, respectively.

By Lemmas 8.1 and 8.2, if we continue this process of decomposition into partial fractions for each of the roots of the polynomial $P(x)$, then we can formulate an important theorem about the partial-fraction decomposition of a proper fraction $R(x)$.

To obtain such a decomposition, we must first factor the denominator $P(x)$ into a product of linear factors of the form $(x - \alpha)$ and irreducible quadratic factors of the form $x^2 + px + q$, $\dfrac{p^2}{4} - q < 0$. This is always possible in principle but may be quite difficult in practice.

Theorem 8.2. Let $R(x) = \dfrac{Q(x)}{P(x)}$ be a proper fraction with real coefficients and

$$P(x) = (x - \alpha_1)^{k_1}(x - \alpha_2)^{k_2} \cdots (x - \alpha_m)^{k_m}(x^2 + p_1 x + q_1)^{l_1} \cdots (x^2 + p_r x + q_r)^{l_r}. \tag{8.13}$$

Then the partial-fraction decomposition of $R(x)$ *has the form*

$$\frac{Q(x)}{P(x)} = \frac{A_{k_1}^{(1)}}{(x-\alpha_1)^{k_1}} + \frac{A_{k_1-1}^{(1)}}{(x-\alpha_1)^{k_1-1}} + \cdots + \frac{A_1^{(1)}}{x-\alpha_1}$$

$$+ \cdots + \frac{A_{k_m}^{(m)}}{(x-\alpha_m)^{k_m}} + \frac{A_{k_m-1}^{(m)}}{(x-\alpha_m)^{k_m-1}} + \cdots + \frac{A_1^{(m)}}{x-\alpha_m}$$

$$+ \frac{M_{l_1}^{(1)}x+N_{l_1}^{(1)}}{(x^2+p_1x+q_1)^{l_1}} + \frac{M_{l_1-1}^{(1)}x+N_{l_1-1}^{(1)}}{(x^2+p_1x+q_1)^{l_1-1}} + \cdots + \frac{M_1^{(1)}x+N_1^{(1)}}{x^2+p_1x+q_1}$$

$$+\cdots+ \frac{M_{l_r}^{(r)}x+N_{l_r}^{(r)}}{(x^2+p_rx+q_r)^{l_r}} + \frac{M_{l_r-1}^{(r)}x+N_{l_r-1}^{(r)}}{(x^2+p_rx+q_r)^{l_r-1}} + \cdots + \frac{M_1^{(r)}x+N_1^{(r)}}{x^2+p_rx+q_r}, \qquad (8.14)$$

where $A_1^{(1)}, A_2^{(1)}, \ldots, A_{k_m}^{(m)}, M_1^{(1)}, N_1^{(1)}, \ldots, M_{l_r}^{(r)}, N_{l_r}^{(r)}$ *are real constants to be determined.*

Example 8.17. Find the partial-fraction decomposition of the proper fraction

$$\frac{x^2+2}{(x+1)^3(x-2)}.$$

By formula (8.14),

$$\frac{x^2+2}{(x+1)^3(x-2)} = \frac{A}{(x+1)^3} + \frac{A_1}{(x+1)^2} + \frac{A_2}{x+1} + \frac{B}{x-2}.$$

To find the constants A, A_1, A_2, B, we multiply both sides of the equality by $(x+1)^3(x-2)$, to obtain

$$x^2+2 = A(x-2)+A_1(x+1)(x-2)+A_2(x+1)^2(x-2)+B(x+1)^3,$$

and hence,

$$x^2+2 = (A_2+B)x^3 + (A_1+3B)x^2 + (A-A_1-3A_2+3B)x + (-2A-2A_1-2A_2+B).$$

Equating coefficients of like terms gives

$$\begin{cases} 0 = A_2+B, \\ 1 = A_1+3B, \\ 0 = A-A_1-3A_2+3B, \\ 2 = -2A-2A_1-2A_2+B \end{cases}$$

and solving this system yields $A = -1, A_1 = \dfrac{1}{3}, A_2 = -\dfrac{2}{9}, B = \dfrac{2}{9}.$

Thus,

$$\frac{x^2+2}{(x+1)^3(x-2)} = -\frac{1}{(x+1)^3} + \frac{1}{3(x+1)^2} - \frac{2}{9(x+1)} + \frac{2}{9(x-2)}.$$

Example 8.18. Find the partial-fraction decomposition of the proper fraction $\dfrac{3x^4 + 2x^3 + 3x^2 - 1}{(x-2)(x^2+1)^2}$.

The quadratic polynomial $x^2 + 1$ is irreducible and so, by (8.14),

$$\frac{3x^4 + 2x^3 + 3x^2 - 1}{(x-2)(x^2+1)^2} = \frac{A}{x-2} + \frac{M_1 x + N_1}{x^2+1} + \frac{M_2 x + N_2}{(x^2+1)^2}.$$

To find the constants A, M_i, N_i $(i = 1, 2)$, we multiply both sides of this equation by the left-hand denominator $(x-2)(x^2+1)^2$ and find that

$$3x^4 + 2x^2 + 3x^2 - 1 = A(x^4 + 2x^2 + 1) + (M_1 x + N_1)(x^3 - 2x^2 + x - 2) + (M_2 x + N_2)(x-2).$$

Again we remember that two polynomials are equal only if the coefficients of the corresponding powers of x are the same, and so we may conclude that

$$\begin{cases} A + M_1 = 3, \\ N_1 - 2M_1 = 2, \\ 2A + M_1 - 2N_1 + M_2 = 3, \\ N_1 - 2M_1 + N_2 - 2M_2 = 0, \\ A - 2N_1 - 2N_2 = -1. \end{cases}$$

We solve this system and find that $A = 3$, $M_1 = 0$, $N_1 = 2$, $M_2 = 1$, $N_2 = 0$.

Substituting these values into the previous partial-fraction decomposition, we have

$$\frac{3x^4 + 2x^3 + 3x^2 - 1}{(x-2)(x^2+1)^2} = \frac{3}{x-2} + \frac{2}{x^2+1} + \frac{x}{(x^2+1)^2}.$$

Example 8.19. Find the partial-fraction decomposition of $\dfrac{x+1}{x(x-1)(x-2)}$.

By Theorem 8.2,

$$\frac{x+1}{x(x-1)(x-2)} = \frac{A_1}{x-1} + \frac{A_2}{x} + \frac{A_3}{x-2}.$$

Thus

$$x + 1 = A_1 x(x-2) + A_2(x-1)(x-2) + A_3 x(x-1).$$

There is an alternative way of finding A_i $(i = 1, 2, 3)$, one that is especially convenient in the case of non-repeated linear factors. Substitution of $x = 1$, 2, and 0 into the last equation immediately gives $A_1 = -2$, $A_3 = \dfrac{3}{2}$, and $A_2 = \dfrac{1}{2}$, respectively.

Above, we saw that by using long division, any rational function can be written as a sum of a polynomial and proper rational function. To see how to integrate an arbitrary rational function, we therefore only need to know how to integrate the partial-fraction decomposition of rational functions. By Theorem 8.2, it turns out that a proper fraction is the sum of partial fractions of the first and second types. We begin, then, with the integration of these.

(1) Integration of the partial fraction of the first type.

$$\int \frac{A}{(x-\alpha)^k}\,dx = A\int \frac{dx}{(x-\alpha)^k}, \quad k=1,2,3,\ldots.$$

We can substitute $x-\alpha = t$, obtaining

$$\int \frac{dx}{(x-\alpha)^k} = \int \frac{dt}{t^k}$$

and

$$\int \frac{dt}{t^k} = \begin{cases} \ln|t|+C, & \text{if } k=1,, \\ \dfrac{t^{1-k}}{1-k}, & \text{if } k>1. \end{cases}$$

Thus,

$$\int \frac{A}{(x-\alpha)^k}\,dx = \begin{cases} A\ln|x-\alpha|+C, & \text{if } k=1, \\ \dfrac{A}{(1-k)(x-\alpha)^{k-1}}, & \text{if } k>1. \end{cases} \tag{8.15}$$

(2) Integration of the partial fraction of the second type, $\dfrac{Mx+N}{(x^2+px+q)^l}$ $(l=1,2,3,\ldots)$.

Since there are no real roots of the quadratic polynomial x^2+px+q, we have that $q-\dfrac{p^2}{4}>0$. Obviously,

$$x^2+px+q = \left(x+\frac{p}{2}\right)^2 + \left(q-\frac{p^2}{4}\right).$$

Then, by substituting $a=\sqrt{q-\dfrac{p^2}{4}}$ and $t=x+\dfrac{p}{2}$, we have

$$\int \frac{Mx+N}{x^2+px+q}\,dx = \int \frac{Mt+\left(N-\dfrac{Mp}{2}\right)}{t^2+a^2}\,dt$$

$$= \frac{M}{2}\int \frac{2t\,dt}{t^2+a^2} + \left(N-\frac{Mp}{2}\right)\int \frac{dt}{t^2+a^2}$$

$$= \frac{M}{2}\int \frac{d(t^2+a^2)}{t^2+a^2} + \left(N-\frac{Mp}{2}\right)\frac{1}{a}\int \frac{d\left(\dfrac{t}{a}\right)}{1+\left(\dfrac{t}{a}\right)^2}$$

$$= \frac{M}{2}\ln(t^2+a^2) + \frac{2N-Mp}{2a}\tan^{-1}\frac{t}{a}+C$$

$$= \frac{M}{2}\ln(x^2+px+q) + \frac{2N-Mp}{2\sqrt{q-\dfrac{p^2}{4}}}\tan^{-1}\frac{x+\dfrac{p}{2}}{\sqrt{q-\dfrac{p^2}{4}}}+C. \tag{8.16}$$

On the other hand, for $l \neq 1$,

$$\int \frac{Mx+N}{(x^2+px+q)^l}\,dx = \int \frac{Mt + \left(N - \dfrac{Mp}{2}\right)}{(t^2+a^2)^l}\,dt$$

$$= \frac{M}{2}\int \frac{d(t^2+a^2)}{(t^2+a^2)^l} + \left(N - \frac{Mp}{2}\right)\int \frac{dt}{(t^2+a^2)^l}. \qquad (8.17)$$

Now, the first of the integrals on the right-hand side of (8.17) can be easily calculated:

$$\int \frac{d(t^2+a^2)}{(t^2+a^2)^l} = \frac{1}{(1-l)(t^2+a^2)^{l-1}} + C = \frac{1}{(1-l)(x^2+px+q)^{l-1}} + C. \qquad (8.18)$$

And the second integral is evaluated as in Formula 8.5 of Example 8.16.

Consequently, using the formulas (8.5) and (8.15)–(8.18), we can integrate any rational function and express the integral in terms of log and arctan.

In particular, it is easy to see from (8.5) and (8.17), that when $l > 1$,

$$\int \frac{Mx+N}{(x^2+px+q)^l}\,dx = \frac{Q_1(x)}{(x^2+px+q)^{l-1}} + K\int \frac{dx}{x^2+px+q}, \qquad (8.19)$$

where the first term of the right-hand side of is a proper fraction and K is a constant.

Example 8.20. Find $\displaystyle\int \frac{x^2+2}{(x+1)^3(x-2)}\,dx$.

By using the partial-fraction decomposition of the integrand given in Example 8.17, we obtain

$$\int \frac{x^2+2}{(x+1)^3(x-2)}\,dx = -\int \frac{dx}{(x+1)^3} + \frac{1}{3}\int \frac{dx}{(x+1)^2} - \frac{2}{9}\int \frac{dx}{x+1} + \frac{2}{9}\int \frac{dx}{x-2}$$

$$= \frac{1}{2}\frac{1}{(x+1)^2} - \frac{1}{3(x+1)} - \frac{2}{9}\ln|x+1| + \frac{2}{9}\ln|x-2| + C$$

$$= -\frac{2x-1}{6(x+1)^2} + \frac{2}{9}\ln\left|\frac{x-2}{x+1}\right| + C.$$

Example 8.21. Find $\displaystyle\int \frac{3x^4+2x^3+3x^2-1}{(x-2)(x^2+1)^2}\,dx$.

By using the partial-fraction decomposition of the integrand given in Example 8.18, we have

$$\int \frac{3x^4+2x^3+3x^2-1}{(x-2)(x^2+1)^2}\,dx = \int \frac{3}{x-2}\,dx + \int \frac{2}{x^2+1}\,dx + \int \frac{x}{(x^2+1)^2}\,dx$$

$$= 3\ln|x-2| + 2\arctan x + \frac{1}{2}\int \frac{d(x^2+1)}{(x^2+1)^2}$$

$$= 3\ln|x-2| + 2\arctan x - \frac{1}{2(x^2+1)} + C.$$

Example 8.22. Evaluate $\int \dfrac{x+1}{x(x-1)(x-2)}\,dx$.

The partial-fraction decomposition of the integrand (see Example 8.19) shows that

$$\int \frac{x+1}{x(x-1)(x-2)}\,dx = -\int \frac{2}{x-1}\,dx + \int \frac{dx}{2x} + \int \frac{3\,dx}{2(x-2)}$$

$$= -2\ln|x-1| + \frac{1}{2}\ln|x| + \frac{3}{2}\ln|x-2| + C.$$

Now, let us state the method of Ostrogradsky[4] for integrating proper rational functions $R(x) = \dfrac{Q(x)}{P(x)}$. For definiteness, suppose the polynomial $P(x)$ has the form (8.13) and its degree is n.

It easy to see (see (8.14), (8.15), and (8.19)) that the integral of $R(x)$ has the representation

$$\int \frac{Q(x)}{P(x)}\,dx = \frac{Q_1(x)}{P_1(x)} + \int \frac{Q_2(x)}{P_2(x)}\,dx, \tag{8.20}$$

where $P_1(x)$ is a polynomial

$$P_1(x) = (x-\alpha_1)^{k_1-1}\cdots(x-\alpha_m)^{k_m-1}(x^2+p_1x+q_1)^{l_1-1}\cdots(x^2+p_rx+q_r)^{l_r-1} \tag{8.21}$$

of degree $n-m-2r$ and

$$P_2(x) = \frac{P(x)}{P_1(x)}. \tag{8.22}$$

Here $Q_1(x)$ and $Q_2(x)$ are polynomials of degree not exceeding $(n-m-2r-1)$ and $(m+2r-1)$, respectively. Remember that $P(x)$ is a polynomial of degree n and $k_1+k_2+\cdots+k_m+2l_1+\cdots+2l_r = n$ (Section 6.3).

Observe that $\dfrac{Q_1(x)}{P_1(x)}$ is the rational part of the integral (8.20) and the integral $\int \dfrac{Q_2(x)}{P_2(x)}\,dx$ is a linear combination of the transcendental functions $\ln|x-\alpha_i|$, $\ln(x^2+px+q)$, and $\tan^{-1}\dfrac{2x+p_i}{2a}$ $\left(a = \sqrt{q_i - \dfrac{p_i^2}{4}}\right)$.

In order to determine the rational part of (8.20), the method of Ostrogradsky consists of the following: first, determine the polynomial $P_1(x)$. Observe that it not necessary for this purpose to have the representation (8.13) of $P(x)$. It can be shown that $P_1(x)$ is the greatest common divisor of $P(x)$ and its derivative $P'(x)$. The greatest common divisor of two polynomials is obtained by the use of the Euclidean Algorithm[5].

In turn, $P_2(x)$ is defined by (8.22). Further differentiating (8.20), where the polynomials $Q_1(x)$ and $Q_2(x)$ are written with n undetermined coefficients, we have

$$\frac{Q(x)}{P(x)} = \frac{Q_1'(x)P_1(x) - Q_1(x)P_1'(x)}{[P_1(x)]^2} + \frac{Q_2(x)}{P_2(x)}.$$

---

[4]M. Ostrogradsky (1801–1862), Russian mathematician.
[5]Euclid (ca. 325 BC– ca. 265 BC), Egyptian mathematician, *Elements* VII.2. See, for example, A. G. Kurosh, *A Course of Advanced Algebra*, Nauka, Moscow, 1975, p. 134.

Multiplying both sides by $P(x)$, we obtain

$$Q(x) = Q_1'(x)P_2(x) - Q_1(x)\frac{P_1'(x)P_2(x)}{P_1(x)} + Q_2(x)P_1(x). \qquad (8.23)$$

It is easy to see that the expression $\dfrac{P_1'(x)P_2(x)}{P_1(x)}$ on the right-hand side of (8.23) is a polynomial. In fact, by (8.22),

$$\frac{P_1'(x)P_2(x)}{P_1(x)} = \frac{[P_1(x)P_2(x)]' - P_1(x)P_2'(x)}{P_1(x)} = \frac{P'(x)}{P_1(x)} - P_2'(x).$$

Then, since $\dfrac{P'(x)}{P_1(x)}$ is a polynomial, the expression under our consideration is a polynomial, too. Now, we equate the coefficients of like powers of x on each side of (8.23) and obtain a system consisting of n equations in n unknowns.

Example 8.23. Use Ostrogradsky's method to find $\displaystyle\int \frac{x^2 + 2x}{(x^2 + 1)^2}\,dx$.

For the given integral, (8.20) has the form:

$$\int \frac{x^2 + 2x}{(x^2 + 1)^2}\,dx = \frac{ax + b}{x^2 + 1} + \int \frac{cx + d}{x^2 + 1}\,dx.$$

Differentiating this relation, we obtain

$$\frac{x^2 + 2x}{(x^2 + 1)^2} = \frac{A(x^2 + 1) - 2x(Ax + B)}{(x^2 + 1)^2} + \frac{Cx + D}{x^2 + 1},$$

or

$$x^2 + 2x = A(x^2 + 1) - 2x(Ax + B) + (Cx + D)(x^2 + 1).$$

The coefficients of like powers of x are equal, and so

$$\begin{cases} C = 0, \\ D - A = 1 \\ C - 2B = 2 \\ A + D = 0 \end{cases}$$

Solving this system, we find $A = -\dfrac{1}{2}, B = -1, C = 0, D = \dfrac{1}{2}$.
Consequently,

$$\int \frac{x^2 + 2x}{(x^2 + 1)^2}\,dx = -\frac{x + 2}{2(x^2 + 1)} + \frac{1}{2}\int \frac{dx}{x^2 + 1} = -\frac{x + 2}{2(x^2 + 1)} + \frac{1}{2}\tan^{-1}x + C.$$

8.6 Integrals Involving Quadratic Polynomials

1 The Evaluation of the Integral $I_0 = \int \dfrac{dx}{ax^2 + bx + c}$.

The object is to convert $ax^2 + bx + c$ into either a sum or a difference of two squares $t^2 \pm k^2$.

Thus,

$$ax^2 + bx + c = a\left[x^2 + \frac{b}{a}x + \frac{c}{a}\right] = a\left[\left(x + \frac{b}{2a}\right)^2 + \left(\frac{c}{a} - \frac{b^2}{4a^2}\right)\right] = a\left[\left(x + \frac{b}{2a}\right)^2 \pm k^2\right],$$

$$\frac{c}{a} - \frac{b^2}{4a^2} = \pm k^2.$$

(As always, the parameter k is assumed to be a real number.) The sign depends on whether the roots of $ax^2 + bx + c$ are real or complex.

Indeed, since

$$\frac{c}{a} - \frac{b^2}{4a^2} = -\frac{b^2 - 4ac}{4a^2}$$

if $b^2 - 4ac > 0$, the roots of the quadratic polynomial are real and one must put a minus sign before the k^2. And if $b^2 - 4ac < 0$, then the roots are complex and the sign before k^2 must be taken to be positive.

Now, if we substitute $x + \dfrac{b}{2a} = t$, then $dx = dt$, and so

$$I_0 = \frac{1}{a}\int \frac{dt}{t^2 \pm k^2}.$$

By the table of integrals, entries 10 and 12, we can write

$$\int \frac{dt}{t^2 + k^2} = \frac{1}{k}\tan^{-1}\frac{t}{k} + C, \quad \int \frac{dt}{t^2 - k^2} = \frac{1}{2k}\ln\left|\frac{t-k}{t+k}\right| + C. \qquad (8.24)$$

2. The Evaluation of the Integral $I_1 = \int \dfrac{Mx + N}{ax^2 + bx + c}dx.$

Clearly,

$$I_1 = \int \frac{\dfrac{M}{2a}(2ax + b) + \left(N - \dfrac{Mb}{2a}\right)}{ax^2 + bx + c}dx$$

$$= \frac{M}{2a}\int \frac{2ax + b}{ax^2 + bx + c}dx + \left(N - \frac{Mb}{2a}\right)\int \frac{dx}{ax^2 + bx + c},$$

and the second integral on the right hand side is I_0, already determined, above. We now find the first integral.

The substitution $ax^2 + bx + c = t$ gives us $(2ax + b)dx = dt$, and so,

$$\int \frac{(2ax + b)dx}{ax^2 + bx + c} = \int \frac{dt}{t} = \ln|t| + C = \ln|ax^2 + bx + c| + C.$$

Therefore,

$$I_1 = \frac{M}{2a} \ln|ax^2 + bx + c| + \left(N - \frac{Mb}{2a}\right)I_0. \tag{8.25}$$

Example 8.24. Find $I_0 = \displaystyle\int \frac{dx}{2x^2 + 8x + 20}$.

Since $I_0 = \dfrac{1}{2} \displaystyle\int \frac{dx}{(x+2)^2 + 6}$, the substitution $x + 2 = t$ gives $dx = dt$, and then

$$I_0 = \frac{1}{2} \int \frac{dt}{t^2 + 6} = \frac{1}{2\sqrt{6}} \tan^{-1} \frac{x+2}{\sqrt{6}} + C. \tag{8.26}$$

Example 8.25. Find $I_1 = \displaystyle\int \frac{x+3}{2x^2 + 8x + 20} \, dx$.

Here $M = 1$, $N = 3$, $a = 2$, $b = 8$, and $c = 20$. Then, by formula (8.25) and the previously accomplished evaluation of the integral (8.26),

$$I_1 = \frac{1}{4} \ln|2x^2 + 8x + 20| + \left(3 - \frac{8}{4}\right)I_0 = \frac{\ln 2}{4} + \frac{1}{4}\ln(x^2 + 4x + 10) + I_0$$

$$= \frac{\ln 2}{4} + \frac{1}{4}\ln(x^2 + 4x + 10) + \frac{\ln 2}{4} + \frac{1}{4}\ln(x^2 + 4x + 10) + \frac{1}{2\sqrt{6}} \tan^{-1}\frac{x+2}{\sqrt{6}} + C.$$

Here, we took into account that the roots of $x^2 + 4x + 10 = 0$ are complex, and so $x^2 + 4x + 10$ is positive for all x.

If, in the last relation, we denote $\dfrac{\ln 2}{4} + C$ by C_1, we have

$$I_1 = \frac{1}{4}\ln(x^2 + 4x + 10) + \frac{1}{2\sqrt{6}} \tan^{-1}\frac{x+2}{\sqrt{6}} + C_1. \tag{8.27}$$

Remark 8.3. In this section we saw that an integral involving a quadratic polynomial can often be simplified by the process of completing the square. On the other hand, it is clear that in this case the integrand is a rational fraction, and so we can also apply the procedure of partial fractions to, for instance, $I_1 = \displaystyle\int \frac{x+3}{2x^2 + 8x + 20} \, dx$ (Example 8.24). Indeed, since by (8.16), $M = 1$, $N = 3$, $p = 4$, and $q = 10$, we can write

$$I_1 = \frac{1}{2} \int \frac{x+3}{x^2 + 4x + 10} \, dx$$

$$= \frac{1}{2}\left[\frac{1}{2}\ln(x^2 + 4x + 10) + \frac{6-4}{2\sqrt{6}} \tan^{-1}\frac{x+2}{\sqrt{6}} + C\right]$$

$$= \frac{1}{4}\ln(x^2 + 4x + 10) + \frac{1}{2\sqrt{6}} \tan^{-1}\frac{x+2}{\sqrt{6}} + \frac{1}{2}C.$$

Denoting $\dfrac{1}{2}C$ by C, we thus derive (8.27).

3. The Determination of the Integral $J_0 = \int \dfrac{dx}{\sqrt{ax^2 + bx + c}}.$

As a result of a simple transformation, we may write

$$\int \frac{dx}{\sqrt{ax^2 + bx + c}} = \int \frac{dx}{\sqrt{a\left[\left(x + \dfrac{b}{2a}\right)^2 - \dfrac{b^2 - 4ac}{4a^2}\right]}}.$$

Now there are two cases.

(1) The case $a < 0$.

Then it is clear that the sign of the square brackets in the square root must be negative. Since $a < 0$, the quadratic polynomial is positive if its roots are real, that is, if $b^2 - 4ac > 0$ and x lies between these roots.

Denoting $\dfrac{b^2 - 4ac}{4a^2}$ by k^2 and $x + \dfrac{b}{2a}$ by t, we obtain

$$\int \frac{dx}{\sqrt{ax^2 + bx + c}} = \frac{1}{\sqrt{|a|}} \int \frac{dx}{\sqrt{k^2 - \left(x + \dfrac{b}{2a}\right)^2}}$$

$$= \frac{1}{\sqrt{|a|}} \int \frac{dt}{\sqrt{k^2 - t^2}} = \frac{1}{\sqrt{|a|}} \sin^{-1} \frac{x + \dfrac{b}{2a}}{k} + C.$$

(2) The case $a > 0$.

In this case, depending on the sign of $b^2 - 4ac$, we denote $\dfrac{b^2 - 4ac}{4a^2} = \pm k^2$. By substituting $x + \dfrac{b}{2a} = t$, we derive

$$\int \frac{dx}{\sqrt{ax^2 + bx + c}} = \frac{1}{\sqrt{a}} \int \frac{dt}{\sqrt{t^2 \pm k^2}} = \frac{1}{\sqrt{a}} \ln\left(t + \sqrt{t^2 \pm k^2}\right) + C.$$

4. The Evaluation of the Integral $J_1 = \int \dfrac{Mx + N}{\sqrt{ax^2 + bx + c}}\, dx.$

Analogously to our previous calculation of the integral I_1,

$$J_1 = \int \frac{\dfrac{M}{2a}(2ax + b) + \left(N - \dfrac{Mb}{2a}\right)}{\sqrt{ax^2 + bx + c}}\, dx$$

$$= \frac{M}{2a} \int \frac{2ax + b}{\sqrt{ax^2 + bx + c}}\, dx + \left(N - \frac{Mb}{2a}\right) \int \frac{dx}{\sqrt{ax^2 + bx + c}}.$$

We next use the substitution $ax^2 + bx + c = t$ for the first integral of the right-hand side. Then $(2ax + b)dx = dt$, and so

$$\int \frac{(2ax + b)\, dx}{\sqrt{ax^2 + bx + c}} = \int \frac{dt}{\sqrt{t}} = 2\sqrt{t} + C = 2\sqrt{ax^2 + bx + c} + C.$$

The second integral is the previously calculated integral J_0.

Thus

$$J_1 = \frac{M}{a}\sqrt{ax^2+bx+c} + \left(N-\frac{Mb}{2a}\right)J_0 + C. \tag{8.28}$$

Example 8.26. (a) Find $J_0 = \displaystyle\int \frac{dx}{\sqrt{x^2+6x+5}}$.

$$J_0 = \int \frac{dx}{\sqrt{(x+3)^2-4}} = \ln\left|x+3+\sqrt{x^2+6x+5}\right| + C. \tag{8.29}$$

(b) Find $J_1 = \displaystyle\int \frac{5x+3}{\sqrt{x^2+6x+5}}\,dx$.

Here $M = 5$, $N = 3$, $a = 1$, $b = 6$, and $c = 5$. Then, by (8.28) and (8.29),

$$J_1 = \frac{5}{1}\sqrt{x^2+6x+5} + \left(3-\frac{5\cdot6}{2}\right)\ln\left|x+3+\sqrt{x^2+6x+5}\right| + C$$

$$= 5\sqrt{x^2+6x+5} - 12\ln\left|x+3+\sqrt{x^2+6x+5}\right| + C.$$

5. The Evaluation of the Integral $\int \sqrt{ax^2+bx+c}\,dx$.

As before, a substitution reduces the expression ax^2+bx+c to one of the cases $t^2\pm k^2$ or k^2-t^2. Thus, in place of the original integral, we obtain

$$\int \sqrt{k^2-t^2}\,dt \quad \text{or} \quad \int \sqrt{t^2\pm k^2}\,dt.$$

For the integral $\int \sqrt{k^2-t^2}\,dt$, we next use the substitution $t = k\sin\theta$. In order to evaluate the second integral, $\int \sqrt{t^2\pm k^2}\,dt$, we use instead the substitution $t = k\tan\theta$ in the case and $t = k\sec\theta$ in the case t^2-k^2. Then we obtain easily computed trigonometric integrals.

Example 8.27. Find $\int \sqrt{15+6x-9x^2}\,dx$.

Since $15+6x-9x^2 = 4^2 - (1-3x)^2$, we should use the substitution $1-3x = 4\sin\theta$. Then $\sqrt{15+6x-9x^2} = 4\cos\theta$ and $-3dx = 4\cos\theta\,d\theta$.

Therefore,

$$\int \sqrt{15+6x-9x^2}\,dx = -\frac{16}{3}\int \cos^2\theta\,d\theta = -\frac{8}{3}\int (1+\cos2\theta)d\theta$$

$$= -\frac{8}{3}\left(\theta+\frac{1}{2}\sin2\theta\right) + C = -\frac{8}{3}\theta - \frac{8}{3}\sin\theta\cos\theta + C.$$

Since

$$\sin\theta = \frac{1-3x}{4}, \quad \theta = \arcsin\frac{1-3x}{4}, \quad \cos\theta = \frac{1}{4}\sqrt{15+6x-9x^2},$$

we obtain for the original integral,

$$\int \sqrt{15+6x-9x^2}\,dx = -\frac{8}{3}\arcsin\frac{1-3x}{4} - \frac{1}{6}(1-3x)\sqrt{15+6x-9x^2} + C.$$

8.7 Integration by Transformation into Rational Functions

Definition 8.4. The following expression is called a *polynomial* in x and y of degree n:

$$P_n(x,y) = \sum_{0 \leqslant i+j \leqslant n} a_{ij} x^i y^j,$$

where a_{ij} are constants.

For instance, $P_2(x,y) = a_{00} + a_{10} + a_{01}y + a_{20}x^2 + a_{11}xy + a_{02}y^2$ is a polynomial of degree two in x and y.

Definition 8.5. A function that is expressible as a ratio of two polynomials $P_n(x,y)$ and $Q_m(x,y)$, of degrees n and m, respectively, is called a *rational function* in x and y:

$$R(x,y) = \frac{P_n(x,y)}{Q_m(x,y)}.$$

1. The Evaluation of the Integral $\int R(\sin x, \cos x)dx$.

We will use the substitution $\tan \dfrac{x}{2} = t$ to convert this integral into the integral of a rational function. Indeed, since

$$\sin x = \frac{2\tan\dfrac{x}{2}}{1 + \tan^2\dfrac{x}{2}} = \frac{2t}{1+t^2},$$

$$\cos x = \frac{1 - \tan^2\dfrac{x}{2}}{1 + \tan^2\dfrac{x}{2}} = \frac{1-t^2}{1+t^2},$$

and

$$x = 2\tan^{-1}t, \quad dx = \frac{2dt}{1+t^2},$$

it is clear that these substitutions will convert the integrand into a rational function of t:

$$\int R(\sin x, \cos x)dx = \int R\left(\frac{2t}{1+t^2}, \frac{1-t^2}{1+t^2}\right) \frac{2dt}{1+t^2}.$$

Example 8.28. Find $\displaystyle\int \frac{dx}{\cos x}$.

Taking into account that $\cos x = \dfrac{1-t^2}{1+t^2}$ and $dx = \dfrac{2dt}{1+t^2}$, we can derive

$$\int \frac{dx}{\cos x} = \int \frac{\dfrac{2dt}{1+t^2}}{\dfrac{1-t^2}{1+t^2}} dt = 2\int \frac{dt}{1-t^2} = \ln\left|\frac{1+t}{1-t}\right| + C = \ln\left|\frac{1 + \tan\dfrac{x}{2}}{1 - \tan\dfrac{x}{2}}\right| + C.$$

There are various special cases of rational functions where simpler substitutions also work.

(a) For the integral $\int R(\sin x)\cos x\,dx$, we substitute $\sin x = t$ and $\cos x\,dx = dt$.

(b) For the integral $\int R(\cos x)\sin x\,dx$, we substitute $\cos x = t$ and $\sin x\,dx = -dt$.

(c) For the integral $\int R(\tan x)\,dx$, we substitute $\tan x = t$, $dx = \dfrac{dt}{1+t^2}$, and then

$$\int R(\tan x)\,dx = \int R(t)\frac{dt}{1+t^2}.$$

(d) If the integrand $R(\sin x, \cos x)$ involves only even powers of $\sin x$ and $\cos x$, then we substitute $\tan x = t$, and

$$\cos^2 x = \frac{1}{1+\tan^2 x} = \frac{1}{1+t^2},$$

$$\sin^2 x = \frac{\tan^2 x}{1+\tan^2 x} = \frac{t^2}{1+t^2}, \quad dx = \frac{dt}{1+t^2},$$

and the original integrand is transformed into a rational function.

Example 8.29. Find $I = \displaystyle\int \frac{dx}{2-\sin^2 x}$.

By substituting $\tan x = t$, we obtain

$$I = \int \frac{dt}{\left(2 - \dfrac{t^2}{1+t^2}\right)(1+t^2)} = \int \frac{dt}{2+t^2} = \frac{1}{\sqrt{2}} \tan^{-1}\frac{t}{\sqrt{2}} + C = \frac{1}{\sqrt{2}} \tan^{-1}\left(\frac{\tan x}{\sqrt{2}}\right) + C.$$

(e) $R(\sin x, \cos x) = \sin^m x \cos^n x$.

We treat two cases separately.

Case 1. At least one of the two numbers m and n is an odd positive integer (say $n = 2p+1$, p a non-negative integer). If so, the other may be any real number.

Case 2. Both m and n are nonnegative even integers, say $m = 2p$, $n = 2q$.

In the first case we split off one $\cos x$ factor and use the identity $\cos^2 x = 1 - \sin^2 x$ to express the remaining factor $\cos^{n-1} x$ in terms of $\sin x$, as follows:

$$\int \sin^m x \cos^{2p+1} x\,dx = \int \sin^m x \cos^{2p} x \cos x\,dx$$

$$= \int \sin^m x (1 - \sin^2 x)^p \cos x\,dx.$$

Then the substitution $\sin x = t$, $\cos x\,dx = dt$ yields

$$\int \sin^m x \cos^n x\,dx = \int t^m (1-t^2)^p\,dt.$$

Observe that the factor $(1-t^2)^p$ of the integrand is a polynomial in t, and so its product with t^m is easy to integrate.

Before treating the second case, we give a concrete example of the first case.

Example 8.30. Find $\int \dfrac{\cos^3 x}{\sin^4 x}\,dx$.

The substitution $\sin x = t$, $\cos x\,dx = dt$ yields

$$\int \frac{\cos^3 x}{\sin^4 x}\,dx = \int \frac{\cos^2 x \cos x\,dx}{\sin^4 x} = \int \frac{(1-\sin^2 x)\cos x\,dx}{\sin^4 x}$$

$$= \int \frac{(1-t^2)\,dt}{t^4} = -\frac{1}{3t^3} + \frac{1}{t} + c = -\frac{1}{3\sin^3 x} + \frac{1}{\sin x} + C.$$

In Case 2, we use the half-angle formulas of elementary trigonometry,

$$\sin^2 x = \frac{1}{2} - \frac{1}{2}\cos 2x \quad\text{and}\quad \cos^2 x = \frac{1}{2} + \frac{1}{2}\cos 2x, \tag{8.30}$$

to rewrite the even powers of $\sin x$ and $\cos x$ as follows:

$$\int \sin^{2p} x \cos^{2q} x\,dx = \int \left(\frac{1}{2} - \frac{1}{2}\cos 2x\right)^p \left(\frac{1}{2} + \frac{1}{2}\cos 2x\right)^q dx.$$

By applying (8.30) repeatedly to the resulting powers of $\cos 2x$—if necessary—we eventually are reduced to integrals involving only odd powers, and we have seen how to handle these in Case 1.

Example 8.31. Evaluate $\int \sin^4 x\,dx$.

With the use of (8.30), we find that

$$\int \sin^4 x\,dx = \frac{1}{2^2}\int (1-\cos 2x)^2 dx = \frac{1}{4}\int (1 - 2\cos 2x + \cos^2 2x)\,dx$$

$$= \frac{1}{4}\left[x - \sin 2x + \frac{1}{2}\int(1+\cos 4x)\,dx\right] = \frac{1}{4}\left[\frac{3}{2}x - \sin 2x + \frac{\sin 4x}{8}\right] + C.$$

Example 8.32. Find $\int \dfrac{\sin^2 x}{\cos^4 x}\,dx$.

We use the substitution $\tan x = t$, and obtain

$$\int \frac{\sin^2 x}{\cos^4 x\,dx} = \int \frac{\sin^2 x(\sin^2 x + \cos^2 x)}{\cos^4 x\,dx} = \int \tan^2 x(1 + \tan^2 x)\,dx$$

$$= \int t^2(1+t^2)\frac{dt}{1+t^2} = \int t^2\,dt = \frac{t^3}{3} + c = \frac{tg^3 x}{3} + C.$$

(f) For the integral $\int \tan^m x \sec^n x\,dx$, the procedure breaks up into two cases.

Case 1. m is an odd positive integer.

Case 2. n is an even positive integer.

In Case 1, we use the substitution $t = \sec x$ and so split off the factor $\sec x\tan x$ to obtain $\sec x\tan x\,dx$, the differential of $\sec x$. Then we use the identity $\tan^2 x = \sec^2 x - 1$ to convert the remaining even power of $\tan x$ into powers of $\sec x$.

In Case 2, we use the substitution $t = \tan x$ and then split off $\sec^2 x$ to obtain the differential of $\tan x$. Use of the identity $\sec^2 x = 1 + \tan^2 x$ to convert the remaining even power of $\sec x$ to powers of $\tan x$ completes the process.

Example 8.33. Find $\int \tan^3 x \sec^3 x \, dx$.

We are in case 1, and so

$$\int \tan^3 x \sec^3 x \, dx = \int (\sec^2 x - 1) \sec^2 x \cdot \sec x \cdot \tan x \, dx = \int (t^4 - t^2) \, dt$$

$$= \frac{1}{5} t^5 - \frac{1}{3} t^3 + C = \frac{1}{5} \sec^5 x - \frac{1}{3} \sec^3 x + C.$$

(g) $R(\sin x, \cos x)$ has one of the forms $\cos mx \cos nx$, $\sin mx \cos nx$, or $\sin mx \sin nx$ $(m \neq n)$.

In this case, we use the representations for products of trigonometric functions obtained from the addition formulas:

$$\cos mx \cos nx = \frac{1}{2}[\cos(m+n)x + \cos(m-n)x],$$

$$\sin mx \cos nx = \frac{1}{2}[\sin(m+n)x + \sin(m-n)x],$$

and

$$\sin mx \sin nx = \frac{1}{2}[\cos(m-n)x - \cos(m+n)x].$$

Example 8.34. Find $\int \sin 5x \sin 3x \, dx$.

$$\int \sin 5x \sin 3x \, dx = \frac{1}{2} \int [\cos 2x - \cos 8x] \, dx = -\frac{\sin 8x}{16} + \frac{\sin 2x}{4} + C.$$

2. The Integration of Irrational Algebraic Functions.

The evaluation of $\int R\left[x, \left(\dfrac{ax+b}{cx+d}\right)^{\frac{p}{q}}\right] dx$.

The substitution $\left(\dfrac{ax+b}{cx+d}\right) = t^q$ yields $\left(\dfrac{ax+b}{cx+d}\right)^{\frac{p}{q}} = t^p$ and $x = \dfrac{t^q d - b}{a - ct^q}$ and the integral $\int R\left[x, \left(\dfrac{ax+b}{cx+d}\right)^{\frac{p}{q}}\right] dx$ is then converted into an integral of rational functions.

Example 8.35. Find $I = \int \sqrt{\dfrac{1+x}{1-x}} \, dx$.

The substitution $\sqrt{\dfrac{1+x}{1-x}} = t$ yields

$$x = \frac{t^2 - 1}{t^2 + 1}, \qquad dx = \frac{4t \, dt}{(t^2 + 1)^2}.$$

Thus,

$$I = \int t \frac{4t \, dt}{(t^2 + 1)^2} = 2 \int t \frac{2t}{(t^2 + 1)^2} \, dt = -\frac{2t}{t^2 + 1} + 2 \int \frac{dt}{t^2 + 1}$$

$$= -\frac{2t}{t^2 + 1} + 2 \tan^{-1} t + C$$

(using integration by parts).

Thus, in term of x, we obtain

$$I = -\sqrt{1-x^2} + 2\arctan\sqrt{\frac{1+x}{1-x}} + C.$$

More generally, for evaluating the integral

$$\int R\left[x, \left(\frac{ax+b}{cx+d}\right)^{\frac{m}{n}}, \ldots, \left(\frac{ax+b}{cx+d}\right)^{\frac{p}{q}}\right]dx,$$

we use the substitution $\dfrac{ax+b}{cx+d} = t^k$ to arrive at an integral of rational functions, where k is the least common multiple of the denominators of the fractions $\dfrac{m}{n}, \ldots, \dfrac{p}{q}$.

3. The Evaluation of the Integral $\int x^m(a+bx^n)^p dx$.

We assume that m, n, and p are rational numbers, and that a and b are real.

We consider the following three cases.

(a) If p is an integer, then the integral has the form $\int R(x^m, x^n)dx$, where R is rational function. We then use the substitution $x = t^k$, where k is the least common multiple of m and n.

(b) If $\dfrac{m+1}{n}$ is an integer, then the substitution $t = x^n$ yields

$$\int x^m(a+bx^n)^p dx = \frac{1}{n}\int t^q(a+bt)^p dt,$$

where $q = \dfrac{m+1}{n} - 1$ is an integer. Let r be the denominator of p. Then using the substitution $u = \sqrt[r]{a+bt}$, the integral is converted into the integral of rational functions of u.

(c) If $\dfrac{m+1}{n} + p$ is an integer, then by substituting $t = \sqrt[r]{ax^{-n}+b}$, where r is the denominator of p, the original integral is converted into the integral of a rational function of t.

Example 8.36. Find $I = \int x^3(1+x^2)^{-1/2}dx$.

Here, $\dfrac{m+1}{n} = \dfrac{3+1}{2} = 2$ and $p = -\dfrac{1}{2}$. Then we use the substitution $1+x^2 = u^2$.

Thus,

$$I = \int \frac{(u^2-1)u\,du}{u^2} = \int(u^2-1)du = \frac{u^3}{3} - u + C$$

$$= \frac{u(u^2-3)}{3} + C = \frac{1}{3}\sqrt{1+x^2}(x^2-2) + C.$$

Example 8.37. Find $I = \int x^{-3}(x^4+2)^{1/2}dx$.

Here, $\dfrac{m+1}{n} = \dfrac{-3+1}{4} = -\dfrac{1}{2}$ is not an integer. But $\dfrac{m+1}{n} + p = -\dfrac{1}{2} + \dfrac{1}{2} = 0$ is an integer. Then the substitution $x^4 + 2 = u^2 x^4$ implies

$$x^4 = \frac{2}{u^2 - 1}, \quad dx = \frac{-u\,du}{x^3(u^2-1)^2}.$$

Thus,

$$I = \int \frac{ux^2}{x^3}\frac{-u\,du}{x^3(u^2-1)} = \int \frac{-u^2\,du}{x^4(u^2-1)^2} = \int \frac{-u^2(u^2-1)}{2(u^2-1)^2}\,du$$

$$= -\frac{1}{2}\int\left(1+\frac{1}{u^2-1}\right)du = -\frac{1}{2}u - \frac{1}{4}\ln\frac{u-1}{u+1} + C.$$

$$= -\frac{\sqrt{x^4+2}}{2x^2} - \frac{1}{4}\ln\frac{\sqrt{x^2+2}-x^2}{\sqrt{x^4+4}+x^2} + C.$$

4. The Evaluation of the Integral $\int R\left(x, \sqrt{ax^2+bx+c}\right)dx$.

We use the trigonometric substitutions. Suppose $a \neq 0$ and $b^2 - 4ac \neq 0$ for a quadratic polynomial $ax^2 + bx + c$. Note that if $a = 0$, then we have a known integral of algebraic functions. On the other hand, if $b^2 - 4ac = 0$, then $ax^2 + bx + c = a\left(x+\dfrac{b}{2a}\right)^2$, and for $a > 0$ we are dealing with an integral of rational functions (if the discriminant is zero and $a < 0$, then the square root $\sqrt{ax^2+bx+c}$ is not defined for any x). Thus if the discriminant is nonzero and $a \neq 0$, then we have

$$ax^2 + bx + c = a\left(x+\frac{b}{2a}\right)^2 + \left(c - \frac{b^2}{4a}\right).$$

Denoting

$$x + \frac{b}{2a} = t, \quad |a| = m^2 \quad \text{and} \quad \left|c - \frac{b^2}{4a}\right| = n^2,$$

the square root $\sqrt{ax^2+bx+c}$ is converted into $\sqrt{m^2t^2 \pm n^2}$ or $\sqrt{n^2 - m^2t^2}$.
Finally, we have the integrals

$$\int R\left(t, \sqrt{m^2t^2+n^2}\right)dt, \tag{8.31}$$

$$\int R\left(t, \sqrt{m^2t^2-n^2}\right)dt, \tag{8.32}$$

and

$$\int R\left(t, \sqrt{n^2-m^2t^2}\right)dt. \tag{8.33}$$

Thus by substituting $t = \dfrac{n}{m}\tan z, t = \dfrac{n}{m}\sec z, t = \dfrac{n}{m}\sin z$ in (8.31)–(8.33), respectively, we obtain a trigonometric integral of the form $\int R_1(\sin z, \cos z)$.

Example 8.38. Find $I = \int \dfrac{dx}{\sqrt{(4-x^2)^3}}$.

This integral has the form (8.33) and so we make the substitution $x = 2\sin z$. Then, since $dx = 2\cos z\, dz$, we can write

$$I = \int \frac{2\cos z\, dz}{\sqrt{(4 - 4\sin^2 z)^3}} = \int \frac{2\cos z\, dz}{8\cos^3 z} = \frac{1}{4}\int \frac{dz}{\cos^2 z}$$

$$= \frac{1}{4}\tan z + C = \frac{1}{4}\frac{\sin z}{\sqrt{1 - \sin^2 z}} + C = \frac{x}{4\sqrt{4-x^2}} + C.$$

Example 8.39. Evaluate $\int \dfrac{\sqrt{x^2 - 25}}{x}\, dx$.

The integral has the form (8.32). We make the substitution $x = 5\sec z$ so that $dx = 5\sec z\tan z\, dz$. Therefore,

$$\int \frac{\sqrt{x^2 - 25}}{x}\, dx = \int \frac{\sqrt{25\sec^2 z - 25}}{5\sec z}(5\sec z\tan z)dz$$

$$= 5\int \tan^2 z\, dz = 5\int (\sec^2 z - 1)dz = 5\tan z - 5z + C.$$

But

$$z = \sec^{-1}\frac{x}{5} \quad \text{and} \quad \tan z = \frac{\sqrt{x^2 - 25}}{5},$$

so that

$$\int \frac{\sqrt{x^2 - 25}}{x}\, dx = \sqrt{x^2 - 25} - 5\sec^{-1}\frac{x}{5} + C.$$

8.8   Problems

In each of problems (1)–(11), use the table of integrals to evaluate the given integral.

(1) $\displaystyle\int (x + \sqrt{x})dx$.

Answer: $\dfrac{x^2}{2} + \dfrac{2x\sqrt{x}}{3} + C.$

(2) $\displaystyle\int \dfrac{x^2 dx}{\sqrt{x}}$.

Answer: $\dfrac{2}{5}x^2\sqrt{x} + C.$

(3) $\displaystyle\int \left(x^2 + \dfrac{1}{\sqrt[3]{x}}\right)^2 dx$.

Answer: $\dfrac{x^5}{5} + \dfrac{3}{4}x^2\sqrt[3]{x^2} + 3\sqrt[3]{x} + C.$

(4) $\displaystyle\int \dfrac{e^{2x} + 1}{e^x + 1}dx$.

Answer: $\dfrac{1}{2}e^{2x} - e^x + x + C.$

(5) $\displaystyle\int \tan^2 x\, dx$.

Answer: $-x + \tan x + C.$

(6) $\displaystyle\int \cot^2 x\, dx$.

Answer: $x - \cot x + C.$

(7) $\displaystyle\int (2^x + 3^x)^2 dx$.

Answer: $\dfrac{4^x}{\ln 4} + 2\dfrac{6^x}{\ln 6} + \dfrac{9^x}{\ln 9} + C.$

(8) $\displaystyle\int \frac{\sqrt{1+x^2}+\sqrt{1-x^2}}{\sqrt{1-x^4}}dx.$ *Answer:* $\arcsin x + \ln\left(x+\sqrt{1+x^2}\right)+C.$

(9) $\displaystyle\int\left(1-\frac{1}{x^2}\right)\sqrt{x\sqrt{x}}dx.$ *Answer:* $\dfrac{4(x^2+7)}{7\sqrt[4]{x}}+C.$

(10) $\displaystyle\int \coth^2 x\,dx.$ *Answer:* $x-\coth x+C.$

(11) $\displaystyle\int \sqrt{1-\sin 2x}\,dx.$ *Answer:* $(\cos x+\sin x)\,\mathrm{sgn}(\cos x+\sin x)+C.$

In problems (12)–(58), evaluate the integrals by making the appropriate substitutions.

(12) $\displaystyle\int \frac{\ln x}{x}dx.$ *Answer:* $\dfrac{1}{2}\ln^2 x+C.$

(13) $\displaystyle\int \frac{dx}{\sin^2 3x}.$ *Answer:* $-\dfrac{\cot 3x}{3}+C.$

(14) $\displaystyle\int \tan x\cdot \sec^2 x\,dx.$ *Answer:* $\dfrac{1}{2}\tan^2 x+C.$

(15) $\displaystyle\int \sin^2 x\cos x\,dx.$ *Answer:* $\dfrac{\sin^3 x}{3}+C.$

(16) $\displaystyle\int \cos^3 x\sin x\,dx.$ *Answer:* $-\dfrac{\cos^4 x}{4}+C.$

(17) $\displaystyle\int \frac{dx}{\cot 3x}.$ *Answer:* $-\dfrac{1}{3}\ln|\cos 3x|+C.$

(18) $\displaystyle\int \frac{x^2 dx}{\sqrt{x^3+1}}.$ *Answer:* $\dfrac{2}{3}\sqrt{x^3+1}+C.$

(19) $\displaystyle\int \frac{\tan x}{\cos^2 x}dx.$ *Answer:* $\dfrac{\tan^2 x}{2}+C.$

(20) $\displaystyle\int \frac{\cot x}{\sin^2 x}dx.$ *Answer:* $-\dfrac{\cot^2 x}{2}+C.$

(21) $\displaystyle\int \frac{\cos x\,dx}{\sqrt{2\sin x+1}}.$ *Answer:* $\sqrt{2\sin x+1}+C.$

(22) $\displaystyle\int \frac{\sin 2x\,dx}{\sqrt{1+\sin^2 x}}.$ *Answer:* $2\sqrt{1+\sin^2 x}+C.$

(23) $\displaystyle\int \frac{\cos 2x\,dx}{(2+3\sin 2x)^3}.$ *Answer:* $-\dfrac{1}{12(2+3\sin 2x)^2}+C.$

(24) $\displaystyle\int \frac{\sin^{-1}x\,dx}{\sqrt{1-x^2}}.$ *Answer:* $\dfrac{(\sin^{-1})^2 x}{2}+C.$

(25) $\displaystyle\int \frac{(\cos^{-1})^2 x\,dx}{\sqrt{1-x^2}}.$ *Answer:* $-\dfrac{(\cos^{-1})^3 x}{3}+C.$

(26) $\displaystyle\int \frac{\cos x\,dx}{2\sin x+3}.$ *Answer:* $\dfrac{1}{2}\ln(2\sin x+3)+C.$

(27) $\displaystyle\int \frac{x+1}{x^2+2x+3}dx.$ *Answer:* $\dfrac{1}{2}\ln(x^2+2x+3)+C.$

(28) $\displaystyle\int \tan^4 x\,dx.$ *Answer:* $\dfrac{\tan^3 x}{3}-\tan x+x+C.$

(29) $\displaystyle\int \frac{dx}{\cos^2 x(3\tan x+1)}.$ *Answer:* $\dfrac{1}{3}\ln|3\tan x+1|+C.$

(30) $\displaystyle\int \frac{\tan^3 x}{\cos^2 x}\,dx.$ *Answer:* $\dfrac{\tan^4 x}{4}+C.$

(31) $\displaystyle\int \frac{\cos 2x}{2+3\sin 3x}\,dx.$ *Answer:* $\dfrac{1}{6}\ln|2+3\sin 2x|+C.$

(32) $\displaystyle\int e^{x/3}\,dx.$ *Answer:* $3e^{x/3}+C.$

(33) $\displaystyle\int 3^x e^x\,dx.$ *Answer:* $\dfrac{3x e^x}{\ln 3+1}+C.$

(34) $\displaystyle\int e^{\sin x}\cos x\,dx.$ *Answer:* $e^{\sin x}+C.$

(35) $\displaystyle\int (e^{5x}+a^{5x})\,dx.$ *Answer:* $\dfrac{1}{5}\left(e^{5x}+\dfrac{a^{5x}}{\ln a}+C\right).$

(36) $\displaystyle\int \frac{(a^x-b^x)^2}{a^x b^x}\,dx.$ *Answer:* $\dfrac{\left(\dfrac{a}{b}\right)^x-\left(\dfrac{b}{a}\right)^e}{\ln a-\ln b}-2x+C.$

(37) $\displaystyle\int x^2\sqrt[3]{1-x}\,dx.$ *Answer:* $-\dfrac{3}{140}(9+12x+14x^2)(1-x^{4/3})+C.$

(38) $\displaystyle\int \frac{\sin x\cos^3 x}{1+\cos^2 x}\,dx.$ *Answer:* $-\dfrac{1}{2}\cos^2 x+\dfrac{1}{2}\ln(1+\cos^2 x)+C.$

(39) $\displaystyle\int \cos^5 x\sqrt{\sin x}\,dx.$ *Answer:* $\left(\dfrac{2}{3}-\dfrac{4}{7}\sin^2 x+\dfrac{2}{11}\sin^4 x\right)\sqrt{\sin^3 x}+C.$

(40) $\displaystyle\int \frac{\sin^2 x}{\cos^6 x}\,dx.$ *Answer:* $\dfrac{1}{3}\tan^3 x+\dfrac{1}{5}\tan^5 x+C.$

(41) $\displaystyle\int \frac{dx}{e^{x/2}+e^x}.$ *Answer:* $-x-2e^{-x/2}+2\ln\left(1+e^{x/2}\right)+C.$

(42) $\displaystyle\int \frac{dx}{\sqrt{1+e^x}}.$ *Answer:* $x-2\ln(1+\sqrt{1+e^x})+C.$

(43) $\displaystyle\int \frac{x^2}{\sqrt{2-x}}\,dx.$ *Answer:* $-\dfrac{2}{15}(32+8x+3x^2)\sqrt{2-x}+C.$

(44) $\displaystyle\int \frac{\ln x\,dx}{x\sqrt{1+\ln x}}.$ *Answer:* $\dfrac{2}{3}(-2+\ln x)\sqrt{1+\ln x}+C.$

(45) $\displaystyle\int \frac{x^5}{\sqrt{1-x^2}}\,dx.$ *Answer:* $-\dfrac{1}{15}(8+4x^2+3x^4)\sqrt{1-x^2}+C.$

(46) $\displaystyle\int \frac{dx}{\sqrt{16-9x^2}}.$ *Answer:* $\dfrac{1}{3}\sin^{-1}\dfrac{3x}{4}+C.$

(47) $\displaystyle\int \frac{dx}{9x^2+4}.$ *Answer:* $\dfrac{1}{6}\tan^{-1}\dfrac{3x}{2}+C.$

(48) $\displaystyle\int \frac{dx}{4-9x^2}.$ *Answer:* $\dfrac{1}{12}\ln\left|\dfrac{2+3x}{2-3x}\right|+C.$

(49) $\displaystyle\int \frac{dx}{\sqrt{4+25x^2}}.$ *Answer:* $\dfrac{1}{5}\ln\left|5x+\sqrt{4+25x^2}\right|+C.$

(50) $\displaystyle\int \frac{x^2\,dx}{5-x^6}.$ *Answer:* $\dfrac{1}{6\sqrt{5}}\ln\left|\dfrac{x^3+\sqrt{5}}{x^3-\sqrt{5}}\right|+C.$

(51) $\displaystyle\int \frac{x\,dx}{x^4+3^4}.$ *Answer:* $\dfrac{1}{18}\tan^{-1}\dfrac{x^2}{9}+C.$

(52) $\displaystyle\int \frac{dx}{x\sqrt{1-\ln^2 x}}.$ *Answer:* $\sin^{-1}(\ln x)+C.$

(53) $\displaystyle\int \frac{x - \tan^{-1} x}{1 + x^2}\, dx.$ *Answer:* $-\dfrac{1}{2}(\cos^{-1} x)^2 + \sqrt{1 - x^2} + C.$

(54) $\displaystyle\int \frac{dx}{\sqrt{x}\sqrt{1 + \sqrt{x}}}.$ *Answer:* $4\sqrt{1 + \sqrt{x}} + C.$

(55) $\displaystyle\int \frac{\sqrt[3]{\tan^2 x}}{\cos^2 x}\, dx.$ *Answer:* $\dfrac{3}{5}\sqrt[3]{\tan^5 x} + C.$

(56) $\displaystyle\int \frac{dx}{2\sin^2 x + 3\cos^2 x}.$ *Answer:* $\dfrac{1}{\sqrt{6}} \tan^{-1}\left(\sqrt{\dfrac{2}{3}}\tan x\right) + C.$

(57) $\displaystyle\int \frac{\sin 2x}{\sqrt{1 + \cos^2 x}}\, dx.$ *Answer:* $-2\sqrt{1 + \cos^2 x} + C.$

(58) $\displaystyle\int \frac{e^x\, dx}{1 + 3e^{2x}}.$ *Answer:* $\tan^{-1} e^x + C.$

In each of problems (59)–(63), evaluate the integrals by making the substitutions $x = a\sin t$, $x = a\tan t$, $x = a\sin^2 t$.

(59) $\displaystyle\int \sqrt{1 - x^2}\, dx.$ *Answer:* $\dfrac{x}{2}\sqrt{1 - x^2} + \dfrac{1}{2}\sin^{-1} x + C.$

(60) $\displaystyle\int \sqrt{\frac{3 + x}{3 - x}}\, dx.$ *Answer:* $-\sqrt{9 - x^2} + 3\sin^{-1}\dfrac{x}{3} + C.$

(61) $\displaystyle\int \frac{dx}{\sqrt{(x - 5)(6 - x)}}.$ *Answer:* $2\sin^{-1}\sqrt{x - 5} + C.$

Suggestion: use the substitution $x - 5 = \sin^2 t$.

(62) $\displaystyle\int \frac{x^2 dx}{\sqrt{x^2 - 2}}.$ *Answer:* $\dfrac{x}{2}\sqrt{x^2 - 2} + \ln\left|x + \sqrt{x^2 - 2}\right| + C.$

(63) $\displaystyle\int \frac{dx}{(x^2 + 4)^{3/2}}.$ *Answer:* $\dfrac{x}{4\sqrt{4 + x^2}} + C.$

In each of problems (64)–(79), use integration by parts to evaluate the integrals.

(64) $\displaystyle\int xe^x\, dx.$ *Answer:* $e^x(x - 1) + C.$

(65) $\displaystyle\int \sin^{-1} x\, dx.$ *Answer:* $x\sin^{-1} x + \sqrt{1 - x^2} + C.$

(66) $\displaystyle\int x\tan^{-1} x\, dx.$ *Answer:* $\dfrac{1}{2}[(x^2 + 1)\tan^{-1} x - x] + C.$

(67) $\displaystyle\int \tan^{-1}\sqrt{x}\, dx.$ *Answer:* $(x + 1)\tan^{-1}\sqrt{x} - \sqrt{x} + C.$

(68) $\displaystyle\int \left(\frac{\ln x}{x}\right)^2 dx.$ *Answer:* $-\dfrac{1}{x}(\ln^2 x + 2\ln x + 2) + C.$

(69) $\displaystyle\int x^2 e^{-2x}\, dx.$ *Answer:* $-\dfrac{e^{-2x}}{2}\left(x^2 + x + \dfrac{1}{2}\right) + C.$

(70) $\displaystyle\int \frac{x\tan^{-1} x}{(x^2 + 1)^2}\, dx.$ *Answer:* $\dfrac{x}{4(1 + x^2)} + \dfrac{1}{4}\tan^{-1} x - \dfrac{1}{2}\dfrac{\tan^{-1} x}{1 + x^2} + C.$

(71) $\displaystyle\int x\cos x\, dx.$ *Answer:* $x\sin x + \cos x + C.$

(72) $\displaystyle\int \frac{\sin^{-1}x}{x^2}\,dx.$ *Answer:* $\ln\left|\dfrac{1-\sqrt{1-x^2}}{x}\right|-\dfrac{1}{x}\sin^{-1}x+C.$

(73) $\displaystyle\int x\tan^{-1}\sqrt{x^2-1}\,dx.$ *Answer:* $\dfrac{1}{2}x^2\tan^{-1}\sqrt{x^2-1}-\dfrac{1}{2}\sqrt{x^2-1}+C.$

(74) $\displaystyle\int \ln\left(x+\sqrt{1+x^2}\right)dx.$ *Answer:* $x\ln\left|x+\sqrt{1+x^2}\right|-\sqrt{1+x^2}+C.$

(75) $\displaystyle\int x^3\cosh 3x\,dx.$ *Answer:* $\left(\dfrac{x^3}{3}+\dfrac{2x}{9}\right)\sinh x-\left(\dfrac{x^2}{3}+\dfrac{2}{27}\right)\cosh 3x+C.$

(76) $\displaystyle\int x^2\cos^{-1}x\,dx.$ *Answer:* $-\dfrac{2+x^2}{9}\sqrt{1-x^2}+\dfrac{x^3}{3}\cos^{-1}3x+C.$

(77) $\displaystyle\int x\ln\dfrac{1+x}{1-x}\,dx.$ *Answer:* $x-\dfrac{1-x^2}{2}\ln\dfrac{1+x}{1-x}+C.$

(78) $\displaystyle\int \sin x\cdot\ln(\tan x)\,dx.$ *Answer:* $\ln\left|\tan\dfrac{x}{2}\right|-\cos x\cdot\ln\tan x+C.$

(79) $\displaystyle\int x^3 e^{2x}\,dx.$ *Answer:* $\left(\dfrac{1}{2}x^3-\dfrac{3}{4}x^2+\dfrac{3}{4}x-\dfrac{3}{8}\right)e^{2x}+C.$

Evaluate the integrals in (80)–(96) of rational fractions:

(80) $\displaystyle\int \frac{2x-1}{(x-1)(x-2)}\,dx.$ *Answer:* $\ln\left|\dfrac{(x-2)^3}{x-1}\right|+C.$

(81) $\displaystyle\int \frac{x^4\,dx}{(x^2-1)(x+2)}.$ *Answer:* $\dfrac{x^2}{2}-2x+\dfrac{1}{6}\ln\dfrac{|x-1|}{|x+1|^3}$

$\qquad\qquad +\dfrac{16}{3}\ln|x+2|+C.$

(82) $\displaystyle\int \frac{x^{10}\,dx}{x^2+x-2}.$ *Answer:* $\dfrac{x^9}{9}-\dfrac{x^8}{8}+\dfrac{3x^7}{7}-\dfrac{5x^6}{6}+\dfrac{11x^5}{5}$

$\qquad\qquad -\dfrac{21x^4}{4}+\dfrac{43x^3}{3}-\dfrac{85x^2}{2}+171x$

$\qquad\qquad +\dfrac{1}{3}\ln\left|\dfrac{x-1}{(x+2)^{1024}}\right|+C.$

(83) $\displaystyle\int \frac{dx}{x(1+x)(1+x+x^2)}.$ *Answer:* $-\dfrac{1}{5(x-1)}+\dfrac{1}{50}\ln\dfrac{(x-1)^2}{x^2+2x+2}$

$\qquad\qquad -\dfrac{8}{25}\tan^{-1}(x+1)+C.$

(84) $\displaystyle\int \frac{x^2+1}{x^3-5x^2+6x}\,dx.$ *Answer:* $x+\dfrac{1}{6}\ln|x|-\dfrac{9}{2}\ln|x-2|$

$\qquad\qquad +\dfrac{28}{3}\ln|x-3|+C.$

(85) $\displaystyle\int \frac{x^3-6}{x^4+6x^2+8}\,dx.$ *Answer:* $\ln\dfrac{x^2+4}{\sqrt{x^2+2}}+\dfrac{3}{2}\tan^{-1}\dfrac{x}{2}$

$\qquad\qquad -\dfrac{3}{\sqrt{2}}\tan^{-1}\dfrac{x}{\sqrt{2}}+C.$

(86) $\displaystyle\int \frac{dx}{(x+1)(x+2)^2(x+3)^3}.$ *Answer:* $\dfrac{9x^2+50x+68}{4(x+2)(x+3)^2}$

$\qquad\qquad +\dfrac{1}{8}\ln\left|\dfrac{(x+1)(x+2)^{16}}{(x+3)^{17}}\right|+C.$

(87) $\int \dfrac{x^2+5x+4}{x^4+5x^2+4}\,dx.$ *Answer:* $\tan^{-1}x+\dfrac{5}{6}\ln\dfrac{x^2+1}{x^2+4}+C.$

(88) $\int \dfrac{3x-7}{x^3+x^2+4x+4}\,dx.$ *Answer:* $\ln\dfrac{x^2+4}{(x+1)^2}+\dfrac{1}{2}\tan^{-1}\dfrac{x}{2}+C.$

(89) $\int \dfrac{dx}{x(x^2+1)}.$ *Answer:* $\ln\dfrac{|x|}{\sqrt{x^2+1}}+C.$

(90) $\int \dfrac{dx}{(x+1)(x^2+1)}.$ *Answer:* $\dfrac{1}{2}\tan^{-1}x+\dfrac{1}{4}\ln\dfrac{(x+1)^2}{x^2+1}+C.$

(91) $\int \dfrac{dx}{(x^2-4x+4)(x^2-4x+5)}.$ *Answer:* $-\dfrac{1}{2}-\tan^{-1}(x-2)+C.$

(92) $\int \dfrac{dx}{(x^3+1)}.$ *Answer:* $\dfrac{1}{6}\ln\dfrac{(x+1)^2}{x^2-x+1}$
$+\dfrac{1}{\sqrt{3}}\tan^{-1}\dfrac{2x-1}{\sqrt{3}}+C.$

(93) $\int \dfrac{dx}{x^5-x^4+x^3-x^2+x-1}.$ *Answer:* $\dfrac{1}{6}\ln\dfrac{(x-1)^2}{x^2+x+1}$
$-\dfrac{1}{\sqrt{3}}\tan^{-1}\dfrac{2x-1}{\sqrt{3}}+C.$

(94) $\int \dfrac{(4x^2-8x)dx}{(x-1)^2(x^2+1)^2}.$ *Answer:* $\dfrac{3x^2-x}{(x-1)(x^2+1)}$
$+\ln\dfrac{(x-1)^2}{x^2+1}\tan^{-1}x+C.$

(95) $\int \dfrac{dx}{(x^2-x)(x^2-x+1)^2}.$ *Answer:* $\ln\left|\dfrac{x-1}{x}\right|-\dfrac{10}{3\sqrt{3}}\tan^{-1}\dfrac{2x-1}{\sqrt{3}}$
$-\dfrac{2x-1}{3(x^2-x+1)}+C.$

(96) $\int \dfrac{dx}{x^4+x^2+1}.$ *Answer:* $\dfrac{1}{4}\ln\dfrac{x^2+x+1}{x^2-x+1}$
$+\dfrac{1}{2\sqrt{3}}\tan^{-1}\dfrac{x^2-1}{x\sqrt{3}}+C.$

Use Ostrogradsky's method to evaluate the integrals in problems (97)–(100).

(97) $\int \dfrac{dx}{(x^4+1)^2}.$ *Answer:* $\dfrac{x}{4(x^4+1)}+\dfrac{3}{16\sqrt{2}}\ln\dfrac{x^2+x\sqrt{2}+1}{x^2-x\sqrt{2}+1}$
$-\dfrac{3}{8\sqrt{2}}\tan^{-1}\dfrac{x\sqrt{2}}{x^2-1}+C.$

(98) $\int \dfrac{x^2dx}{(x^2+2x+2)^2}.$ *Answer:* $\dfrac{1}{x^2+2x+2}+\tan^{-1}(x+1)+C.$

(99) $\int \dfrac{xdx}{(x-1)^2(x+1)^2}.$ *Answer:* $-\dfrac{x^2+x+2}{8(x-1)(x+1)^2}+\dfrac{1}{16}\ln\left|\dfrac{x+1}{x-1}\right|+C.$

(100) $\int \dfrac{dx}{(x^3+x+1)^3}.$ *Answer:* $-\dfrac{8x^4+8x^2+4x-1}{28(x^3+x+1)^2}+C.$

Find the integrals of the irrational functions in problems (101)–(106).

(101) $\int \dfrac{\sqrt{x}}{\sqrt[4]{x^3}+1}\,dx.$ *Answer:* $\dfrac{4}{3}\left[\sqrt[4]{x^3}-\ln\left(\sqrt[4]{x^3}+1\right)\right]+C.$

(102) $\int \dfrac{\sqrt[6]{x}+1}{\sqrt[6]{x^7}+\sqrt[4]{x^5}}\,dx.$ *Answer:* $-\dfrac{6}{\sqrt[6]{x}}+\dfrac{12}{\sqrt[12]{x}}+2\ln x$

$-24\ln\left(\sqrt[12]{x}+1\right)+C.$

(103) $\int \sqrt{\dfrac{1-x}{1+x}}\dfrac{dx}{x^2}.$ *Answer:* $\ln\left|\dfrac{\sqrt{1-x}+\sqrt{1+x}}{\sqrt{1-x}-\sqrt{1+x}}\right|-\dfrac{\sqrt{1-x^2}}{x}+C.$

(104) $\int \sqrt{\dfrac{2+3x}{x-3}}\,dx.$ *Answer:* $\sqrt{3x^2-7x-6}$

$+\dfrac{11}{2\sqrt{3}}\ln\left(x-\dfrac{7}{6}+\sqrt{x^2-\dfrac{7}{3}x-2}\right)+C.$

(105) $\int \dfrac{\sqrt{x^3}-\sqrt[3]{x}}{6\sqrt[4]{x}}\,dx.$ *Answer:* $\dfrac{2}{27}\sqrt[4]{x^9}-\dfrac{2}{13}\sqrt[12]{x^{13}}+C.$

(106) $\int \dfrac{dx}{1+\sqrt{x}+\sqrt{1+x}}.$ *Answer:* $\dfrac{x}{2}+\sqrt{x}-\dfrac{1}{2}\sqrt{x(1+x)}$

$-\dfrac{1}{2}\ln\left(\sqrt{x}+\sqrt{1+x}\right)+C.$

Suggestion: Use the substitution $x=\left(\dfrac{4^2-1}{24}\right)^2.$

Find the integrals of the trigonometric functions in problems (107)–(117).

(107) $\int \sin^5 x\,dx.$ *Answer:* $-\cos x+\dfrac{2}{3}\cos^2 x-\dfrac{\cos^5 x}{5}+C.$

(108) $\int \cos^6 x\,dx.$ *Answer:* $\dfrac{1}{16}\left(5x+4\sin 2x-\dfrac{\sin^3 2x}{3}+\dfrac{3}{4}\sin 4x\right)+C.$

(109) $\int \cos^4 x\sin^3 x\,dx.$ *Answer:* $-\dfrac{1}{5}\cos^5 x\dfrac{1}{7}\cos^7 x+C.$

(110) $\int \sin^4 x\cos^4 x\,dx.$ *Answer:* $\dfrac{1}{128}\left(3x-\sin 4x-\dfrac{\sin 8x}{8}\right)+C.$

(111) $\int \cot^3 x\,dx.$ *Answer:* $-\dfrac{\cot^2 x}{2}-\ln|\sin x|+C.$

(112) $\int \tan^4 x\sec^4 x\,dx.$ *Answer:* $\dfrac{\tan^7 x}{7}+\dfrac{\tan^5 x}{5}+C.$

(113) $\int \dfrac{dx}{\cos^4 x}.$ *Answer:* $\tan x+\dfrac{1}{3}\tan^3 x+C.$

(114) $\int \sin\dfrac{1}{4}x\cos\dfrac{3}{4}x\,dx.$ *Answer:* $-\dfrac{\cos x}{2}+\cos\dfrac{1}{2}x+C.$

(115) $\int \dfrac{\sin x\,dx}{1+\sin x}.$ *Answer:* $\dfrac{2}{1+\tan\dfrac{x}{2}}+x+C.$

(116) $\int \dfrac{\sin^2 x}{1+\cos^2 x}\,dx.$ *Answer:* $\sqrt{2}\tan^{-1}\left(\dfrac{\tan x}{\sqrt{2}}\right)+C.$

(117) $\int \dfrac{\sin 2x}{\cos^4 x+\sin^4 x}\,dx.$ *Answer:* $\tan^{-1}(2\sin^2 x-1)+C.$

In problems (118)–(121), evaluate the integrals of the type $\int x^m (a+bx^n)^p dx$.

(118) $\int \sqrt{x^3 + x^4} dx.$ *Answer:* $\dfrac{1}{3}\sqrt{(x+x^2)^3} - \dfrac{1+2x}{8}\sqrt{x+x^2}$

$$+\frac{1}{8}\ln\left(\sqrt{x}+\sqrt{1+x}\right) + C, x > 0.$$

(119) $\int \dfrac{dx}{x\sqrt[6]{1+x^6}}.$ *Answer:* $\dfrac{1}{6}\ln\dfrac{z-1}{z+1} + \dfrac{1}{12}\ln\dfrac{z^2+z+1}{z^2-z+1}$

$$+\frac{1}{2\sqrt{3}}\tan^{-1}\frac{z^2-1}{z\sqrt{3}} + C, z = \sqrt[6]{1+x^6}.$$

(120) $\int \dfrac{x^5\, dx}{\sqrt{1-x^2}}.$ *Answer:* $-z + \dfrac{2}{3}z^3 - \dfrac{z^5}{5} + C, z = \sqrt{1-x^2}.$

(121) $\int \sqrt{3x - x^3} dx.$ *Answer:* $\dfrac{3z}{2(z^3+1)} - \dfrac{1}{4}\ln\dfrac{(z+1)^2}{z^2-z+1}$

$$-\frac{\sqrt{3}}{2}\tan^{-1}\frac{2z-1}{\sqrt{3}} + C, z = \frac{\sqrt[3]{3x-x^3}}{x}.$$

Find the integrals of the form $\int R\left(x, \sqrt{ax^2+bx+c}\right) dx.$

(122) $\int \dfrac{dx}{x\sqrt{2+x-x^2}}.$ *Answer:* $-\dfrac{1}{\sqrt{2}}\ln\left|\dfrac{\sqrt{2+x-x^2}}{x}\right| + \dfrac{1}{2\sqrt{2}}\right| + C.$

(123) $\int \dfrac{\sqrt{x^2+x}}{x} dx.$ *Answer:* $\sqrt{x^2+2x} + \ln\left|x+1+\sqrt{x^2+2x}\right| + C.$

(124) $\int \dfrac{dx}{x-\sqrt{x^2-1}}.$ *Answer:* $\dfrac{x^2}{2} + \dfrac{x}{2}\sqrt{x^2-1} - \dfrac{1}{2}\ln\left|x+\sqrt{x^2-1}\right| + C.$

(125) $\int \dfrac{dx}{\sqrt{(2x-x^2)^3}}.$ *Answer:* $\dfrac{x-1}{\sqrt{2x-x^2}} + C.$

(126) $\int \dfrac{dx}{(1+x)\sqrt{1+x+x^2}}.$ *Answer:* $\ln\left|\dfrac{x+\sqrt{1+x+x^2}}{2+x+\sqrt{1+x+x^2}}\right| + C.$

(127) $\int \dfrac{dx}{x\sqrt{x^2-x+3}}.$ *Answer:* $\dfrac{1}{\sqrt{3}}\ln\left|\dfrac{\sqrt{x^2-x+3}-\sqrt{3}}{3}\right| + \dfrac{1}{2\sqrt{3}}\right| + C.$

Chapter 9

The Definite Integral

In this chapter we define Newton's and Riemann's integrals. We investigate the conditions under which they are equal. Necessary and sufficient conditions for the existence of the Riemann integral are formulated in terms of a partition of a given interval of the real line. We prove mean value theorems in the context of integrals which are analogous to the ones studied in Chapter 5. The Fundamental Theorem of Calculus is then proved. The Lebesgue criterion, which uses the concept of measure zero, for the existence of the integral is given. In particular, it is shown that the measure of the Cantor set is zero, even though it has the cardinality of the continuum. Next, we introduce the Lebesgue integral, and use this to examine the integrability of Dirichlet's and Riemann's functions. The topics of functional series, the improper integral, the Cauchy principal value, and numerical integration are developed. Finally, we introduce the concept of a function with bounded variation in order to introduce the Stieltjes integral.

9.1 Newton's Integral

Chapter 8 already showed that if f is an antiderivative (Definition 8.1) for a function f on $[a, b]$, then $F(x) + C$, for any constant C, is another such antiderivative.

Observe that the difference $F(b) - F(a)$ does not depend on C, because

$$(F(b) + C) - (F(a) + C) = F(b) - F(a).$$

Definition 9.1. The difference $F(b) - F(a)$ is said to be the *Newton integral of the function* f on $[a, b]$, and is denoted by $\int_a^b f(x)dx$:

$$\int_a^b f(x)dx = F(b) - F(a). \tag{9.1}$$

Sometimes $F(b) - F(a)$ is denoted by $F(x)\big|_a^b$.

The numbers a and b are called the *lower limit* and the *upper limit*, respectively.

Formula 9.1 is called the *Newton–Leibniz formula*.

Observe that in (9.1) it may be the case that $a \geqslant b$. In particular, if $a = b$, the integral is zero, i.e., $\int_a^a f(x)dx = 0$.

E. Mahmudov, *Single Variable Differential and Integral Calculus*,
DOI: 10.2991/978-94-91216-86-2_9, © Atlantis Press 2013

Using the results of Section 8.2 we note the following properties of Newton's integral:

(1) $\int_a^b f(x)dx = -\int_b^a f(x)dx.$

(2) $\int_a^b kf(x)dx = k\int_a^b f(x)dx.$

(3) $\int_a^b [f(x) + g(x)]dx = \int_a^b f(x)dx + \int_a^b g(x)dx.$

(4) $\int_a^b f(x)dx = \int_a^c f(x)dx + \int_c^b f(x)dx.$

It is assumed that the antiderivatives of the functions f and g exist.

(5) Suppose the functions u and v are differentiable on $[a,b]$ and there exists an antiderivative of the product function $v \cdot u'$. Then the integral $\int_a^b u(x)v'(x)dx$ is defined, and

$$\int_a^b u(x)v'(x)dx = u(x)v(x)\Big|_a^b - \int_a^b v(x)u'(x)dx.$$

(6) Suppose that the function f has an antiderivative on $[a,b]$ and φ is differentiable on $[\alpha,\beta]$. Suppose further that $f[\varphi(t)]$ is defined and continuous on $[\alpha,\beta]$, and $\varphi(\alpha) = a$, $\varphi(\beta) = b$. Then,

$$\int_a^b f(x)dx = \int_\alpha^\beta f[\varphi(t)]\varphi'(t)dt.$$

Note that the mechanical meaning of Newton's integral is the net distance between two points. Indeed, consider a particle traveling along a straight line according to the equation $s = f(t)$. We want to compute the net distance s it travels between time $t = t_0$ and time $t = t_1$. At time t its velocity is $f'(t) = v(t)$, and so

$$\int_{t_0}^{t_1} v(t)dt = f(t_1) - f(t_0) \quad (t_1 > t_0).$$

Remark 9.1. It is easy to see that

$$\int_a^x f(t)dt = F(x) - F(a) \quad (x, a \in A),$$

and so $\int_a^x f(t)dt$ itself is an antiderivative.

If a function f is an antiderivative of f on the half-interval $[a,b)$ and the limit $\lim_{x \to b-0} F(x)$ exists, then we make the further definition,

$$\int_a^b f(x)dx = \lim_{x \to b-0} F(x) - F(a) = \lim_{x \to b-0} \int_a^x f(t)dt.$$

Analogously, if a function f is an antiderivative of f on the half-interval $[a, +\infty)$ and $\lim_{x \to +\infty} F(x)$ exists, then we make the further definition that

$$\int_a^{+\infty} f(x)dx = \lim_{x \to +\infty} F(x) - F(a) = \lim_{x \to +\infty} \int_a^x f(t)dt.$$

The definite integral of f on the half-intervals $(a,b]$ and $(-\infty,b]$ is defined similarly:

$$\int_{-\infty}^{+\infty} f(x)dx = \int_{-\infty}^{c} f(x)dx + \int_{c}^{+\infty} f(x)dx.$$

Such integrals are called improper integrals.

Example 9.1. Evaluate $I_n = \int_0^{\frac{\pi}{2}} \sin^n x\, dx.$

Using integration by parts, we have

$$I_n = \int_0^{\frac{\pi}{2}} \sin^n x\, dx = \int_0^{\frac{\pi}{2}} \sin^{n-1} x \sin x\, dx = -\int_0^{\frac{\pi}{2}} \sin^{n-1} x\, d(\cos x)$$

$$= -\sin^{n-1} x \cos x \Big|_0^{\frac{\pi}{2}} + (n-1)\int_0^{\frac{\pi}{2}} \sin^{n-2} x \cos x \cos x\, dx$$

$$= (n-1)\int_0^{\frac{\pi}{2}} \sin^{n-2} x \cos^2 x\, dx = (n-1)\int_0^{\frac{\pi}{2}} \sin^{n-2} x(1-\sin^2 x)\, dx$$

$$= (n-1)\int_0^{\frac{\pi}{2}} \sin^{n-2} x\, dx - (n-1)\int_0^{\frac{\pi}{2}} \sin^n x\, dx.$$

Therefore,

$$I_n = (n-1)I_{n-2} - (n-1)I_n.$$

This is a reduction formula and it is easy to see that

$$I_n = \frac{n-1}{n}I_{n-2}.$$

Likewise, if we express I_{n-2} by I_{n-4}, and then I_{n-4} by I_{n-6}, and so on, we have

$$I_{2m} = \frac{2m-1}{2m}\cdot\frac{2m-3}{2m-2}\cdots\frac{1}{2}I_0, \quad \text{if } n = 2m,$$

$$I_{2m+1} = \frac{2m}{2m+1}\cdot\frac{2m-2}{2m-1}\cdots\frac{2}{3}I_1, \quad \text{if } n = 2m+1.$$

Here,

$$I_0 = \int_0^{\frac{\pi}{2}} \sin^0 x\, dx = \frac{\pi}{2},$$

$$I_1 = \int_0^{\frac{\pi}{2}} \sin x\, dx = 1.$$

Example 9.2. Find $\int_0^1 \frac{dx}{\sqrt{1-x}}$.

The integrand is not defined at $x = 1$, and so, as an improper integral, we can write

$$\int_0^1 \frac{dx}{\sqrt{1-x}} = \lim_{x\to 1-0} \int_0^x \frac{dt}{\sqrt{1-t}} = \lim_{x\to 1-0}\left(-2\sqrt{1-x}\right) - (-2) = 2.$$

Example 9.3. Find $\int_1^{+\infty} \frac{dx}{x^2}$.

The interval of integration is the infinite interval $[1,+\infty)$. Using the definition of an improper integral, we obtain

$$\int_1^{+\infty} \frac{dx}{x^2} = \lim_{x\to +\infty}\int_1^x \frac{dt}{t^2} = \lim_{x\to +\infty}\left(-\frac{1}{x}\right) - (-1) = 1.$$

9.2   Mean Value Theorems for the Newton Integral

Theorem 9.1. *If a function f defined on an interval $[a,b]$ has an antiderivative, then there is at least one number ξ in (a,b) such that*

$$\int_a^b f(x)dx = f(\xi)(b-a). \tag{9.2}$$

☐   Since, by the definition of antiderivative, $F'(x) = f(x)$, the function $F(x)$ is necessarily continuous on the closed interval $[a,b]$.

Then, by Theorem 5.4, there exists at least one $\xi \in (a,b)$ such that

$$F(b) - F(a) = F'(\xi)(b-a) = f(\xi)(b-a). \tag{9.3}$$

Now, (9.1) and (9.3) imply (9.2).   ∎

Remark 9.2. If f is integrable on $[a,b]$, then the average value of f on $[a,b]$ is defined to be $\dfrac{1}{b-a}\int_a^b f(x)\,dx$. Then, by Theorem 9.1, there exists a point $\xi \in (a,b)$ such that $f(\xi)$ is the average value of $f(x)$ on $[a,b]$. Note that $f(\xi)$ is closely related to the familiar notion of the arithmetic average.

Corollary 9.1. *Suppose that there exists an antiderivative of a function f on $[a,b]$ such that $f(x) \geqslant 0$, $x \in (a,b)$. Then*

$$\int_a^b f(x)dx \geqslant 0. \tag{9.4}$$

☐   Since $f(x) \geqslant 0$ for all $x \in (a,b)$, the right hand side of (9.2) is non-negative.   ∎

Corollary 9.2. *Suppose that for all x in $[a,b]$, $f(x) \geqslant g(x)$, and that the functions f and g possess antiderivatives on $[a,b]$. Then*

$$\int_a^b f(x)dx \geqslant \int_a^b g(x)dx. \tag{9.5}$$

☐   It suffices to put $\varphi(x) = f(x) - g(x) \geqslant 0$, $x \in [a,b]$ for $f(x)$ in (9.4).   ∎

Corollary 9.3. *Suppose that the functions g and $f \cdot g$ have antiderivatives on $[a,b]$ and $m \leqslant f(x) \leqslant M$, $g(x) \geqslant 0$, $x \in [a,b]$. Then*

$$m\int_a^b g(x)dx \leqslant \int_a^b f(x)g(x)dx \leqslant M\int_a^b g(x)dx. \tag{9.6}$$

☐   The proof follows from the inequalities $mg(x) \leqslant f(x)g(x) \leqslant Mg(x)$, $x \in [a,b]$, and (9.5).   ∎

In particular, if $g(x) \equiv 1$ in (9.6), then

$$m(b-a) \leqslant \int_a^b f(x)dx \leqslant M(b-a).$$

Corollary 9.4. *If there exist antiderivatives of $f(x)$ and $|f(x)|$ on $[a,b]$, then*

$$\left| \int_a^b f(x)dx \right| \leqslant \int_a^b |f(x)|dx. \tag{9.7}$$

☐ Taking into account the inequality $-|f(x)| \leqslant f(x) \leqslant |f(x)|$, it follows from (9.5) that

$$-\int_a^b |f(x)|dx \leqslant \int_a^b f(x)dx \leqslant \int_a^b |f(x)|dx.$$

Clearly, this inequality is equivalent to (9.7).   ∎

Theorem 9.2. *Suppose that a function f satisfying $f(x) \geqslant 0$ is not identically zero on $[a,b]$ and has an antiderivative on this interval.*
Then

$$\int_a^b f(x)dx > 0. \tag{9.8}$$

☐ It is obvious that (9.4) is true. We must show that the inequality is strict. On the contrary, suppose $\int_a^b f(x)dx = 0$. By hypothesis, $F'(x) = f(x) \geqslant 0$, and so $F(x)$ is nondecreasing on $[a,b]$. On the other hand, $F(b) = F(a)$ because $F(b) - F(a) = \int_a^b f(x)dx = 0$. It follows that $F(x)$ is constant. Hence, $f(x)$ is identically zero. This contradiction completes the proof of the theorem.   ∎

Theorem 9.3 (General Mean Value Theorem). *Suppose that the functions $f(x)$, $g(x)$, and possess antiderivatives on $[a,b]$, and that $g(x) \geqslant 0$, $x \in [a,b]$. Then there exists a point ξ in $[a,b]$ such that*

$$\int_a^b f(x)g(x)dx = f(\xi) \int_a^b g(x)dx. \tag{9.9}$$

☐ Note that for a constant function f the equality (9.9) is true for every $\xi \in (a,b)$. Suppose that f is not constant. By Theorem 9.2, $\int_a^b g(x)dx > 0$. Put

$$f^0 = \frac{\int_a^b f(x)g(x)dx}{\int_a^b g(x)dx}. \tag{9.10}$$

We show that there exist points $x_1, x_2 \in [a,b]$ such that $f(x_1) < f^0 < f(x_2)$.
On the contrary, suppose that for all $x \in [a,b]$, either $f^0 \leqslant f(x)$ or $f^0 \geqslant f(x)$. If $f^0 \leqslant f(x)$,

$x \in [a,b]$, then $f(x) - f^0$ is not identically zero, and so by Theorem 9.2, $\int_a^b (f(x) - f^0)dx > 0$. Therefore,

$$f^0 = \frac{\displaystyle\int_a^b (f(x) - f^0)g(x)dx + \int_a^b f^0 g(x)dx}{\displaystyle\int_a^b g(x)dx} > f^0.$$

Thus we have the contradiction, $f^0 > f^0$. Hence, there is a point x_1 in $[a,b]$ such that $f(x_1) < f^0$. Similarly, we can show that there exists a point x_2 in $[a,b]$ such that $f^0 < f(x_2)$. Thus, $f(x_1) < f^0 < f(x_2)$, for $x_1, x_2 \in [a,b]$. Now, by Theorem 5.6, for any $f(x) = F'(x)$, there exists at least one point $\xi \in (a,b)$ such that

$$f(\xi) = f^0. \tag{9.11}$$

Now, (9.9) follows from (9.10) and (9.11). ∎

9.3   The Riemann Integral

Newton's integral cannot exist for a function having a point of discontinuity of the first kind. Indeed, if the Newton integral $\int_a^b f(x)dx$ exists, then $f(x)$ has an antiderivative, $F(x)$, for $F'(x) = f(x)$. But, by Corollary 5.4, a point satisfying $F'(x) = f(x)$ is either a point of continuity of $f(x)$ or a point of discontinuity of the second kind.

In this section we introduce the concept of the Riemann integral, which remedies this problem.

Definition 9.2. A *partition* of $[a,b]$ is a set of points $P = \{x_0, x_1, \ldots, x_n\}$ satisfying $a = x_0 < x_1 < \cdots < x_n = b$.

Definition 9.3. A partition P_2 of $[a,b]$ is said to be a *refinement* of P_1 if $P_1 \subset P_2$.

Definition 9.4. If P is any partition of $[a,b]$ and $\Delta x_i = x_{i+1} - x_i$ $(i = 0, 1, \ldots, n-1)$, then

$$\mu(P) = \max_{0 \leqslant i \leqslant n-1} \Delta x_i$$

is said to be the *mesh of the partition P*. It is clear that $n \geqslant \dfrac{b-a}{\mu(P)}$.

Definition 9.5. Let f be a function defined on the interval $[a,b]$. If P is a partition of $[a,b]$ and $\{\xi_0, \xi_1, \ldots, \xi_{n-1}\}$, $\xi_i \in [x_i, x_{i+1}]$ $(i = 0, 1, \ldots, n-1)$ is a selection of points for P, then the Riemann sum for f determined by P and the selection $\{\xi_0, \xi_1, \ldots, \xi_{n-1}\}$ is

$$\sigma(x_i, \xi_i) = \sum_{i=0}^{n-1} f(\xi_i)\Delta x_i.$$

We also say that the Riemann sum is associated with the partition P.

Definition 9.6. Let P be a partition of $[a,b]$ and $\{\xi_0, \xi_1, \ldots, \xi_{n-1}\}$ be a selection for P. We say that I is the limit of the Riemann sums $\sigma(x_i, \xi_i)$ as the mesh of the partition tends to zero, if for every $\varepsilon > 0$, there is a $\delta = \delta(\varepsilon)$ such that $\mu(P) < \delta$ implies $|I - \sigma| < \varepsilon$.

Definition 9.7. The Riemann integral of the function f over the interval from a to b is the number I

$$\lim_{\mu(P)\to 0} \sigma(x_i, \xi_i) = \lim_{\mu(P)\to 0} \sum_{i=0}^{n-1} f(\xi_i)\Delta x_i = I. \tag{9.12}$$

provided that this limit exists, in which case we say that f is integrable on $[a,b]$.

In this case we write

$$\int_a^b f(x)dx = \lim_{\mu(P)\to 0} \sum_{i=0}^{n-1} f(\xi_i)\Delta x_i.$$

The definition of the Riemann integral is motivated by the problem of defining and calculating the area of the region lying between the graph of a non-negative function f and the x-axis over a closed interval $[a,b]$. Thus, if $f(x) \geqslant 0$, then the Riemann sum $\sigma(x_i, \xi_i)$ is the sum of rectangular areas with base Δx_i and height $f(\xi_i)$, as is shown in Figure 9.1:

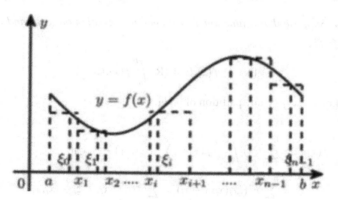

Fig. 9.1   The Riemann sum $\sigma(x_i, \xi_i)$ for $y = f(x)$.

Consequently, the Riemann integral $\lim_{\mu(P)\to 0} \sigma(x_i, \xi_i) = \int_a^b f(x)dx$ of f on $[a,b]$ is the area of the region $A = \{(x,y) : a \leqslant x \leqslant b, 0 \leqslant y \leqslant f(x)\}$.

Example 9.4. Evaluate $\int_0^1 x^2 dx$.

By the geometrical meaning of the Riemann integral, we must compute the area of $A = \{(x,y) : 0 \leqslant x \leqslant 1, 0 \leqslant y \leqslant x^2\}$. We take a so-called regular partition of $[0,1]$, that is, $x_0 = 0$, $x_1 = \Delta x, \ldots, x_n = 1 = n\Delta x$. Here, $\Delta x = 1/n$, and $\mu(P) = \Delta x = 1/n$ is the mesh of the partition $P = \{x_0, x_1, \ldots, x_n\}$. As the, we take the right-hand points x_{i+1} of each i-th subinterval, $[x_i, x_{i+1}]$.

Then

$$\sigma(x_i, \xi_i) = \sum_{i=0}^{n-1} \xi_i^2 \Delta x = \sum_{i=0}^{n-1} x_{i+1}^2 \Delta x$$

$$= [(\Delta x)^2 \Delta x + (2\Delta x)^2 \Delta x + \cdots + (n\Delta x)^2 \Delta x] = (\Delta x)^3 [1^2 + 2^2 + \cdots + n^2].$$

Since $1^2 + 2^2 + \cdots + n^2 = \dfrac{n(n+1)(2n+1)}{6}$, we have

$$\sigma(x_i, \xi_i) = \frac{1}{n^3} \cdot \frac{n(n+1)(2n+1)}{6} = \frac{1}{6}\left(1 + \frac{1}{n}\right)\left(2 + \frac{1}{n}\right).$$

Thus, see (9.12),

$$A = \int_0^1 x^2 dx = \lim_{\mu(P) \to 0} \sigma = \lim_{n \to \infty} \sigma = \frac{1}{3}.$$

Newton's and Riemann's integrals of a function f on $[a, b]$ will be denoted by $(\text{N}) \int_a^b f(x)$ and $(\text{R}) \int f(x) dx$, respectively.

Theorem 9.4. *Suppose that a function f is Riemann integrable on $[a, b]$ and has an antiderivative, f. Then*

$$(\text{N}) \int_a^b f(x)dx = (\text{R}) \int_a^b f(x)dx. \tag{9.13}$$

☐ Let $P = \{x_0, \ldots, x_n\}$ be any partition of $[a, b]$.

Then

$$(\text{N}) \int_a^b f(x)dx = \sum_{i=0}^{n-1} (\text{N}) \int_{x_i}^{x_{i+1}} f(x) = \sum_{i=0}^{n-1} [F(x_{i+1}) - F(x_i)]$$

$$= \sum_{i=0}^{n-1} F'(\xi_i)(x_{i+1} - x_i) = \sum_{i=0}^{n-1} f(\xi_i)\Delta x_i, \quad \xi_i \in (x_i, x_{i+1}). \tag{9.14}$$

Using Theorem 5.4 in (9.14), then passing to the limit (see (9.12)) as $\mu(P) \to 0$, we obtain the desired formula, (9.13). ∎

Theorem 9.5. *If the function f is Riemann integrable on $[a, b]$, then f is bounded on this interval.*

☐ On the contrary, suppose that f is integrable on $[a, b]$ but unbounded on this interval. By Definition 5.6, for every $\varepsilon > 0$ there is a positive number $\delta = \delta(\varepsilon)$ such that $\mu(P) < \delta(\varepsilon)$ implies $|I - \sigma| < \varepsilon$. It follows that if $\mu(P) < \delta(\varepsilon)$, then the Riemann sum σ is bounded.

Let P be a partition such that $\mu(P) < \delta(\varepsilon)$. Since, by assumption, f is unbounded on $[a,b]$, there exists a subinterval $[x_k, x_{k+1}]$ such that the function f is unbounded on $[x_k, x_{k+1}]$. Then it is possible, by selecting an appropriate $\xi_k \in [x_k, x_{k+1}]$, to arrange that $f(\xi_k)\Delta x_k$ is larger than any pre-assigned number. This implies that the Riemann sums are unbounded, which contradiction proves the theorem. ∎

Remark 9.3. For the existence of the Riemann integral of f, the boundedness of f is necessary, but not sufficient. For instance, consider the Dirichlet function,

$$D(x) = \begin{cases} 1, & \text{if } x \text{ is rational,} \\ 0, & \text{if } x \text{ is irrational,} \end{cases}$$

on the interval $[a,b]$. This function is bounded, but not integrable on $[a,b]$. For let P be a partition with sufficiently small mesh $\mu(P)$. Clearly, if all points ξ_i of a selection for the partition P are irrational, then the corresponding Riemann sum $\sigma_1 = 0$. If the ξ_i are rational, then the Riemann sum is $\sigma_2 = b - a$. Then for the existence of the limit I of the Riemann sum as the mesh μ tends to zero, for $\varepsilon = \dfrac{b-a}{2}$ and sufficiently small μ, we would have to have that

$$|\sigma_1 - I| < \frac{b-a}{2}, \quad |\sigma_2 - I| < \frac{b-a}{2}.$$

But from this, it would follow that $|\sigma_2 - \sigma_1| = |(\sigma_2 - I) + (I - \sigma_1)| \leqslant |\sigma_2 - I| + |\sigma_1 - I| \leqslant b - a$.

On the other hand, it is clear that $\sigma_2 - \sigma_1 = b - a$. This contradiction proves that the integral in the sense of Riemann does not exist.

Example 9.5. Let $f(x) = \begin{cases} 2x\sin\dfrac{1}{x^2} - \dfrac{2}{x}\cos\dfrac{1}{x^2}, & \text{if } x \neq 0, \\ 0, & \text{if } x = 0, \end{cases}$

It is easy to see that

$$F(x) = \begin{cases} x^2 \sin\dfrac{1}{x^2}, & \text{if } x \neq 0, \\ 0, & \text{if } x = 0, \end{cases}$$

is an antiderivative for f. Hence the Newton integral of f, (N) $\int_a^b f(x)\,dx$, exists. But the Riemann integral (R) $\int_a^b f(x)\,dx$ does not exist, because f is unbounded on the interval $[0,1]$.

Example 9.6. Show that $f(x) = \operatorname{sgn} x$ is Riemann integrable. Recall that the *signum* function, sgn x, also called the *sign function*, is defined by

$$\operatorname{sgn} x = \begin{cases} -1, & \text{if } x < 0, \\ 0, & \text{if } x = 0, \\ +1, & \text{if } x > 0, \end{cases}$$

Note that is a point of discontinuity of the first kind. The Riemann sum of on for a selection $\xi_i \in [-1, 0]$ $(i = 0, \ldots, m)$ and $\xi_i \in [0, 1]$ $(i = m+1, \ldots, n-1)$ is

$$\sum_{i=0}^{n-1} f(\xi_i)\Delta x_i = \sum_{i=0}^{m} f(\xi_i)\Delta x_i + \sum_{i=m+1}^{n-1} f(\xi_i)\Delta x_i = -1(0+1) + 1(1-0) = 0.$$

Then

$$(R) \int_{-1}^{1} \operatorname{sgn} x\, dx = \lim_{\mu(P) \to 0} \sum_{i=0}^{n-1} f(\xi_i)\Delta x_i = 0.$$

But the Newton integral (N) $\int_{-1}^{1} \operatorname{sgn} x\, dx$ does not exist, because the derivative $F'(x) = f(x)$ cannot have a point of discontinuity of the first kind (Corollary 5.4).

Consequently, if a function f integrable on $[a, b]$ both in the sense of Newton and in the sense of Riemann, then both definite integrals are equal.

9.4 Upper and Lower Riemann Sums and Their Properties

Suppose a function f is bounded on the closed interval $[a, b]$, and that $P = \{x_0, x_1, \ldots, x_n\}$ is any partition of $[a, b]$. We adopt the following notation:

$$\begin{aligned}
M &= \sup_{x \in [a,b]} f(x), & m_i &= \sup_{x \in [x_i, x_{i+1}]} f(x), \\
m &= \inf_{x \in [a,b]} f(x), & m_i &= \inf_{x \in [x_i, x_{i+1}]} f(x), \\
S(P) &= \sum_{i=0}^{n-1} M_i \Delta x_i, & s(P) &= \sum_{i=0}^{n-1} M_i \Delta x_i.
\end{aligned} \tag{9.15}$$

Definition 9.8. The $S(P)$ and $s(P)$ are called the *upper Riemann sum* and the *lower Riemann sum*, respectively, for f, associated with the partition P of $[a, b]$.

For each $i = 1, \ldots, n$ consider the rectangle with base $[x_{i-1}, x_i]$, height M_i, and area $M_i \Delta x_i$. The union of these n rectangles contains the region $A = \{(x, y) : a \leqslant x \leqslant b, 0 \leqslant y \leqslant f(x)\}$ under the graph; it is a circumscribed rectangular polygon associated with the partition of $[a, b]$. Its area is $S(P)$. Similarly, $s(P)$ is the area of an inscribed rectangular polygon, that is, the sum of the areas of the n rectangles each with base length Δx_i and height m_i (Figure 9.2).

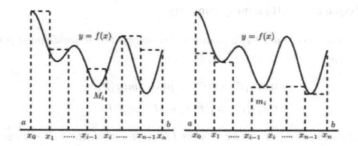

Fig. 9.2   The upper and the lower Riemann sums of $y = f(x)$.

Lemma 9.1. *Let $\sigma(x_i, \xi_i)$ be the Riemann sum for f determined by an arbitrary partition P of $[a, b]$ and any selection $\{\xi_0, \xi_1, \ldots, \xi_{n-1}\}$ for the partition P.*
Then

$$s(P) \leqslant \sigma(x_i, \xi_i) \leqslant S(P).$$

☐   Indeed, by the definition of M_i and m_i, the inequality $m_i \leqslant f(\xi_i) \leqslant M_i$ holds for any $\xi_i \in [x_i, x_{i+1}]$. Hence, by multiplying by Δx_i, $i = 0, \ldots, n-1$, and then summing these inequalities, we have the desired inequality.   ■

Lemma 9.2. *For any partition P of $[a, b]$,*

$$S(P) = \sup_{\xi_i \in [x_i, x_{i+1}]} \sigma(x_i, \xi_i); \quad s(P) = \inf_{\xi_i \in [x_i, x_{i+1}]} \sigma(x_i, \xi_i). \tag{9.16}$$

☐   Let $\varepsilon > 0$ be any positive real number. By definition (see (9.15)), M_i is the least upper bound, that is $M_i = \sup_{x \in [x_i, x_{i+1}]} f(x)$. Then there exist numbers ξ_i such that

$$M_i - \frac{\varepsilon}{b-a} < f(\xi_i) \leqslant M_i \quad (i = 0, 1, \ldots, n-1).$$

Thus,

$$\sum_{i=0}^{n-1} \left(M_i - \frac{\varepsilon}{b-a} \right) \Delta x_i < \sum_{i=0}^{n-1} f(\xi_i) \Delta x_i \leqslant \sum_{i=0}^{n-1} M_i \Delta x_i,$$

or

$$S(P) - \varepsilon < \sigma(x_i, \xi_i) \leqslant S(P).$$

Consequently,

$$S(P) = \sup_{\xi_i \in [x_i, x_{i+1}]} \sigma(x_i, \xi_i).$$

The second equality in (9.16) is proved similarly. ■

Lemma 9.3. *If a refinement P^* of a partition P is obtained by the addition of p selection points to P, then*

$$0 \leqslant S(P) - S(P^*) \leqslant p(M - m)\mu(P),$$
$$0 \leqslant s(P^*) - s(P) \leqslant p(M - m)\mu(P),$$

i.e., upper Riemann sums do not increase and lower Riemann sums do not decrease by adding more selection points, that is, by refining P.

□ For simplicity, we take $p = 1$. Then $P^* = P \cup \{x^*\}$ $(x_i < x^* < x_{i+1})$ for some new point x^*. It is clear that

$$0 \leqslant M_i - \sup_{x \in [x_i, x^*]} f(x) \leqslant M - m,$$
$$0 \leqslant M_i - \sup_{x \in [x^*, x_{i+1}]} f(x) \leqslant M - m, \tag{9.17}$$

and

$$S(P^*) = \sum_{k=0}^{i-1} M_k \Delta x_k + \left(\sup_{x \in [x_i, x^*]} f(x) \right)(x^* - x_i) + \left(\sup_{x \in [x^*, x_{i+1}]} f(x) \right)(x_{i+1} - x^*) + \sum_{k=i+1}^{n-1} M_k \Delta x_k.$$

Then

$$S(P) - S(P^*) = M_i(x_{i+1} - x_i) - \left(\sup_{x \in [x_i, x^*]} f(x) \right)(x^* - x_i) - \left(\sup_{x \in [x^*, x_{i+1}]} f(x) \right)(x_{i+1} - x^*)$$

$$= \left(M_i - \sup_{x \in [x_i, x^*]} f(x) \right)(x^* - x_i) + \left(M_i - \sup_{x \in [x^*, x_{i+1}]} f(x) \right)(x_{i+1} - x^*).$$

Thus, taking into account inequality (9.17), we have

$$0 \leqslant S(P) - S(P^*) \leqslant (M - m)(x_{i+1} - x_i) \leqslant (M - m)\mu(P).$$

The second inequality of the lemma is proved analogously. ■

Lemma 9.4. *If P_1 and P_2 are any partitions of $[a, b]$, then*

$$s(P_1) \leqslant S(P_2),$$

i.e., a lower Riemann sum is never greater than any upper Riemann sum, regardless of the partition used.

□ Indeed, since $P_1 \subset P_1 \cup P_2$ and $P_2 \subset P_1 \cup P_2$, a partition $P_1 \cup P_2$ is a refinement both of P_1 and P_2. Then, by Lemma 9.3,

$$s(P_1) \leqslant s(P_1 \cup P_2) \quad \text{and}$$
$$S(P_1 \cup P_2) \leqslant S(P_2). \tag{9.18}$$

On the other hand, by Lemma 9.1, $s(P_1 \cup P_2) \leqslant S(P_1 \cup P_2)$. Then it follows from (9.18) that $s(P_1) \leqslant S(P_2)$.

Moreover, by Lemma 9.4, the lower Riemann sums are bounded above and the upper Riemann sums are bounded below. Thus, in the set of all possible partitions of $[a,b]$, there exists a least upper bound of the upper Riemann sums and a greatest lower bound of the lower Riemann sums.

Then, denoting

$$I^* = \inf_P S(P) \quad \text{and} \quad I_* = \sup_P s(P),$$

we have

$$s(P) \leqslant I_* \leqslant I^* \leqslant S(P). \tag{9.19}$$

∎

The bounds I^* and I_* are called the upper and lower Riemann integrals, respectively.

Lemma 9.5. *For every* $\varepsilon > 0$, *there is a* $\delta = \delta(\varepsilon) > 0$ *such that for every partition P of* $[a,b]$, *the inequality* $\mu(P) \leqslant \Delta(\varepsilon)$ *implies*

$$I_* - \varepsilon < s(P) \leqslant S(P) < I^* + \varepsilon. \tag{9.20}$$

□ Choose any $\varepsilon > 0$. By the definition of the least upper bound, it follows from $I_* = \sup_P s(P)$ that there is a partition, such that

$$s(P_1) > I_* - \frac{\varepsilon}{2}. \tag{9.21}$$

Let p be the number of selection points of P_1 belonging to (a,b). Further, let P be a partition of $[a,b]$ such that

$$\mu(P) < \frac{\varepsilon}{2p(M-m)} = \delta_1(\varepsilon).$$

Without loss of generality, we may assume $M \neq m$, because if $M = m$, that is $f(x) = c-$ constant, $s(P) = S(P) = c(b-a)$ for all partitions P. This means that $I_* = I^* = c(b-a)$. Now, we choose a refinement of P and P_1, and partition $P^* = P \cup P_1$. By Lemma 9.3,

$$S(P_1) \leqslant s(P^*) = s(P) + p(M-m)\mu(P) < s(P) + p(M-m)\frac{\varepsilon}{2p(M-m)} = s(P) + \frac{\varepsilon}{2},$$

or

$$s(P_1) < s(P) + \frac{\varepsilon}{2}.$$

Then, using (9.21), we have

$$I_* - \frac{\varepsilon}{2} < s(P) + \frac{\varepsilon}{2} \text{ and}$$
$$I_* - \varepsilon < s(P). \tag{9.22}$$

Thus, for a partition P of $[a,b]$ with mesh satisfying $\mu(P) < \delta_1(\varepsilon)$, the inequalities (9.22) hold. Similarly, $\delta_2 = \delta_2(\varepsilon)$ can be found such that for every partition P of $[a,b]$, the inequality $\mu(P) < \delta_2$ implies

$$S(P) < I^* + \varepsilon.$$

Now, if we denote the smallest of the numbers $\delta_1(\varepsilon)$ and $\delta_2(\varepsilon)$ by $\delta(\varepsilon)$, then we see that $\mu(P) < \delta(\varepsilon)$ implies (9.20).　　　　　　　　　　　　　　　　　　　　　■

9.5　Necessary and Sufficient Conditions for the Existence of the Riemann Integral; Basic Properties of the Riemann Integral

Theorem 9.6. *For the existence of the Riemann integral of a bounded function f defined on a closed interval $[a,b]$, it is necessary and sufficient that for every $\varepsilon > 0$, there exists a partition P of $[a,b]$ such that*

$$S(P) - s(P) < \varepsilon. \tag{9.23}$$

☐　(*Necessity*). Suppose that the Riemann integral of f on $[a,b]$ exists. Then for each $\varepsilon > 0$, there is a $\delta = \delta(\varepsilon) > 0$ such that

$$|\sigma(x_i, \xi_i) - I| < \frac{\varepsilon}{3}$$

for every partition P of $[a,b]$ with mesh $\mu(P) < \delta$ and every $\xi_i \in [x_i, x_{i+1}]$.
Hence,

$$I - \frac{\varepsilon}{3} < \sigma(x_i, \xi_i) < I + \frac{\varepsilon}{3}.$$

Then, by Lemmas 9.1 and 9.2,

$$I - \frac{\varepsilon}{3} < s(P) \leqslant S(P) \leqslant I + \frac{\varepsilon}{3}.$$

Thus, for every mesh P with $\mu(P) < \delta(\varepsilon)$,

$$S(P) - s(P) \leqslant 2 \cdot \frac{\varepsilon}{3} < \varepsilon.$$

(*Sufficiency*). Let ε be any positive number and P any partition of $[a,b]$ satisfying the inequality (9.23). We have, from (9.19) and (9.23), that $0 \leqslant I^* - I_* < \varepsilon$. Since ε is an arbitrary positive number, it follows that $I^* = I_*$. Denote $I_* = I^* = I$. By Lemma 9.5, there is a number $\delta = \delta(\varepsilon) > 0$ such that for every P of $[a,b]$, the inequality $\mu(P) < \delta(\varepsilon)$ implies

$$I - \varepsilon < s(P) \leqslant S(P) < I + \varepsilon,$$

or

$$|\sigma(x_i, \xi_i) - I| < \varepsilon.$$

Consequently, f is integrable on $[a,b]$.   ∎

Remark 9.4. Using (9.15), we can write (9.23) in the form

$$\sum_{i=0}^{n-1} (M_i - m_i)\Delta x_i < \varepsilon.$$

Now, if we introduce the oscillation $\omega_i = M_i - m_i$ of a function f on the interval $[x_i, x_{i+1}]$, then, in terms of the oscillation, the necessary and sufficient condition (9.23) can be expressed as

$$\sum_{i=0}^{n-1} \omega_i \Delta x_i < \varepsilon.$$

Remark 9.5. Theorem 9.6 can be reformulated as follows: For the existence of the Riemann integral of a bounded function f defined on $[a,b]$, it is necessary and sufficient that the upper and lower integrals are equal.

Property 9.1. *If f is Riemann integrable on $[a,b]$, then f is Riemann integrable on any $[c,d] \subset [a,b]$.*

☐ Let ε be any positive number. By Theorem 9.6, there is a partition of $[a,b]$ such that

$$S(P) - s(P) < \varepsilon. \tag{9.24}$$

Put $P^* = P \cup \{c,d\}$. Let P_1 be a partition of $[c,d]$ consisting of the points of P^* belonging to the interval $[c,d]$.

Then

$$\sum_{P_1} \omega_i \Delta x_i \leqslant \sum_{P^*} \omega_i \Delta x_i = S(P^*) - s(P^*). \tag{9.25}$$

Here the first sum corresponds to a partition P_1, and the second sum corresponds to P^* and consist of nonnegative terms, so that the second sum includes all the terms of the first sum. Since P^* is a refinement of P, then, by Lemma 9.3 and the inequality (9.24),

$$S(P^*) - s(P^*) \leqslant S(P) - s(P) < \varepsilon. \tag{9.26}$$

From (9.25) and (9.26) we have

$$\sum_{P_1} \omega_i \Delta x_i < \varepsilon.$$

Thus, by Remark 9.4, f is Riemann integrable on $[c,d]$.   ∎

Property 9.2. *If f is Riemann integrable on $[a,c]$ and $[c,b]$ $(a < c < b)$, then f is Riemann integrable on $[a,b]$, and moreover,*

$$\int_a^b f(x)dx = \int_a^c f(x)dx + \int_c^b f(x)dx.$$

☐ Let $\varepsilon > 0$ be any positive number. Further, let P_1 and P_2 be partitions of $[a,c]$ and $[c,b]$, respectively. By Theorem 9.6, there exist partitions P_1 and P_2 such that

$$\sum_{P_1} \omega_i \Delta x_i < \frac{\varepsilon}{2} \quad \text{and} \quad \sum_{P_2} \omega_i \Delta x_i < \frac{\varepsilon}{2}.$$

For a partition $P = P_1 \cup P_2$, it is clear that

$$\sum_{P} \omega_i \Delta x_i = \sum_{P_1} \omega_i \Delta x_i + \sum_{P_2} \omega_i \Delta x_i < \varepsilon.$$

By Theorem 9.6, f is integrable on $[a,b]$. The Riemann sum associated with P can be written as follows:

$$\sum_{P} f(\xi_i) \Delta x_i = \sum_{P_1} f(\xi_i) \Delta x_i + \sum_{P_2} f(\xi_i) \Delta x_i.$$

The partition P is a refinement of P_1 and P_2, and so $\mu(P_1) \to 0$ and $\mu(P_2) \to 0$ as $\mu(P) \to 0$. Thus, in the last relation, by passing to the limit as $\mu(P) \to 0$, we have

$$\int_a^b f(x)dx = \int_a^c f(x)dx + \int_c^b f(x)dx.$$

                                                                                                    ∎

Property 9.3. *If f and g are Riemann integrable on $[a,b]$, then $(f+g)(x)$ is also, and*

$$\int_a^b [f(x) + g(x)]dx = \int_a^b f(x)dx + \int_a^b g(x)dx. \tag{9.27}$$

☐ For any partition P of $[a,b]$ and any selection of points $\xi_i \in [x_i, x_{i+1}]$, we have

$$\sum_{i=0}^{n-1} [f(\xi_i) + g(\xi_i)] \Delta x_i = \sum_{i=0}^{n-1} f(\xi_i) \Delta x_i + \sum_{i=0}^{n-1} g(\xi_i) \Delta x_i. \tag{9.28}$$

Since the integrals of f and g exist, the limit, as $\mu(P) \to 0$, of the right hand side of (9.28) exists. Thus the limit of the left hand side of (9.28) as $\mu(P) \to 0$ also exists, and so, according to the definition of the Riemann integral, (9.27) is true.   ∎

Property 9.4. *If f is Riemann integrable on $[a,b]$ and c is constant, then $c \cdot f$ is Riemann integrable on $[a,b]$, and*

$$\int_a^b c \cdot f(x)dx = c \int_a^b f(x)dx.$$

☐ Indeed,

$$\int_a^b cf(x)dx = \lim_{\mu(P)\to 0}\sum_{i=0}^{n-1}cf(\xi_i)\Delta x_i = c\lim_{\mu(P)\to 0}\sum_{i=0}^{n-1}f(\xi_i)\Delta x_i = c\int_a^b f(x)dx.$$

■

Property 9.5. *Suppose f and g are defined on $[a,b]$. Suppose further that f is Riemann integrable on $[a,b]$ and with the exception of finitely many points. Then the function g is Riemann integrable on $[a,b]$, and*

$$\int_a^b g(x)dx = \int_a^b f(x)dx.$$

☐ Put $u(x) = g(x) - f(x)$. Suppose $\bar{x}_1, \bar{x}_2, \ldots, \bar{x}_p$ are all the points at which the functions f and g are not equal. We have $u(x) = 0$ except at these points. Let $k = \max\{|u(\bar{x}_1)|, |u(\bar{x}_2)|, \ldots, |u(\bar{x}_p)|\}$. It is easy to see that each of the points $\bar{x}_1, \ldots, \bar{x}_p$ belong to subintervals $[x_i, x_{i+1}]$ $(i = 0, 1, \ldots, n-1)$ not more than twice. For all other points x, we have $u(x) = 0$. Then for every partition P of $[a,b]$ and selection of points $\xi_i \in [x_i, x_{i+1}]$, we can write

$$\left|\sum_{i=0}^{n-1}u(\xi_i)\Delta x_i\right| \leqslant \sum_{i=0}^{n-1}|u(\xi_i)|\Delta x_i \leqslant 2pk\mu(P).$$

By passing to the limit as $\mu(P) \to 0$, we have

$$\int_a^b u(x)dx = 0.$$

By Property 9.3, the function is integrable on $[a,b]$, and then

$$\int_a^b g(x)dx = \int_a^b u(x)dx + \int_a^b f(x)dx = \int_a^b f(x)dx.$$

■

Remark 9.6. As is seen from Property 9.5, the Riemann integral can be defined for a function f defined on $[a,b]$ with the exception of a finite number points, because if we define f at such points arbitrarily, then the integral of the new function g and f are the same.

Property 9.6. *If the functions f and g are Riemann integrable on $[a,b]$, then their product $f \cdot g$ is Riemann integrable on $[a,b]$.*

☐ The functions f and g are integrable on $[a,b]$ and so are bounded: $|f(x)| \leqslant C_1$, $|g(x)| \leqslant C_2|$ on $[a,b]$. By Theorem 9.6, for every $\varepsilon > 0$ there are partitions P_1 and P_2 of $[a,b]$ such that

$$S'(P_1) - s'(P_1) < \frac{\varepsilon}{2C_2},$$
$$S''(P_2) - s''(P_2) < \frac{\varepsilon}{2C_1}.$$

Here $S'(P_1)$ and $S''(P_2)$ are the upper Riemann sums for f and g, respectively. Similarly, $s'(P_1)$ and $s''(P_2)$ are their lower Riemann sums.

Thus, for a partition $P = P_1 \cup P_2$, we have

$$S'(P) - s'(P) < \frac{\varepsilon}{2C_2},$$
$$S''(P) - s''(P) < \frac{\varepsilon}{2C_1},$$

or

$$\sum_{i=0}^{n-1} \omega_i' \Delta x_i < \frac{\varepsilon}{2C_2},$$
$$\sum_{i=0}^{n-1} \omega_i'' \Delta x_i < \frac{\varepsilon}{2C_1}. \tag{9.29}$$

Here, ω_i', ω_i'' denote the oscillations of f and g respectively on the subintervals $[x_i, x_{i+1}]$.

It is easy to see that for the oscillation ω_i of a product function $f \cdot g$ on $[x_i, x_{i+1}]$, we have

$$\omega_i \leqslant C_1 \omega_i'' + C_2 \omega_i'. \tag{9.30}$$

In fact, for every $\xi, \eta \in [x_i, x_{i+1}]$, we have the identity

$$f(\xi)g(\xi) - f(\eta)g(\eta) = f(\xi)[g(\xi) - g(\eta)] + g(\eta)[f(\xi) - f(\eta)].$$

Obviously,

$$|f(\xi)g(\xi) - f(\eta)g(\eta)| \leqslant C_1|g(\xi) - g(\eta)| + C_2|f(\xi) - f(\eta)|,$$

and so

$$\sup_{\xi, \eta \in [x_i, x_{i+1}]} |f(\xi)g(\xi) - f(\eta)g(\eta)|$$
$$\leqslant C_1 \sup_{\xi, \eta \in [x_i, x_{i+1}]} |g(\xi) - g(\eta)| + C_2 \sup_{\xi, \eta \in [x_i, x_{i+1}]} |f(\xi) - f(\eta)|. \tag{9.31}$$

If we put $\overline{k}_i = [x_i, x_{i+1}]$, then

$$\sup_{\xi, \eta \in \overline{k}_i} |f(\xi) - f(\eta)| = \sup_{\xi, \eta \in \overline{k}_i} [f(\xi) - f(\eta)] = \sup_{x \in \overline{k}_i} f(x) - \inf_{x \in \overline{k}_i} f(x) = M_i - m_i = \omega_i'. \tag{9.32}$$

Similarly,

$$\sup_{\xi, \eta \in \overline{k}_i} |g(\xi) - g(\eta)| = \omega_i'' \text{ and}$$
$$\sup_{\xi, \eta \in \overline{k}_i} |f(\xi)g(\xi) - f(\eta)g(\eta)| = \omega_i. \tag{9.33}$$

Thus, it follows from (9.31)–(9.33) that (9.30) is true. Now, by (9.29) and (9.30), for a partition $P = P_1 \cup P_2$ we have

$$\sum_{i=0}^{n-1} \omega_i \Delta x_i \leqslant C_1 \sum_{i=0}^{n-1} \omega_i'' \Delta x_i + C_2 \sum_{i=0}^{n-1} \omega_i' \Delta x_i < \varepsilon,$$

which means that $f \cdot g$ is Riemann integrable on $[a,b]$.   ∎

Property 9.7. *Let f be Riemann integrable on $[a,b]$ and $f(x) \geqslant 0$ on $[a,b]$.*
Then

$$\int_a^b f(x)dx \geqslant 0.$$

☐   Since $f(x) \geqslant 0$ on $[a,b]$, for every partition P of $[a,b]$ and selection of points $\xi_i \in [x_i, x_{i+1}]$, we have

$$\sigma(x_i, \xi_i) \geqslant 0.$$

By passing to the limit as $\mu(P) \to 0$, we obtain the desired inequality.   ∎

Corollary 9.5. *If f and g are Riemann integrable on $[a,b]$ and if $f(x) \geqslant g(x)$ for all x in $[a,b]$, then*

$$\int_a^b f(x)dx \geqslant \int_a^b g(x)dx. \tag{9.34}$$

☐   By Property 9.7,

$$\int_a^b [f(x) - g(x)]dx \geqslant 0.$$

Then Properties 9.3 and 9.4 imply (9.34).   ∎

Property 9.8. *If f is Riemann integrable on $[a,b]$, then $|f|$ also is Riemann integrable on this interval, and*

$$\left| \int_a^b f(x)dx \right| \leqslant \int_a^b |f(x)|dx. \tag{9.35}$$

☐   Denote the oscillations of the functions $|f|$ and f on the subinterval $[x_i, x_{i+1}]$ by $\overline{\Omega}_i$ and ω_i, respectively. Then it follows from the familiar inequality $\left| |f(\xi)| - |f(\eta)| \right| \leqslant |f(\xi) - f(\eta)|$ (see (9.32)), that

$$\overline{\Omega}_i \leqslant \omega_i.$$

On the other hand, since f is Riemann integrable on $[a,b]$, for any given $\varepsilon > 0$ there is a partition P of $[a,b]$ such that

$$\sum_{i=0}^{n-1} \omega_i \Delta x_i < \varepsilon.$$

Thus, for this same P, we have

$$\sum_{i=0}^{n-1} \overline{\Omega}_i \Delta x_i < \varepsilon.$$

Therefore, $|f(x)|$ is Riemann integrable on $[a,b]$.

Let now P be an arbitrary partition of $[a,b]$. It is clear that

$$\left| \sum_{i=0}^{n-1} f(\xi_i)\Delta x_i \right| \leqslant \sum_{i=0}^{n-1} |f(\xi_i)|\Delta x_i.$$

Finally, by passing to the limit as $\mu(P) \to 0$, we obtain the inequality (9.35). ∎

Remark 9.7. In general, for a Riemann integrable $|f|$ on $[a,b]$, the function f might not be Riemann integrable on $[a,b]$. For instance, let f be defined as

$$f(x) = \begin{cases} 1, & \text{if } x \text{ is rational,} \\ -1, & \text{if } x \text{ is irrational.} \end{cases}$$

We already showed that this, the Dirichlet function, is not Riemann integrable (see Remark 9.3). Nevertheless, $|f(x)| = 1$ is Riemann integrable on $[a,b]$.

If a function f is Riemann integrable on and $a > b$, we define that

$$\int_a^b f(x)dx = -\int_b^a f(x)dx.$$

Clearly, Properties 9.3 and 9.4 remain true for this extended definition.

In particular, if $a = b$, then

$$\int_a^a f(x)dx = 0.$$

9.6   The Class of Riemann Integrable Functions; Mean Value Theorems

Theorem 9.7. *A continuous function f on $[a,b]$ is Riemann integrable on $[a,b]$.*

□   By Cantor's theorem (Theorem 3.13), for every $\varepsilon > 0$ there is a $\delta = \delta(\varepsilon) > 0$ such that the oscillation of f on any interval $[a_1,b_1] \subset [a,b]$ with $b_1 - a_1 < \delta$ is less than $\dfrac{\varepsilon}{b-a}$.

Let P be a partition of $[a,b]$ with mesh $\mu(P) < \delta$. Then the oscillation ω_i of f on any subinterval $[x_i,x_{i+1}]$ satisfies the inequality $\omega_i < \dfrac{\varepsilon}{b-a}$.

For such a partition P,

$$\sum_{i=0}^{n-1} \omega_i \Delta x_i < \sum_{i=0}^{n-1} \frac{\varepsilon}{b-a}\Delta x_i = \frac{\varepsilon}{b-a}\sum_{i=0}^{n-1}\Delta x_i = \varepsilon,$$

which, by Theorem 9.6, means that f is Riemann integrable on $[a,b]$. ∎

Theorem 9.8. *A bounded function f on $[a,b]$ having only finitely many points of discontinuity on $[a,b]$ is Riemann integrable on this interval.*

☐ At first, consider the case when f has only one point of discontinuity in $[a,b]$, which is either a or b. For definiteness, suppose that a is the point of discontinuity. Denote by ω denote the oscillation of f on $[a,b]$. For every $\varepsilon > 0$, choose a point x_1 ($a < x_1 < b$) such that

$$x_1 - a < \frac{\varepsilon}{2\omega} . \qquad (9.36)$$

Now, f is continuous on $[x_1,b]$, so Theorem 9.7 implies it is Riemann integrable on this interval. On the other hand, by Theorem 9.6, there is a partition $P_1 = \{x_1, x_2, \ldots, x_n\}$ of $[x_1,b]$ such that

$$\sum_{i=1}^{n-1} \omega_i \Delta x_i < \frac{\varepsilon}{2} . \qquad (9.37)$$

For a partition $P = \{x_0, x_1, \ldots, x_n\}$ ($x_0 = a$) of $[x_1,b]$, with the use of (9.36) and (9.37) we can write

$$\sum_{i=0}^{n-1} \omega_i \Delta x_i = \omega_0 (x_1 - a) + \sum_{i=1}^{n-1} \omega_i \Delta x_i < \omega \cdot \frac{\varepsilon}{2\omega} + \frac{\varepsilon}{2} = \varepsilon.$$

Thus, by Theorem 9.6, the function f is Riemann integrable on $[a,b]$.

We now investigate the general case. Assume that ξ_i ($i = 1, \ldots, p$) are the points of discontinuity of f contained in (a,b) and $\xi_1 < \xi_2 < \cdots < \xi_p$. Choose points η_i, $i = 1, \ldots, p+1$ such that

$$a < \eta_1 < \xi_1 < \eta_2 < \xi_2 < \cdots < \eta_p < \xi_p < \eta_{p+1} < b.$$

There is no more than one point of discontinuity of f in each of the subintervals $[a, \eta_1]$, $[\eta_1, \xi_1]$, $[\xi_1, \eta_2], \ldots, [\xi_p, \eta_{p+1}], [\eta_{p+1}, b]$, and so f is Riemann integrable on each subinterval. Therefore, by Property 9.2, f is integrable on all of $[a,b]$.   ∎

Definition 9.9. A function f is said to be *piecewise continuous* on $[a,b]$ if it has only finitely many points of discontinuity, and provided that each of them is either removable (Definition 3.8) or of the first kind (Definition 3.9).

Corollary 9.6. *A piecewise continuous function f on $[a,b]$ is Riemann integrable on this interval.*

Theorem 9.9. *Monotone functions on $[a,b]$ are Riemann integrable on this interval.*

☐ For definiteness, let f be nondecreasing on $[a,b]$. Take an arbitrary $\varepsilon > 0$. Let P be a partition of $[a,b]$ with mesh satisfying the inequality $\mu(P) < \dfrac{\varepsilon}{f(b) - f(a)}$. Obviously, the oscillation ω_i on the subinterval $[x_i, x_{i+1}]$ is equal to $f(x_{i+1}) - f(x_i)$. Therefore,

$$\sum_{i=0}^{n-1} \omega_i \Delta x_i \leqslant \mu(P) \sum_{i=0}^{n-1} \omega_i = \mu(P) \sum_{i=0}^{n-1} [f(x_{i+1}) - f(x_i)] = \mu(P)[f(b) - f(a)],$$

or

$$\sum_{i=0}^{n-1} \omega_i \Delta x_i < \varepsilon.$$

Thus, by Theorem 9.6, f is Riemann integrable on $[a, b]$.   ∎

Definition 9.10. A set A of real numbers is said to *have Lebesgue*[1] *measure zero* (or, put more briefly, *measure zero*) if for any given $\varepsilon > 0$, there exists a countable set of open intervals I_k, $k = 1, 2, 3, \ldots$, such that $A \subset \bigcup_k I_k$, and such that the sum of the lengths of the I_k is less than ε.

For instance, we will see in Section 9.12 that a set of finitely many points, a set of countably many points, and hence the set of rational numbers contained in some interval, are sets with zero measure.

Theorem 9.10 (The Lebesgue Criterion). *A necessary and sufficient condition that a bounded function f be Riemann integrable on $[a, b]$ is that the set of discontinuities of f in $[a, b]$ have measure zero.*

Thus, a bounded function defined on $[a, b]$ with countably many discontinuity points can be a Riemann integrable function.

At the end of this section we will discuss the mean value theorems. The mean value theorems proved in Section 9.2 for the Newton integral are true also for the Riemann integral if instead of assuming the existence of antiderivatives we assume Riemann integrability. Below, we formulate the general mean value theorem.

Theorem 9.11. *Suppose f is continuous on $[a, b]$ and g is Riemann integrable on $[a, b]$, and that $g(x) \geqslant 0$. Then there exists a number $\xi \in [a, b]$ such that*

$$\int_a^b f(x)g(x)dx = f(\xi) \cdot \int_a^b g(x)dx. \tag{9.38}$$

☐ Let $m = \inf_{x \in [a,b]} f(x)$ and $M = \sup_{x \in [a,b]} f(x)$. Since f is continuous on $[a, b]$, the inequality (9.6) is true:

$$m \int_a^b g(x)dx \leqslant \int_a^b f(x)g(x)dx \leqslant M \int_a^b g(x)dx. \tag{9.39}$$

If $\int_a^b g(x)dx = 0$ in (9.39), then $\int_a^b f(x)g(x)dx = 0$. Therefore, (9.38) is valid for every $\xi \in [a, b]$.

---

[1]H. Lebesgue (1875–1941), French mathematician

If $\int_a^b g(x)dx \neq 0$, then, by virtue of (9.39), we have

$$m \leqslant \frac{\int_a^b f(x)g(x)dx}{\int_a^b g(x)dx} \leqslant M. \tag{9.40}$$

By Weierstrass's theorem (Theorem 3.12), there are points $x_0 \in [a,b]$ and $x_1 \in [a,b]$ such that $f(x_0) = m$ and $f(x_1) = M$. On the other hand, by Theorem 3.10, f assumes every intermediate value in $[m, M]$. Hence, by (9.40), there is a $\xi \in [a,b]$ such that

$$f(\xi) = \frac{\int_a^b f(x)g(x)dx}{\int_a^b g(x)dx}.$$

That is, Formula 9.38 is valid. ∎

Theorem 9.12 (The Mean Value Theorem). *Suppose f is continuous on $[a,b]$. Then, for some point ξ of $[a,b]$,*

$$\int_a^b f(x)dx = f(\xi)(b-a).$$

☐ The proof of the theorem follows immediately from Formula 9.38, provided that $g(x) \equiv 1, x \in [a,b]$. ∎

Example 9.7. If a function is not continuous, then Formula 9.38 may not be correct. For instance, consider the function

$$f(x) = \begin{cases} \dfrac{1}{2}, & \text{if } x \in \left[0, \dfrac{1}{2}\right], \\ \dfrac{3}{4}, & \text{if } x \in \left(\dfrac{1}{2}, 1\right], \end{cases}$$

Observe that $\int_0^1 f(x)dx = \dfrac{5}{8}$. On the other hand, for every point ξ in $[0,1]$, the function f omits the value $\dfrac{5}{8}$. Consequently, there does not exist any ξ in $[0,1]$ such that $\int_0^1 f(x)dx = f(\xi)$.

Example 9.8. Consider the function defined by $f(x) = \begin{cases} x^x, & \text{if } x \in (0,1], \\ 1, & \text{if } x = 0. \end{cases}$
This function is continuous on $[0,1]$. It is easy to see that $f'(x) = x^x(\ln x + 1)$ (Example 4.12). Equating $f'(x) = 0$ to find a critical point gives us $x_0 = e^{-1}$. Thus, $f(e^{-1}) = e^{-1/e}$ is the global minimum value of f on $[0,1]$. It follows that

$$e^{-1/e} \leqslant \int_0^1 x^x dx \leqslant 1.$$

It is easy to compute that $e^{-1/e} \approx 0.692\ldots.$ On the other hand, note that this integral cannot be evaluated in terms of finite combinations of the familiar algebraic and elementary transcendental functions.

9.7   The Fundamental Theorem of Calculus; Term-wise Integration of Series

Consider a Riemann integrable function f on $[a,b]$. Then, for any $x \in [a,b]$, f is Riemann integrable on $[a,x]$ also, so $\int_a^x f(t)dt$ is a well-defined function of x.

Theorem 9.13. *If a function f is Riemann integrable on $[a,b]$, then the function $F(x) = \int_a^x f(t)dt$ is differentiable at every point of continuity of f. Furthermore, $F'(x_0) = f(x_0)$. (If x_0 is either a or b, then $F'(x_0)$ is to be understood as the appropriate one-sided derivative.)*

□   The function f is continuous at x_0, and so for every $\varepsilon > 0$ there is a positive number $\delta = \delta(\varepsilon)$ such that $|x - x_0| < \delta$ implies $f(x_0) - \varepsilon < f(x) < f(x_0) + \varepsilon$. Since $|t - x_0| \leqslant |x - x_0| < \delta$ for all $t \in [x_0, x]$, it follows that $f(x_0) - \varepsilon \leqslant f(t) \leqslant f(x_0) + \varepsilon$ for every $t \in [x_0, x]$. Thus, by Corollary 9.5,

$$f(x_0) - \varepsilon \leqslant \frac{1}{x - x_0} \int_{x_0}^x f(t)dt \leqslant f(x_0) + \varepsilon, \quad |x - x_0| < \delta.$$

But, since $\frac{1}{x - x_0} \int_{x_0}^x f(t)dt = \frac{f(x) - f(x_0)}{x - x_0}$ for all x satisfying $|x - x_0| < \delta$, we have

$$f(x_0) - \varepsilon \leqslant \frac{f(x) - f(x_0)}{x - x_0} \leqslant f(x_0) + \varepsilon,$$

that is, $F'(x_0)$ exists and $F'(x_0) = f(x_0)$.   ■

Corollary 9.7. *Every function f which is continuous on $[a,b]$ has an antiderivative on this interval. The function $F(x) = \int_a^x f(t)dt$ is one of the antiderivatives. Moreover, Theorem 9.12 is valid if f is continuous on the open interval (a,b).*

Corollary 9.8. *Let f be continuous on $[a,b]$. Then $\int_a^b f(x)dx$ exists both in the sense of Riemann and in that of Newton, and the two integrals are equal.*

By Corollary 9.8 and Properties 9.5 and 9.6 of Newton's integral (Section 9.1) we derive the following assertions for Riemann integrable functions.

(1) If the functions f and g and their derivatives are continuous on $[a,b]$, then the formula for integration by parts is true:

$$\int_a^b f(x)g'(x)dx = f(x)g(x)\Big|_a^b - \int_a^b g(x)f'(x)dx.$$

(2) Suppose f is continuous on $[a,b]$ and that φ and φ' are continuous on $[\alpha,\beta]$. Suppose further that $f[\varphi(t)]$ is defined and continuous on $[\alpha,\beta]$, and that $\varphi(\alpha) = a$, $\varphi(\beta) = b$. Then

$$\int_a^b f(x)dx = \int_\alpha^\beta f[\varphi(t)]\varphi'(t)dt.$$

Remark 9.8. The definite integrals with variable upper (lower) bound can be used to define new functions that cannot be expressed in terms of finite combinations of the familiar elementary functions. For instance (Section 8.2), $\int_0^x e^{-t^2} dt$ (Poisson's integral) and

$$\int_0^x \frac{dx}{\sqrt{1 - k^2 \sin^2 x}}$$ (Legendre's elliptic integral of the first kind).

As we proved earlier, the difference of two antiderivatives of a function defined on $[a,b]$ is a constant. Therefore, if $F(x) = \int_a^x f(t)dt$ and $\varphi(x)$ are two such antiderivatives, then their difference $\varphi(x) - F(x) = C$ is a constant, or, $\varphi(x) = \int_a^x f(t)dt + C$. In particular, if $x = a$ and $x = b$, then

$$\int_a^b f(x)dx = \varphi(b) - \varphi(a). \tag{9.41}$$

Theorem 9.14 (The Fundamental Theorem of Calculus). *Let f be a continuous function defined on $[a,b]$. If φ is any antiderivative of f on $[a,b]$, then (9.41) holds.*

☐ Sometimes the difference on the right hand side of (9.41) is denoted by $\varphi(x)\big|_a^b$.

Suppose that the first $n+1$ derivatives of the function f exist for x contained in an ε-neighborhood of the point a. Consider the equality

$$f(x) - f(a) = \int_a^x f'(t)dt.$$

Let $u(t) = f'(t)$ and $v(t) = -(x-t)$, and apply to $\int_a^x f'(t)dt = \int_a^x u(t)dv(t)$ the rule for integration by parts. Then we find that

$$f(x) - f(a) = \int_a^x f'(t)dt = -[f'(t)(x-t)]\big|_a^x + \int_a^x f''(t)(x-t)dt$$

$$= f'(a)(x-a) + \int_a^x f''(t)(x-t)dt.$$

With the repeated use of integration by parts, we can derive the following formula:

$$f(x) - f(a) = f'(a)(x-a) + \frac{1}{2}f''(a)(x-a)^2 + \frac{1}{2}\int_a^x f'''(t)(x-t)^2 dt$$

$$= \cdots = f'(a)(x-a) + \frac{1}{2}f''(a)(x-a)^2 + \cdots + \frac{1}{n!}f^{(n)}(a)(x-a)^n + \frac{1}{n!}\int_a^x f^{(n+1)}(t)(x-t)^n dt$$

$$\sum_{k=1}^n \frac{1}{k!}f^k(a)(x-a)^k + R_{n+1}(x),$$

$$R_{n+1}(x) = \frac{1}{n!} \int_a^x f^{n+1}(t)(x-t)^n dt.$$

Here, $R_{n+1}(x)$ is called the *remainder term of Taylor's series* in integral form. By using Formula 9.38, we can obtain Lagrange's form for the remainder term. Indeed,

$$R_{n+1}(x) = \frac{1}{n!} \int_a^x f^{n+1}(t)(x-t)^n dt = \frac{f^{(n+1)}(\xi)}{n!} \int_a^x (x-a)^n dt$$

$$= -\frac{f^{n+1}(\xi)}{n!} \frac{(x-t)^{n+1}}{n+1} \Big|_a^x = \frac{f^{n+1}(\xi)}{(n+1)!}(x-a)^{n+1}$$

where ξ is a point contained in $[a,x]$.   ■ By applying Theorem 5.11, we can prove the following theorem.

Theorem 9.15. (1) *Suppose that a sequence of functions $\{g_n(x)\}$ converges uniformly on $[a,b]$ to $g(x)$ and that for each n, $g_n(x)$ has an antiderivative on $[a,b]$. Then*

$$\int_{x_0}^x g_n(t)dt \Longrightarrow \int_{x_0}^x g(t)dt, \quad x_0 \in [a,b]. \tag{9.42}$$

(2) *Suppose that the series $\sum\limits_{k=1}^{\infty} u_k(x)$ of functions converges uniformly on $[a,b]$ and for each k, $u_k(x)$ has an antiderivative on $[a,b]$. Then $\sum\limits_{k=1}^{\infty} \int_{x_0}^x u_k(t)dt$ converges uniformly on $[a,b]$ and the following formula for term-wise integration is valid:*

$$\int_{x_0}^x \left\{ \sum_{k=1}^{\infty} u_k(t) \right\} dt = \sum_{k=1}^{\infty} \int_{x_0}^x u_k(t)dt.$$

☐   It is sufficient to prove case (1). Setting

$$f_n(x) = \int_{x_0}^x g_n(t)dtx, \quad x_0 \in [a,b],$$

we have $f_n'(x) = g_n(x)$. It is clear that the sequence $\{f_n(x)\}$ ($f_n(x_0) = 0$) converges and that the sequence $\{f_n'(x)\} = \{g_n(x)\}$ converges uniformly on $[a,b]$ to the function $g(x)$. By Theorem 5.11, the sequence

$$\{f_n(x)\} = \left\{ \int_{x_0}^x g_n(t)dt \right\}$$

converges uniformly on $[a,b]$ to a function $f(x)$. Furthermore,

$$f'(x) = \lim_{n \to \infty} f_n'(x) = \lim_{n \to \infty} g_n(x) = g(x).$$

It follows that

$$f(x) = \int_{x_0}^x g(t)dt + C.$$

Thus,

$$\int_{x_0}^x g_n(t)dt \implies \int_{x_0}^x g(t)dt + C$$

on $[a,b]$. Setting $x = x_0$ in the last relation, we obtain $C = 0$. Consequently, (9.42) is proved. ∎

Example 9.9. Setting $\alpha = -1$ in (5.35), we have

$$\frac{1}{1+x} = 1 - x + x^2 - \cdots + (-1)^n x^n + \cdots . \tag{9.43}$$

This series converges uniformly on $[-a,a]$ $(0 < a < 1)$, because, by Theorem 3.18, $|x^n| \leqslant a^n$ and so the numerical series $\sum_{k=0}^{\infty} a^k$ is convergent. Then, by Theorem 9.15, the series (9.43) is term-wise integrable:

$$\int_0^x \frac{dt}{1+t} = \int_0^x dt - \int_0^x t\,dt + \cdots + (-1)^n \int_0^x t^n dt + \cdots \quad (|x| \leqslant a).$$

Thus,

$$\ln(1+x) = x - \frac{x^2}{2} + \cdots + (-1)^n \frac{x^{n+1}}{n+1} + \cdots . \tag{9.44}$$

Because $a \in (0,1)$, the series (9.44) is valid for all $x \in (-1,1)$.

Example 9.10. Evaluate $\int_0^a e^{-x^2} dx$.

It is known that the integrand e^{-x^2} cannot be integrated in terms of elementary functions. Any attempt to find such an expression will, therefore, inevitably be unsuccessful. In order to evaluate this integral we use the Maclaurin series of e^x (Section 5.6), and get

$$e^{-x^2} = 1 - \frac{x^2}{1!} + \frac{x^4}{2!} - \frac{x^6}{3!} + \cdots + (-1)^n \frac{x^{2n}}{n!} + \cdots .$$

Term-wise integration give us

$$\int_0^a e^{-x^2} dx = \frac{a}{1} - \frac{a^3}{1! \cdot 3} + \frac{a^5}{2! \cdot 5} - \frac{a^7}{3! \cdot 7} + \cdots .$$

It is easy to see that we can approximate the integral with any desired degree of accuracy, for a given a, simply by choosing a sufficiently large number of terms of the series.

Example 9.11. Evaluate $\int_0^{\pi/2} \sqrt{1 - k^2 \sin^2 x}\,dx$ $(k < 1)$.

Using the series (5.35) with $\alpha = \frac{1}{2}$, we have

$$\sqrt{1 - k^2 \sin^2 x} = 1 - \frac{1}{2}k^2 \sin^2 x - \frac{1}{2} \cdot \frac{1}{4}k^4 \sin^4 x - \frac{1}{2} \cdot \frac{1}{4} \cdot \frac{3}{6}k^6 \sin^6 x - \cdots .$$

This series converges for all value of x and is majorized on any interval. Thus,

$$\int_0^x \sqrt{1 - k^2 \sin^2 x}\,dx = x - \frac{1}{2}k^2 \int_0^x \sin^2 x\,dx - \frac{1}{2}\frac{1}{4}k^2 \int_0^x \sin^4 x\,dx - \frac{1}{2}\frac{1}{4}\frac{3}{6} \int_0^x \sin^6 x\,dx - \cdots .$$

Now, because

$$\int_0^{\pi/2} \sin^{2n} x\, dx = \frac{1\cdot 3\cdots(2n-1)}{2\cdot 4\cdots 2n}\,\frac{\pi}{2}$$

at $x = \frac{\pi}{2}$ (see Example 9.1), we have

$$\int_0^{\pi/2}\sqrt{1-k^2\sin^2 x}\,dx = \frac{\pi}{2}\left[1-\left(\frac{1}{2}\right)^2 k^2 - \left(\frac{1\cdot 3}{2\cdot 4}\right)^2\frac{k^4}{3} - \left(\frac{1\cdot 3\cdot 5}{2\cdot 4\cdot 6}\right)^2\frac{k^6}{5} - \cdots\right].$$

Example 9.12. (1) Replacement of x by x^2 in the series (9.43) gives us

By term-wise integration (Theorem 9.15), we can write

$$\frac{1}{1+x^2} = 1 - x^2 + \cdots + (-1)^k x^{2k} + \cdots \quad (|x| < 1).$$

$$\int_0^x \frac{dt}{1+t^2} = \int_0^x dt - \int_0^x t^2\,dt + \cdots + (-1)^k \int_0^x t^{2k}\,dt + \cdots.$$

Therefore,

$$\tan^{-1} x = x - \frac{x^3}{x} + \cdots + (-1)^k \frac{x^{2k+1}}{2k+1} + \cdots \quad (|x| < 1)$$

(2) If we replace x by $-x$ and take $\alpha = -\frac{1}{2}$, we get the series

$$\frac{1}{\sqrt{1-x}} = 1 + \sum_{n=1}^{\infty}\frac{1\cdot 3\cdot 5\cdots(2n-1)}{2\cdot 4\cdot 6\cdots(2n)}\, x^n.$$

Substituting t^2 for x here yields

$$\frac{1}{\sqrt{1-t^2}} = 1 + \sum_{n=1}^{\infty}\frac{1\cdot 3\cdot 5\cdots(2n-1)}{2\cdot 4\cdot 6\cdots(2n)}\, t^{2n}, \quad |t| < 1.$$

Term-wise integration of this series on $[0,x]$ gives

$$\sin^{-1} x = x + \sum_{n=1}^{\infty}\frac{1\cdot 3\cdot 5\cdots(2n-1)}{2\cdot 4\cdot 6\cdots(2n)}\cdot\frac{x^{2n+1}}{2n+1}.$$

9.8   Improper Integrals

1. The Interval of Integration is Unbounded.   We now extend the definition of the integral to include the possibility that the interval of integration has one of the forms $[a,+\infty)$, $(-\infty,a]$, $(-\infty,+\infty)$. Suppose a function f is Riemann integrable on each subinterval $[a,b] \subset [a,+\infty)$.

Definition 9.11. If the limit

$$\lim_{b \to +\infty} \int_a^b f(x)dx \qquad (9.45)$$

exists and is finite, then we say that the improper integral

$$\int_a^{+\infty} f(x)dx \qquad (9.46)$$

converges, and write

$$\int_a^{+\infty} f(x)dx = \lim_{b \to +\infty} \int_a^b f(x)dx; \qquad (9.47)$$

otherwise, we say that it diverges.

If f is nonnegative on $[a, +\infty)$, then the limit (9.45) either exists or is infinite, and in the latter case we say that the improper integral (9.46) diverges to infinity.

Sometimes the integral (9.46) is called an improper integral of the first kind.

It is clear that if $c > a$, then either both integrals (9.46) and $\int_c^{+\infty} f(x)dx$ converge or both diverge. Indeed, since

$$\int_a^b f(x)dx = \int_a^c f(x)dx + \int_c^b f(x)dx,$$

either both limits (9.45) and $\lim_{b \to +\infty} \int_c^b f(x)dx$ exist, or both limits do not exist.

If there is an antiderivative F of f on $[a, +\infty)$, then

$$\int_a^{+\infty} f(x)dx = \lim_{b \to +\infty} F(b) - F(a).$$

The improper integral $\int_{-\infty}^b f(x)dx$ is defined similarly, and

$$\int_{-\infty}^{+\infty} f(x)dx = \int_{-\infty}^d f(x)dx + \int_d^{+\infty} f(x)dx$$

for any convenient choice of d, provided that both improper integrals on the right converge.
Note that $\int_{-\infty}^{+\infty} f(x)dx$ is not necessarily equal to $\lim_{A \to \infty} \int_{-A}^A f(x)dx$.

Example 9.13. Investigate the improper integral $\int_0^{+\infty} \frac{dx}{1+x^2}$.

Because, for any $b > 0$,

$$F(b) = \int_0^b \frac{dx}{1+x^2} = \tan^{-1} x \Big|_0^b = \tan^{-1} b,$$

we have

$$\int_0^{+\infty} \frac{dx}{1+x^2} = \lim_{b \to +\infty} F(b) - F(0) = \frac{\pi}{2}.$$

Thus, the integral converges.

Example 9.14. Investigate the improper integral $\int_1^{+\infty} \frac{dx}{x^\alpha}$ as the parameter α varies.

If $\alpha \neq 1$, then clearly

$$\int_1^b \frac{dx}{x^\alpha} = \frac{1}{1-\alpha}(b^{1-\alpha}-1),$$

and

$$\int_1^{+\infty} \frac{dx}{x^\alpha} = \lim_{b \to +\infty} \frac{1}{1-\alpha}(b^{1-\alpha}-1).$$

If $\alpha = 1$, then

$$\int_1^b \frac{dx}{x^\alpha} = \ln b - \ln 1 = \ln b.$$

Finally, we can write

$$\int_1^{+\infty} \frac{dx}{x^\alpha} = \begin{cases} \dfrac{1}{\alpha-1}, & \alpha > 1, \\ \infty, & \alpha \leqslant 1. \end{cases}$$

Thus the improper integral converges if $\alpha > 1$, and diverges if $\alpha \leqslant 1$.

Theorem 9.16 (Cauchy's Criterion). *For the convergence of the improper integral* $\int_a^{+\infty} f(x)dx$, *it is necessary and sufficient that for every* $\varepsilon > 0$ *there is a number* $B \geqslant a$ *such that*

$$\left| \int_{b'}^{b''} f(x)dx \right| < \varepsilon$$

for all $b', b'' > B$.

☐ The convergence of $\int_a^{+\infty} f(x)dx$ is equivalent to the existence of the limit of

$$F(b) = \int_a^b f(x)dx \qquad (9.48)$$

as $b \to +\infty$. By Theorem 3.3 (Cauchy's criterion) for the existence of $\lim\limits_{b \to +\infty} F(b)$, it is necessary and sufficient that for any $\varepsilon > 0$ there exists a $B \geqslant a$ such that $|F(b') - F(b'')| < \varepsilon$ for all $b', b'' > B$. Taking into account (9.48) in the inequality $|F(b') - F(b'')| < \varepsilon$, we have the required inequality of the theorem.   ∎

Suppose that the functions f and g are defined on $[a, +\infty)$ and are Riemann integrable on every interval $[a, b] \subset [a, +\infty)$.

Theorem 9.17 (Comparison Test). *Suppose that* $|f(x)| \leqslant g(x)$ *for all* $x \in [a, +\infty)$. *Then the integral* $\int_a^{+\infty} f(x)dx$ *converges if* $\int_a^{+\infty} g(x)dx$ *converges.*

☐ By Theorem 9.16, for every $\varepsilon > 0$ there can be found a $B \geq a$ such that

$$\left| \int_{b'}^{b''} g(x)dx \right| < \varepsilon$$

for all $b', b'' > B$.

On the other hand, because $|f(x)| \leq g(x)$, we have the inequalities

$$\left| \int_{b'}^{b''} f(x)dx \right| \leq \left| \int_{b'}^{b''} |f(x)|dx \right| \leq \left| \int_{b'}^{b} g(x)dx \right| < \varepsilon.$$

By Theorem 9.16, the improper integral $\int_a^{+\infty} f(x)dx$ converges.   ■

Example 9.15. Determine whether the improper integral $\int_1^{+\infty} \dfrac{dx}{x^2(1+e^x)}$ converges or diverges.

Because $\dfrac{1}{x^2(1+e^x)} < \dfrac{1}{x^2}$ for all $x \geq 1$, the improper integral $\int_1^{+\infty} \dfrac{dx}{x^2}$ converges (Example 9.14). Then, by Theorem 9.17, the given integral converges.

Definition 9.12. If the integral $\int_a^{+\infty} |f(x)|dx$ converges, then we say that the improper integral $\int_a^{+\infty} f(x)dx$ converges absolutely.

If $\int_a^{+\infty} f(x)dx$ converges, but the integral $\int_a^{+\infty} |f(x)|dx$ diverges, then $\int_a^{+\infty} f(x)dx$ is said to be *conditionally convergent*.

If we apply Theorem 9.17 to the case $f(x) \leq |f(x)|$, then we see that the absolute convergence of an improper integral implies its convergence.

Remark 9.9. If $f(x) \geq 0$, then for the convergence of the improper integral $\int_a^{+\infty} f(x)dx$, it is necessary and sufficient that

$$\int_a^b f(x)dx \leq C \qquad\qquad (9.49)$$

for some $C > 0$ and every $b \geq a$.

Indeed, if $f(x) \geq 0$, then the function (9.48) is nondecreasing for all $b \geq a$. Hence for the convergence of the improper integral $\int_a^{+\infty} f(x)dx$, it is necessary and sufficient that there exists a constant $C > 0$ such that $F(b) \leq C$. Thus, for every $b \geq a$, (9.49) holds.

Theorem 9.18. *Suppose that $f(x) \geq 0$ $(x \geq a)$, and that $\{a_n\}$ $(a_0 = a)$, $n = 0, 1, 2, \ldots$ is a increasing sequence having the limit $+\infty$. Then the improper integral $\int_a^{+\infty} f(x)dx$ converges if and only if the series*

$$\sum_{n=1}^{\infty} \left\{ \int_{a_{n-1}}^{a_n} f(x)dx \right\} \qquad\qquad (9.50)$$

converges.

☐ Let

$$S_n = \int_a^{a_1} f(x)dx + \int_{a_1}^{a_2} f(x)dx + \cdots + \int_{a_{n-1}}^{a_n} f(x)dx = \int_a^{a_n} f(x)dx$$

be the n-th partial sum of the series (9.50). Because $\{a_n\}$ is increasing and $f(x) \geqslant 0$ for $x \geqslant a$, the sequence of n-th partial sums $\{S_n\}$ is nondecreasing. On the other hand, since $a_n \to +\infty$, for each $b \geqslant a$ there is an integer n such that $a_n \leqslant b \leqslant a_{n+1}$. Therefore,

$$S_n = \int_a^{a_n} f(x)dx \leqslant \int_a^b f(x)dx \leqslant \int_a^{a_{n+1}} f(x)dx = S_{n+1}. \qquad (9.51)$$

Now, if the integral $\int_a^{+\infty} f(x)dx$ converges, then the inequality (9.49) holds. Thus, (9.49) and (9.51) imply $S_n \leqslant C \ (n \geqslant 1)$. Moreover, $\{S_n\}$ is nondecreasing and so converges. Then (9.50) is convergent.

Conversely, if the series (9.50) converges, then its n-th partial sum is bounded above $S_n \leqslant C \ (n \geqslant 1)$. Then it follows from (9.51) that (9.49) is satisfied. Consequently, the improper integral $\int_a^{+\infty} f(x)dx$ converges.   ∎

Theorem 9.19 (The Integral Test). *Suppose that $f(x)$ is a nondecreasing positive-valued function for $x \geqslant n_0$ (where n_0 is an integer). Then the series $\sum\limits_{n=n_0}^{\infty} f(n)$ and the improper integral $\int_{n_0}^{+\infty} f(x)dx$ either both converge or both diverge.*

☐ By hypothesis, $f(n-1) \geqslant f(x) \geqslant f(n)$, $n \geqslant x \geqslant n-1$.
By integrating over the interval $[n-1, n]$, we obtain

$$f(n-1) \geqslant \int_{n-1}^n f(x)dx \geqslant f(n).$$

Then it follows from Theorem 2.19 (Comparison Test) that the series $\sum\limits_{n=0}^{\infty} f(n)$ and $\sum\limits_{n=n_0+1}^{\infty} \left\{ \int_{n-1}^n f(x)dx \right\}$ either both converge or both diverge. On the other hand, by Theorem 9.18, the latter series and the improper integral $\int_{n_0}^{+\infty} f(x)dx$ either both converge or both diverge. This ends the proof of the theorem.   ∎

Example 9.16. Investigate whether the hyperharmonic (or α-series) series $\dfrac{1}{1^\alpha} + \dfrac{1}{2^\alpha} + \cdots + \dfrac{1}{n^\alpha} + \cdots$ converges or diverges.

If $\alpha > 0$, then the function $f(x) = \dfrac{1}{x^\alpha}$, for $x \geqslant 1$, satisfies the conditions of the integral test. As we saw in Example 9.14,

$$\int_1^b \frac{dx}{x^\alpha} = \begin{cases} \left. \dfrac{1}{1-\alpha} x^{1-\alpha} \right|_1^b = \dfrac{1}{1-\alpha}\left(b^{1-\alpha} - 1\right), & \text{if } \alpha \neq 1, \\[2mm] \left. \ln x \right|_1^b = \ln b, & \text{if } \alpha = 1, \end{cases}$$

and the improper integral $\displaystyle\int_1^{+\infty} \frac{dx}{x^\alpha} = \lim_{b \to +\infty} \int_1^b \frac{dx}{x^\alpha}$ converges if $\alpha > 1$, and diverges if $\alpha \leqslant 1$. Then, by Theorem 9.19, it follows that the series converges if $\alpha > 1$. But it is not hard to see that the tests of Cauchy and D'Alembert are inconclusive in this case. Indeed, denoting $u_n = \dfrac{1}{n^\alpha}$, we have

$$\lim_{n \to \infty} \frac{u_n + 1}{u_n} = \lim_{n \to \infty} \left(\frac{n}{n+1}\right)^\alpha = 1 \quad \text{(D'Alembert's test)}$$

$$\lim_{n \to \infty} \sqrt[n]{u_n} = \lim_{n \to \infty} \sqrt[n]{\frac{1}{n^\alpha}} = \lim_{n \to \infty} \left(\sqrt[n]{\frac{1}{n}}\right)^\alpha = 1^\alpha = 1 \quad \text{(Cauchy's test)}$$

(recall that the convergence of the hyperharmonic series was investigated in Theorem 2.29, too).

Theorem 9.20 (The Dirichlet Test). *Suppose that the functions f and g satisfy the following conditions.*

(1) *The function f is Riemann integrable on any interval $[a,b] \subset [a,+\infty)$, and there exists a constant $C > 0$ such that*

$$\left| \int_a^b f(x)dx \right| \leqslant C \quad (b > a).$$

(2) *The function g is continuously differentiable on $[a,+\infty)$, and $g(x) \to 0$ monotonically as $x \to +\infty$.*

Then the integral

$$\int_a^{+\infty} f(x)g(x)dx \tag{9.52}$$

converges.

☐ By hypothesis (1), the function $F(x) = \int_a^x f(t)dt$ is bounded, i.e., $|F(x)| \leqslant C$ for all $x \geqslant a$. Moreover, by hypothesis (2), $g'(x) \leqslant 0$. Then integration by parts gives us

$$\int_a^b |F(x)g'(x)|dx \leqslant C \int_a^b |g'(x)|dx = -C \int_a^b g'(x)dx = C(g(a) - g(b)) \leqslant Cg(a).$$

Thus, by Remark 9.9, the improper integral $\int_0^{+\infty} F(x)g'(x)dx$ is absolutely convergent and therefore converges. Furthermore, since $|F(b)| \leqslant C$ and $g(b) \to 0$ (as $x \to +\infty$), we have $\lim\limits_{b\to\infty} F(b)g(b) = 0$. On the other hand, as a result of integration by parts we conclude that

$$\int_a^b f(x)g(x)dx = F(b)g(b) - F(a)g(a) - \int_a^b F(x)g'(x)dx.$$

We have already proved that all expressions on the right hand side have a finite limit as $b \to +\infty$. Consequently, the improper integral (9.52) converges. ∎

Example 9.17. Investigate the improper integral $\int_0^\infty \sin x^2 dx$.

This integral is often applied to physical problems in optics.

Because $\sin x^2 \equiv x \sin x^2 \left(\dfrac{1}{x}\right)$, we denote $f(x) = x \sin x^2$ and $g(x) = \dfrac{1}{x}$. Obviously, the substitution $x^2 = t$ yields

$$\int_0^b x \sin x^2 dx = \frac{1}{2} \int_0^b \sin x^2 dx^2 = \frac{1}{2} \int_0^{b^2} \sin t \, dt = \frac{1 - \cos b^2}{2}.$$

Then it follows that

$$\left| \int_a^b x \sin x^2 dx \right| \leqslant 1 \quad \text{for all } b > 0.$$

Therefore, the functions $f(x) = x \sin x^2$ and $g(x) = \dfrac{1}{x}$ satisfy the conditions of Theorem 9.20. Hence, the improper integral $\int_0^{+\infty} f(x)g(x)dx = \int_0^{+\infty} \sin x^2 dx$ converges.

Example 9.18. Investigate the improper integral $\int_0^{+\infty} \dfrac{\sin x}{x} dx$ (where the definition of the function $f(x) = \dfrac{\sin x}{x}$ is extended to the point $x = 0$ by declaring that $f(0) = 1$).

First, we consider the improper integral

$$\int_\pi^{+\infty} \frac{\sin x}{x} dx. \tag{9.53}$$

Clearly,

$$\left| \int_\pi^b \sin x \, dx \right| = 1 + \cos b \leqslant 2.$$

On the other hand, $g(x) = \dfrac{1}{x}$ $(x \geqslant \pi)$ satisfies hypothesis (2) of Theorem 9.20. Thus, by Dirichlet's test, the improper integral $\int_\pi^{+\infty} \dfrac{\sin x}{x} dx$ converges. This convergence is only conditional. In fact,

$$S_n = \int_{n\pi}^{(n+1)\pi} \left| \frac{\sin x}{x} \right| dx \geqslant \frac{1}{\pi n} \left| \int_{n\pi}^{(n+1)\pi} \sin x \, dx \right| = \frac{2}{\pi n}$$

and the numerical series $\sum\limits_{n=1}^{\infty} \dfrac{2}{\pi n}$ diverges. It follows that the series $\sum\limits_{n=1}^{\infty} S_n$ diverges, too.

Then, by Theorem 9.18, the series $\displaystyle\int_{\pi}^{+\infty} \left| \dfrac{\sin x}{x} \right| dx$ diverges. Consequently, the improper integral $\displaystyle\int_{\pi}^{+\infty} \dfrac{\sin x}{x} dx$ converges conditionally.

Let f be a function defined on the half-open interval $[a,b)$, and suppose that $f(x) \to \infty$ as $x \to b-0$. Moreover, assume that f is Riemann integrable on every closed interval $[a, \alpha] \subset [a, b)$.

Definition 9.13. If $\displaystyle\lim_{\alpha \to b-0} \int_a^\alpha f(x)dx$ exists and is finite, then we say that the improper integral of the second kind, defined as

$$\int_a^b f(x)dx = \lim_{\alpha \to b-0} \int_a^\alpha f(x)dx, \qquad (9.54)$$

converges; otherwise we say that it diverges.

Definition 9.13 can be extended to the case where the function is unbounded at finitely many points.

Sometimes it is convenient to rewrite (9.54) as follows:

$$\int_a^b f(x)dx = \lim_{\varepsilon \to 0+0} \int_a^{b-\varepsilon} f(x)dx.$$

The improper integral of the second kind is defined similarly: for a function f which is Riemann integrable on any $[\alpha, b] \subset (a, b]$ and unbounded on $(a, b]$ ($f(x) \to \infty$ as $x \to a+0$):

$$\int_a^b f(x)dx = \lim_{\alpha \to a+0} \int_\alpha^b f(x)dx.$$

If both improper integrals $\displaystyle\int_a^c f(x)dx$ and $\displaystyle\int_c^b f(x)dx$ $(a < c < b)$ converge and one or both one-sided limits at c are infinite, then we define

$$\int_a^b f(x)dx = \int_a^c f(x)dx + \int_c^b f(x)dx.$$

Remark 9.10. If $a < \eta < b$, then, by Definition 9.13, the improper integrals $\displaystyle\int_a^b f(x)dx$ and $\displaystyle\int_\eta^b f(x)dx$ either both converge or both diverge. Indeed, since

$$\int_a^b f(x)dx = \int_a^\eta f(x)dx + \int_\eta^b f(x)dx,$$

the limits

$$\lim_{\alpha \to b-0} \int_a^\alpha f(x)dx \quad \text{and} \quad \lim_{\alpha \to b-0} \int_\eta^\alpha f(x)dx$$

either both exist or both fail to exist.

Definition 9.14. If the improper integral

$$\int_a^b |f(x)|dx \tag{9.55}$$

converges, then we say that the *integral* (9.54) *converges absolutely.*

An improper integral that converges but does not converge absolutely is said to be *conditionally convergent.*

The theorems formulated below are proved similarly to previous theorems, and so their proofs are omitted.

Theorem 9.21 (The Cauchy Criterion). *For the convergence of the improper integral* $\lim\limits_{\alpha \to b-0} \int_a^\alpha f(x)dx$, *it is necessary and sufficient that for every* $\varepsilon > 0$ *there exists a* $\delta = \delta(\varepsilon) > 0$ *such that for all* α', $\alpha'' \in [a,b)$ *satisfying* $b - \delta < \alpha' < b$ *and* $b - \delta < \alpha'' < b$,

$$\left| \int_{\alpha'}^{\alpha''} f(x)dx \right| < \varepsilon.$$

Theorem 9.22 (The Comparison Test). *If* $|f(x)| \leqslant g(x)$ *for all x in the half-interval* $[a,b)$, *and if the improper integral* $\int_a^b g(x)dx$ *converges, then* $\int_a^b f(x)dx$ *is convergent.*

Theorem 9.23. *If* $f(x) \geqslant 0$ *for any x in* $[a,b)$, *then the improper integral* $\lim\limits_{\alpha \to b-0} \int_a^\alpha f(x)dx$ *converges if and only if there exists a constant* $C > 0$ *such that*

$$\int_a^\alpha f(x)dx \leqslant C$$

for all $\alpha \in (a,b)$.

Example 9.19. Evaluate $\int_0^1 \frac{dx}{\sqrt{1-x}}$.

The integrand becomes infinite as $x \to 1 - 0$, but

$$\int_0^1 \frac{dx}{\sqrt{1-x}} = \lim_{\alpha \to 1-0} \int_0^\alpha \frac{dx}{\sqrt{1-x}} = -\lim_{\alpha \to 1-0} 2\sqrt{1-x}\,\Big|_0^\alpha = -\lim_{\alpha \to 1-0} 2\left(\sqrt{1-\alpha} - 1\right) = 2,$$

so the integral converges.

Example 9.20. Evaluate the improper integral $\int_{-1}^1 \frac{dx}{x^2}$.

The point $x = 0$ is a point of discontinuity of the second type for the integrand $\frac{1}{x^2}$. By definition,

$$\int_{-1}^1 \frac{dx}{x^2} = \lim_{\alpha_1 \to 0-0} \int_{-1}^{\alpha_1} \frac{dx}{x^2} + \lim_{\alpha_2 \to 0+0} \int_{\alpha_2}^1 \frac{dx}{x^2},$$

but

$$\lim_{\alpha_1 \to 0-0} \int_{-1}^{\alpha_1} \frac{dx}{x^2} = -\lim_{\alpha_1 \to 0-0} \frac{1}{x} \Big|_{-1}^{\alpha_1} = \infty \text{ and}$$

$$\lim_{\alpha_2 \to 0+0} \int_{\alpha_2}^{1} \frac{dx}{x^2} = -\lim_{\alpha_2 \to 0+0} \left(1 - \frac{1}{\alpha_2}\right) = \infty.$$

Thus the integral in question diverges.

Note that if we blindly applied the fundamental theorem of calculus, we would get

$$\int_{-1}^{1} \frac{dx}{x^2} = -\frac{1}{x} \Big|_{-1}^{1} = -2.$$

This answer is obviously incorrect and emphasizes that we cannot ignore the hypotheses of continuity and boundedness of the Fundamental Theorem of Calculus. ■

Suppose that a function f defined on the interval $(-\infty, +\infty)$ is Riemann integrable on every interval $[a, b] \subset (-\infty, +\infty)$. If the limit

$$\lim_{A \to +\infty} \int_{-A}^{A} f(x)dx$$

exists, we say that the value of the limit is the Cauchy principal value (p.v.) of this improper integral of the first kind on the whole real line, and denote it by

$$\text{p.v.} \int_{-\infty}^{+\infty} f(x)dx.$$

In such a case, we say that the Riemann integral of f exists in the sense of the Cauchy principal value. Thus,

$$\text{p.v.} \int_{-\infty}^{+\infty} f(x)dx = \lim_{A \to +\infty} \int_{-A}^{A} f(x)dx.$$

Example 9.21. Investigate the improper integral $\int_{-\infty}^{+\infty} \frac{1+x}{1+x^2} dx$.

It is not hard to see that, as an improper integral of the first kind, this integral diverges. But at the same time the integral of f exists in the Cauchy principal value sense. Indeed,

$$\text{p.v.} \int_{-\infty}^{+\infty} \frac{1+x}{1+x^2} dx = \lim_{A \to +\infty} \left[\tan^{-1} x + \frac{1}{2} \ln(1+x^2) \right] \Big|_{-A}^{A} = \lim_{A \to +\infty} 2\tan^{-1} A = 2 \cdot \frac{\pi}{2} = \pi.$$

Suppose a function f, defined on $[a, b]$ with the possible exception of the point c, $a < c < b$, is integrable on every interval $[a, \eta] \subset [a, c)$ and $[\xi, b] \subset (c, b]$. If the limit

$$\lim_{\varepsilon \to 0+0} \left[\int_{a}^{c-\varepsilon} f(x)dx + \int_{c+\varepsilon}^{b} f(x)dx \right]$$

exists, we say that the integral of f exists in the Cauchy principal value sense and denote this limit by

$$\text{p.v.} \int_a^b f(x)dx.$$

Thus,

$$\text{p.v.} \int_a^b f(x)dx = \lim_{\varepsilon \to 0+0} \left[\int_a^{c-\varepsilon} f(x)dx + \int_{c+\varepsilon}^b f(x)dx \right].$$

Example 9.22. Investigate the improper integral $\int_a^b \dfrac{dx}{x-c}$ $(a < c < b)$. It is easy to verify that this integral is divergent. We show that this integral exists in the Cauchy principal value sense. Indeed,

$$\text{p.v.} \int_a^b \frac{dx}{x-c} = \lim_{\varepsilon \to 0+0} \left[\int_a^{c-\varepsilon} \frac{dx}{x-c} + \int_{c+\varepsilon}^b \frac{dx}{x-c} \right] = \ln \frac{b-c}{c-a}.$$

9.9 Numerical Integration

Note that the Fundamental Theorem of Calculus can be used to evaluate the integral if a convenient formula for the antiderivative can be found. On the other hand, there are many simple functions whose antiderivatives are not elementary functions. For instance, it is known that the elementary function e^{-x^2} has no elementary antiderivative.

1. Rectangular Approximation. Suppose the function f is Riemann integrable on $[a,b]$, and that $P = \{x_0, x_1, \ldots, x_n\}$ $(x_0 = a, x_n = b)$ is a partition of $[a,b]$. For each $i = 0, 1, \ldots, n-1$, we choose points $\xi_i \in [x_i, x_{i+1}]$ and form the Riemann sum

$$\sum_{i=0}^{n-1} f(\xi_i)(x_{i+1} - x_i).$$

Because f is Riemann integrable, we can write

$$\lim_{\mu(P) \to 0} \sum_{i=0}^{n-1} f(\xi_i)(x_{i+1} - x_i) = \int_a^b f(x)dx.$$

Thus for any partition P of sufficiently small mesh $\mu(P)$,

$$\int_a^b f(x)dx \approx \sum_{i=0}^{n-1} f(\xi_i)(x_{i+1} - x_i). \tag{9.56}$$

The right hand side of (9.56) is the sum of the areas of n rectangles with base-length $x_{i+1} - x_i$ and height $f(\xi_i)$, $i = 0, 1, \ldots, n-1$, and so the approximation is called the *rectangular approximation*. In particular, if $\xi_i = x_i$ or $\xi_i = x_{i+1}$, we have the left end-point approximation or the right end-point approximation, respectively. If $\xi_i = \dfrac{x_i + x_{i+1}}{2}$, we have the midpoint approximation.

Suppose that $f'(x)$ is continuous on $[a,b]$ and $M = \max\limits_{x\in[a,b]} |f'(x)|$.

Denote by R_n the error made in replacing the Riemann integral with the Riemann sum of the areas of the n rectangles:

$$R_n = \int_a^b f(x)dx - \sum_{i=0}^{n-1} f(\xi_i)(x_{i+1} - x_i). \tag{9.57}$$

In order to estimate R_n, we use Theorem 5.4:

$$f(x) - f(\xi_i) = f'(p_i)(x - \xi_i), \quad x_i < p_i < x_{i+1}.$$

Hence, for $x \in [x_i, x_{i+1}]$, the inequality $|f(x) - f(\xi_i)| \leqslant \max\limits_{x\in[a,b]} |f'(x)||x - \xi_i| \leqslant M\Delta x_i$ holds, and so

$$\left| \int_{x_i}^{x_{i+1}} f(x)dx - f(\xi_i)(x_{i+1} - x_i) \right| = \left| \int_{x_i}^{x_{i+1}} [f(x) - f(\xi_i)]dx \right| \leqslant \int_{x_i}^{x_{i+1}} M\Delta x_i dx = M\Delta x_i^2.$$

Then it follows from (9.57) that

$$|R_n| \leqslant \sum_{i=0}^{n-1} \left| \int_{x_i}^{x_{i+1}} f(x)dx - f(\xi_i)(x_{i+1} - x_i) \right| \leqslant \sum_{i=0}^{n-1} M\Delta x_i^2 \leqslant M\mu(P) \sum_{i=0}^{n-1} \Delta x_i = M(b-a)\mu(P).$$

Finally,

$$|R_n| \leqslant M(b-a)\mu(P), \quad M = \max\limits_{x\in[a,b]} |f'(x)|.$$

Obviously, $R_n \to 0$ as $\mu(P) \to 0$.

Now, suppose that a continuous second derivative $f''(x)$ of $f(x)$ exists on $[a,b]$. Put $M = \max\limits_{x\in[a,b]} |f''(x)|$ and choose $\xi_i = \dfrac{x_i + x_{i+1}}{2}$. Then (9.57) implies that

$$\int_a^b f(x)dx \approx \sum_{i=0}^{n-1} f\left(\frac{x_i + x_{i+1}}{2} \right) (x_{i+1} - x_i).$$

We wish to estimate the error,

$$R_n = \int_a^b f(x)dx - \sum_{i=0}^{n-1} f\left(\frac{x_i + x_{i+1}}{2} \right) (x_{i+1} - x_i). \tag{9.58}$$

To do this, we first form the following difference:

$$\int_{x_i}^{x_{i+1}} f(x)dx - f(\xi_i)\Delta x_i = \int_{x_i}^{x_{i+1}} [f(x) - f(\xi_i)]dx. \tag{9.59}$$

By Taylor's formula,

$$f(x) = f(\xi_i) + f'(\xi_i)(x - \xi_i) + \frac{1}{2} f''(\eta_i)(x - \xi_i)^2, \quad x_i < \eta_i < x_{i+1}. \tag{9.60}$$

Then the substitution (9.60) in (9.59) gives us

$$\int_{x_i}^{x_{i+1}} f(x)dx - f(\xi_i)\Delta x_i = \frac{1}{2} \int_{x_i}^{x_{i+1}} f''(\eta_i)(x - \xi_i)^2 dx.$$

Thus, by (9.58),

$$|R_n| \leqslant \sum_{i=0}^{n-1} \left| \int_{x_i}^{x_{i+1}} f(x)dx - f(\xi_i)\Delta x_i \right| \leqslant \frac{M}{2} \sum_{i=0}^{n-1} \int_{x_i}^{x_{i+1}} (x - \xi_i)^2 dx$$

$$= \frac{M}{24} \sum_{i=0}^{n-1} \Delta x_i^3 \leqslant \frac{M[\mu(P)]^2}{24} \sum_{i=0}^{n-1} \Delta x_i = \frac{M}{24}(b-a)[\mu(P)]^2. \qquad (9.61)$$

Consequently,

$$|R_n| \leqslant \frac{M}{24}(b-a)[\mu(P)]^2.$$

2. The Trapezoidal Approximation. For numerical calculation of Riemann integrals, we can use the following formula:

$$\int_a^b f(x) \approx \sum_{i=0}^{n-1} \frac{f(x_i) + f(x_{i+1})}{2}(x_{i+1} - x_i).$$

It is easy to see that each term here is the area of the trapezoid with altitude $x_{i+1} - x_i$ and lengths $f(x_1), f(x_{i+1})$, and so this is called the trapezoidal approximation.

By using (9.60), we obtain the following inequality:

$$\left| f\left(\frac{x_i + x_{i+1}}{2}\right) - \frac{f(x_i) + f(x_{i+1})}{2} \right| \leqslant M \frac{\Delta x_i^2}{8}, \qquad M = \max_{x \in [a,b]} f''(x).$$

Then,

$$|R_n| = \left| \sum_{i=0}^{n-1} \left[\int_{x_i}^{x_{i+1}} f(x)dx - \frac{f(x_i) + f(x_{i+1})}{2}(x_{i+1} - x_i) \right] \right|$$

$$\leqslant \sum_{i=0}^{n-1} \left| \int_{x_i}^{x_{i+1}} f(x)dx - f\left(\frac{x_i + x_{i+1}}{2}\right)(x_{i+1} - x_i) \right|$$

$$+ \sum_{i=0}^{n-1} \left| f\left(\frac{x_i + x_{i+1}}{2}\right) - \frac{f(x_i) + f(x_{i+1})}{2} \right| |x_{i+1} - x_i|$$

$$\leqslant \sum_{i=0}^{n-1} \left(\frac{M}{24}\Delta x_i^3 + \frac{M}{8}\Delta x_i^3 \right) = \frac{M}{6}[\mu(P)]^2 \sum_{i=0}^{n-1} \Delta x_i = \frac{M}{6}(b-a)[\mu(P)]^2.$$

In the latter inequality, we have used the inequality (9.61):

$$\sum_{i=0}^{n-1} \left| \int_{x_i}^{x_{i+1}} f(x)dx - f(\xi_i)\Delta x_i \right| \leqslant \frac{M}{24} \sum_{i=0}^{n-1} \Delta x_i^3.$$

Thus,

$$|R_n| \leqslant \frac{M}{6}(b-a)[\mu(P)]^2.$$

Remark 9.11. In general, in numerical integration using the rectangular or trapezoidal approximations, the selection of points within the given partition P has the form $x_i = a + ih$ $(i = 0, 1, \ldots, n)$, where $h = \dfrac{b-a}{n}$.

3. Simpson's Approximation. The rectangular and trapezoidal approximations used rectangles and trapezoids. Simpson's Rule (as Simpson's approximation is often called) uses parabolic approximations to the curve $y = f(x)$. Given $n = 2m$, subdivide the interval $[a, b]$ into $n = 2m$ equal subintervals. We consider a parabolic function

$$y = ax^2 + Bx + C$$

on $[x_0, x_2]$. We choose the coefficients A, B, and C so that $Ax^2 + Bx + C$ agrees with $f(x)$ at the three points x_0, x_1, and x_2. This can be done by solving the three equations

$$Ax_0^2 + Bx_0 + C = f(x_0),$$
$$Ax_1^2 + Bx_1 + C = f(x_1),$$
$$Ax_2^2 + Bx_2 + C = f(x_2).$$

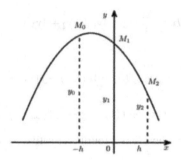

Fig. 9.3   The area under the parabola from $-h$ to h.

We will calculate the area of the region under the parabola $y = ax^2 + Bx + C$ from $-h$ to h (Figure 9.3). This area is

$$S = \frac{h}{3}(y_0 + 4y_1 + y_2) \tag{9.62}$$

where y_0, y_1, y_2 are the ordinates of the points M_0, m_1, m_2, respectively:

$$y_0 = f(-h), \quad y_1 = f(0), \quad y_2 = f(h).$$

Then the coefficients A, B, C can be found by solving the equations

$$y_0 = Ah^2 - Bh + C,$$
$$y_1 = C, \tag{9.63}$$
$$y_2 = Ah^2 + Bh + C.$$

The area of the region under the graph of parabola $y = Ax^2 + Bx + C$ from $-h$ to h is

$$S = \int_{-h}^{h} (Ax^2 + Bx + C)\,dx = \frac{h}{3}(2Ah^2 + 6C).$$

On the other hand, it follows from (9.63) that

$$y_0 + 4y_1 + y_2 = 2Ah^2 + 6C.$$

Finally, from the last two relations we have the formula (9.62). Thus, by using (9.62), we can write $(h = \Delta x)$:

$$\int_{x_{2k-2}}^{x_{2k}} f(x)dx \approx \frac{\Delta x}{3}(y_{2k-2} + 4y_{2k-1} + y_{2k}) \quad (k = 1, 2, \ldots, m).$$

We now approximate $\int_a^b f(x)dx$ by replacing $f(x)$ with the parabola of the form $A_i x^2 + B_i x + C_i$ on the interval $[x_{2i-2}, x_{2i}]$, $k = 1, 2, \ldots, m$. This gives

$$\int_a^b f(x)dx = \sum_{k=1}^{m} \int_{x_{2k-2}}^{x_{2k}} f(x)dx \approx \frac{\Delta x}{3}(y_0 + 4y_1 + 2y_2 + 4y_3 + \cdots + 2y_{2m-2} + 4y_{2m-1} + y_{2m}),$$

or

$$\int_a^b f(x)dx \approx \frac{\Delta x}{3}\left[f(a) + f(b) + 4\big(f(x_1) + f(x_3) + \cdots + f(x_{2m-1})\big)\right.$$

$$\left. + 2\big(f(x_2) + f(x_4) + \cdots + f(x_{2m-2})\big)\right].$$

It can be shown that if a function f has a fourth derivative on $[a, b]$, then Simpson's error estimate is

$$|R_n| \leqslant \frac{M}{180}(b-a)(\Delta x)^4, \quad M = \max_{x \in [a,b]} \left|f^{(4)}(x)\right|.$$

Example 9.23. Use numerical integration to show that $\ln 2 = \int_1^2 \frac{dx}{x}$.

We subdivide the interval $[1, 2]$ into 10 equal subintervals. Since $\Delta x = 0.1$ and $f(x) = \frac{1}{x}$ we can form the table

x	1.0	1.1	1.2	1.3	1.4	1.5
$f(x)$	1.00000	0.90909	0.83333	0.76923	0.71429	0.66667
x	1.6	1.7	1.8	1.9	2.0	
$f(x)$	0.62500	0.58824	0.55556	0.52632	0.50000	

(i) By the rectangular approximation,

$$\int_1^2 \frac{dx}{x} \approx 0.1(y_0 + y_1 + \cdots + y_9) = (0.1)(7.18773) = 0.71877.$$

(ii) By the trapezoidal approximation,

$$\int_1^2 \frac{dx}{x} \approx 0.1 \left(\frac{1+0.5}{2} + 6.18773 \right) = 0.69377.$$

(iii) By Simpson's approximation,

$$\int_1^2 \frac{dx}{x} \approx \frac{0.1}{3} [f(1) + f(2) + 2(f(x_2) + f(x_4) + f(x_6) + f(x_8))$$

$$+ 4(f(x_1) + f(x_3) + f(x_5) + f(x_7) + f(x_9))]$$

$$= \frac{0.1}{3} [1 + 0.5 + 2(2.72818) + 4(3.45955)] = 0.69315.$$

Comparing Simpson's result 0.69315 with the real value $\left(\ln 2 = \int_1^2 \frac{dx}{x} = 0.6931472 \right)$, we see that this method has achieved five-place accuracy.

9.10   Functions of Bounded Variation

In this section we study a family of functions, the functions of bounded variation, which plays an important role in the theory of Stieltjes[2] integrals, of arclengths, etc.

Definition 9.15. Let f be a real-valued function defined on the closed interval $[a, b]$ and let

$$a = x_0 < x_1 < \cdots < x_{n-1} < x_n = b$$

be any subdivision of $[a, b]$. If, for all partitions $P = \{x_0, \ldots, x_n\}$ of $[a, b]$, there exists a positive number M such that

$$\sum_{i=1}^{n} |f(x_i) - f(x_{i-1})| \leqslant M,$$

we say that f is *of bounded variation* over $[a, b]$. The least upper bound of all the sums for all possible partitions of $[a, b]$, that is, $\sup \sum_{i=1}^{n} |f(x_i) - f(x_{i-1})|$, is called the *total variation* of f in $[a, b]$, and will be denoted by $V_a^b f$.

Theorem 9.24. *A monotone function f on $[a, b]$ is of bounded variation in $[a, b]$.*

□   For any partition P of $[a, b]$, we can write

$$\sum_{i=1}^{n} |f(x_i) - f(x_{i-1})| = |f(x_1) - f(x_0)| + |f(x_2) - f(x_1)| + \cdots + |f(x_n) - f(x_{n-1})|.$$

---

[2]T. J. Stieltjes (1856–1894), Dutch mathematician.

Moreover, for a nondecreasing function, $f(x_i) - f(x_{i-1}) \geqslant 0$, and for a nonincreasing function, $f(x_i) - f(x_{i-1}) \leqslant 0$. In the first case, we have

$$\sum_{i=1}^{n} |f(x_i) - f(x_{i-1})| = f(x_1) - f(x_0) + f(x_2) - f(x_1) + \cdots + f(x_n) - f(x_{n-1}) = f(b) - f(a).$$

Similarly, for the second case we obtain

$$\sum_{i=1}^{n} |f(x_i) - f(x_{i-1})| = f(a) - f(b).$$

∎

Theorem 9.25. *If a function f is of bounded variation in $[a,b]$, then it is bounded on $[a,b]$.*

□ Take an arbitrary point x in $[a,b]$. Since f is of bounded variation in $[a,b]$, we can write $|f(x) - f(a)| + |f(b) - f(x)| \leqslant V_a^b f$, and so $|f(x) - f(a)| \leqslant V_a^b f$. Thus, $f(a) - V_a^b f \leqslant f(x) \leqslant V_a^b f + f(a)$.

Now, if we put $C = \max[|f(a) - V_a^b f|, |V_a^b f + f(a)|]$, then $|f(x)| < C$. ∎

Theorem 9.26. *If f and g are of bounded variation in $[a,b]$, then the functions $f \pm g$, $f \cdot g$, f/g are also of bounded variation in $[a,b]$, in the last case, provided that $|g(x)| \geqslant m > 0$ for some constant m.*

□ Put $\varphi_1 = f + g$, $\varphi_2 = f \cdot g$, $\varphi_3 = f/g$. For any partition $P = \{x_0, \ldots, x_n\}$ of $[a,b]$, we can write

(i)
$$\sum_{i=1}^{n} |\varphi_1(x_i) - \varphi_1(x_{i-1})| = \sum_{i=1}^{n} |[f(x_i) + g(x_i)] - [f(x_{i-1}) + g(x_{i-1})]|$$
$$\leqslant \sum_{i=1}^{n} |f(x_i) - f(x_{i-1})| + \sum_{i=1}^{n} |g(x_i) - g(x_{i-1})| \leqslant V_a^b f + V_a^b g;$$

(ii)
$$\sum_{i=1}^{n} |\varphi_2(x_i) - \varphi_2(x_{i-1})| = \sum_{i=1}^{n} |f(x_i)g(x_i) - f(x_{i-1})g(x_{i-1})|$$
$$= \sum_{i=1}^{n} |f(x_i)g(x_i) - f(x_i)g(x_{i-1}) + f(x_i)g(x_{i-1}) - f(x_{i-1})g(x_{i-1})|$$
$$\leqslant \sum_{i=1}^{n} |f(x_i)| \cdot |g(x_i) - g(x_{i-1})| + \sum_{i=1}^{n} |g(x_{i-1})| \cdot |f(x_i) - f(x_{i-1})|$$
$$\leqslant C_1 \cdot V_a^b g + C_2 \cdot V_a^b f.$$

Here, C_1 and C_2 are constants such that $|f(x)| \leqslant C_1$ and $|g(x)| \leqslant C_2$ (by Theorem 9.25 the constants C_1 and C_2 exist).

(iii)
$$\sum_{i=1}^{n} |\varphi_3(x_i) - \varphi_3(x_{i-1})| = \sum_{i=1}^{n} \left| \frac{f(x_i)}{g(x_i)} \frac{f(x_{i-1})}{g(x_{i-1})} \right|$$
$$= \sum_{i=1}^{n} \frac{|f(x_i)g(x_{i-1}) - f(x_{i-1})g(x_i)|}{|g(x_i)| \cdot |g(x_{i-1})|}$$

$$\leqslant \frac{1}{m^2} \Sigma_{i=1}^n |f(x_i)g(x_{i-1}) - f(x_{i-1})g(x_i)|$$

$$= \frac{1}{m^2} \Sigma_{i=1}^n |f(x_i) \cdot g(x_{i-1}) - f(x_{i-1}) \cdot g(x_{i-1}) + f(x_{i-1}) \cdot g(x_{i-1}) - f(x_{i-1}) \cdot g(x_i)|$$

$$\leqslant \frac{1}{m^2} C_2 \cdot V_a^b f + \frac{1}{m^2} C_1 \cdot V_a^b g.$$

∎

Theorem 9.27. *Suppose the function f is of bounded variation in $[a,b]$, with $a < c < b$. Then f is of bounded variation both in $[a,c]$ and $[c,b]$ and*

$$V_a^b f = V_a^c f + V_c^b f.$$

□ Let $P_0 = \{x_1^0, \ldots, x_n^0\}$ and $P_1 = \{x_1^1, \ldots, x_m^1\}$ be any partitions of $[a,c]$ and $[c,b]$, respectively. Let $c = x_n^0 = x_1^1$. Denote by S_1 and S_2 the following sums:

$$S_1 = \sum_{i=1}^n |f(x_i^0) - f(x_{i-1}^0)|,$$

$$S_2 = \sum_{i=1}^n |f(x_i^1) - f(x_{i-1}^1)|.$$

Clearly, the sum $S_1 + S_2$, corresponding to the partition $P = P_0 \cup P_1$ of $[a,b]$, satisfies the inequality $S_1 + S_2 \leqslant V_a^b f$. On the other hand, since P_0 and P_1 are arbitrary, we deduce from the last inequality that

$$V_a^c f + V_c^b f \leqslant V_a^b f. \tag{9.64}$$

It follows that f is of bounded variation in $[a,c]$ and $[c,b]$. Moreover, since $S_1 \leqslant V_a^c f$, $S_2 \leqslant V_c^b f$, it follows that $S_1 + S_2 \leqslant V_a^c f + V_c^b f$. Recall that c ($x_n^0 = x_1^1 = c$) is contained in a partition of $[a,b]$. If $P = \{x_0, \ldots, x_n\}$ is an arbitrary partition of $[a,b]$, that is,

$$a = x_0 < x_1 < \cdots < x_{n-1} < x_n = b,$$

then by adding the point c ($x_i < c < x_{i+1}$) to the partition $P = \{x_0, \ldots, x_n\}$, we have the case considered above:

$$a = x_0 < x_1 < \cdots < x_i < c < x_{i+1} < \cdots < x_{n-1} < x_n = b.$$

Hence the sum S assigned to the partition P is not greater than the sum $S_1 + S_2$ corresponding to $P \cup \{c\}$, that is,

$$S \leqslant S_1 + S_2 \leqslant V_a^c f + V_c^b f.$$

Thus, if we take the supremum over all possible partitions of $[a,b]$ ($V_a^b f$), we clearly have

$$V_a^b f \leqslant V_a^c f + V_c^b f. \tag{9.65}$$

Consequently, comparing (9.64) and (9.65) we have the desired result. ∎

Remark 9.12. (a) If f is continuous on $[a,b]$, it need not be of bounded variation.

For example, consider the continuous function on $[0,2]$ defined by

$$f(x) = \begin{cases} x\sin\dfrac{\pi}{x}, & \text{if } x \neq 0, \\ 0, & \text{if } x = 0. \end{cases}$$

Take the partition $P = \{x_0, \ldots, x_n\}$ consisting of the points $x_0 = 0$, $x_i = \dfrac{2}{2n - (2i-1)}$ ($i = 1, 2, \ldots, n$). For integers $i \geqslant 2$ we have

$$f(x_i) - f(x_{i-1}) = \frac{2}{2n - (2i-1)} \sin\left[\frac{\pi}{2}(2n - 2i + 1)\right] - \frac{2}{2n - (2i-3)} \sin\left[\frac{\pi}{2}(2n - 2i + 3)\right]$$

$$= \frac{2}{2n - 2i + 1} \sin\frac{\pi}{2} - \frac{2}{2n - 2i + 3} \sin\frac{3\pi}{2} = \frac{2}{2n - 2i + 1} + \frac{2}{2n - 2i + 3} > 0.$$

Thus,

$$\sum_{i=1}^{n} |f(x_i) - f(x_{i-1})| = \frac{2}{2n-1} + \sum_{i=2}^{n}\left[\frac{2}{2n - 2i + 1} + \frac{2}{2n - 2i + 3}\right]$$

$$= \frac{2}{2n-1} + \left(\frac{2}{2n-3} + \frac{2}{2n-1}\right) + \left(\frac{2}{2n-5} + \frac{2}{2n-3}\right) + \cdots + \left(\frac{2}{3} + \frac{2}{5}\right) + \left(2 + \frac{2}{3}\right).$$

On the other hand, because

$$\frac{2}{2(n-i+1)-1} + \frac{2}{2(n-i+2)-1} > \frac{1}{n-i+1},$$

we obtain

$$\sum_{i=1}^{n} |f(x_i) - f(x_{i-1})| > \sum_{k=2}^{n} \frac{1}{k}.$$

Since the harmonic series is divergent, by choosing n sufficiently large, the sums $\sum_{i=1}^{n} |f(x_i) - f(x_{i-1})|$ increase without bound. Therefore f is not of bounded variation in $[0,2]$.

(b) Obviously, the total variation of a constant function f over $[a,b]$ is zero. So if we define the class of functions satisfying the condition $f(a) = 0$, this class will be a linear normed space with norm $\|f\| = V_a^b f$.

Theorem 9.28. *A function f is of bounded variation over $[a,b]$ if and only if it can be expressed as the difference of two bounded monotone increasing functions on $[a,b]$.*

□ Suppose f is of bounded variation in $[a,b]$. Let us introduce two auxiliary functions $\Phi(x) = V_a^x f$ and $\varphi(x) = V_a^x f - f(x)$, $x \in [a,b]$. Clearly, $\Phi(x) - \varphi(x) = f(x)$. We will show that $\Phi(x)$ and $\varphi(x)$ are monotone increasing on $[a,b]$. Since $V_x^{x+\delta} f \geqslant 0$ ($\delta > 0$), Theorem 9.27 implies

$$V_a^{x+\delta} f = V_a^x f + V_x^{x+\delta} f \geqslant V_a^x f,$$

from which it follows that $\Phi(x+\delta) \geqslant \Phi(x)$. Thus, $\Phi(x)(x \in [a,b])$ is monotone increasing. On the other hand,

$$\begin{aligned}
\varphi(x+\delta) - \varphi(x) &= V_a^{x+\delta} f - f(x+\delta) - V_a^x f + f(x) \\
&= V_a^x f + V_x^{x+\delta} f - V_a^x f - [f(x+\delta) - f(x)] \\
&= V_x^{x+\delta} f - [f(x+\delta) - f(x)] \geqslant 0,
\end{aligned}$$

that is, $\varphi(x+\delta) - \varphi(x) \geqslant 0$. Therefore, $\varphi(x)$ is monotone increasing. The necessity is proved. Sufficiency follows immediately from Theorem 9.24.   ∎

We have seen a relationship between monotone increasing functions and functions of bounded variation. However, it should be pointed out that even though the set of functions of bounded variation is a linear space, the set of monotone increasing functions is not, in general, a linear space. For example, despite the fact that the functions defined by $f(x) = x$, $g(x) = x^2$, $x \in [0,1]$ are monotone increasing, their difference, $f(x) - g(x) = x - x^2$, is not monotone increasing.

Corollary 9.9. *By Theorem* 3.15 *and Theorem* 9.28, *it is easy to conclude that if f is of bounded variation in $[a,b]$, then the set of discontinuities of f can at most be denumerable. Besides, each point of discontinuity is of the first kind.*

Example 9.24. Let f be a function with bounded derivative on $[a,b]$, that is $|f'(x)| \leqslant M$ (M is constant). Show that f is of bounded variation in $[a,b]$.
Take any partition $P = \{x_0, \ldots, x_n\}$ of $[a,b]$. By Theorem 5.4,

$$f(x_i) - f(x_{i-1}) = f'(\xi)(x_i - x_{i-1}), \quad \xi \in (x_{i-1}, x_i)$$

and so

$$\sum_{i=1}^n |f(x_i) - f(x_{i-1})| = \sum_{i=1}^n |f'(\xi)| \cdot |x_i - x_{i-1}| \leqslant M \sum_{i=1}^n |x_i - x_{i-1}| = M(b-a).$$

Thus taking the supremum over all possible partitions P, we have

$$V_a^b f \leqslant M(b-a).$$

Example 9.25. Suppose that the function f, defined on $[a,b]$, satisfies the following condition, called a Lipschitz[3] condition, viz., there exists a constant L such that for any $x_{i-1}, x_i \in [a,b]$ we have $|f(x_i) - f(x_{i-1})| \leq L|x_i - x_{i-1}|$.

We show that such f are of bounded variation in $[a,b]$.

Indeed, for an arbitrary partition $P = \{x_0, \ldots, x_n\}$ of $[a,b]$, we have

$$\sum_{i=1}^{n} |f(x_i) - f(x_{i-1})| \leq \sum_{i=1}^{n} |x_i - x_{i-1}| = L(b-a),$$

which means that f is of bounded variation in $[a,b]$.

Example 9.26. Show that the function

$$f(x) = \begin{cases} x^2 \sin \dfrac{1}{x}, & \text{if } x \neq 0, \\ 0, & \text{if } x = 0, \end{cases}$$

is of bounded variation in $[-M, M]$ ($M > 0$). As was shown in Example 5.1, the derivative is

$$f'(x) = \begin{cases} 2x \sin \dfrac{1}{x} - \cos \dfrac{1}{x}, & \text{if } x \neq 0, \\ 0, & \text{if } x = 0. \end{cases}$$

Thus, $|f'(x)| \leq |2x| + 1 \leq 2M + 1$, $x \in [-M, M]$. By example 9.24, it follows that the given function is of bounded variation in $[-M, M]$.

Example 9.27. Show that the Dirichlet function,

$$D(x) = \begin{cases} 1, & \text{if } x \in \mathbb{Q}, \\ 0, & \text{if } x \in \mathbb{R} \setminus \mathbb{Q}, \end{cases}$$

is not of bounded variation in $[a,b]$, where $\mathbb{Q}$ is the set of rationals. Take a partition $P = \{x_0, x_1, \ldots, x_n\}$ of $[a,b]$, where x_{2k} is irrational and x_{2k+1} is rational.

It is easy to see that

$$D(x_i) - D(x_{i-1}) = \begin{cases} 1, & \text{if } i = 2k+1, \\ -1, & \text{if } i = 2k, \end{cases}$$

Thus $|D(x_i) - D(x_{i-1})| = 1$ $(i = 1, \ldots, n)$, and so

$$\sum_{i=1}^{n} |D(x_i) - D(x_{i-1})| = n.$$

Consequently, if $n \to \infty$, the sum on the left hand side becomes arbitrarily large, that is, $D(x)$ is not of bounded variation in $[a,b]$.

---

[3] R. Lipschitz (1832–1903), German mathematician.

9.11 Stieltjes Integrals

In this section the concept of the Riemann integral is extended to the concept of the Stieltjes integral. We study the existence of the Stieltjes integral and its main properties.

1. Definition and Existence of Stieltjes Integrals. In fact, the definition of the Stieltjes integral of a function f on $[a,b]$ depends on the choice of some other, fixed, function α on $[a,b]$. For this reason, it is more precise to say, the Stieltjes integral of f with respect to α over $[a,b]$.

Let f and α be bounded functions defined on $[a,b]$ $(a < b)$, and let $P = \{x_0, \ldots, x_n\}$ be any partition of $[a,b]$: $a = x_0 < x_1 < \cdots < x_{i-1} < x_i < x_n = b$.

The sum,

$$\sigma(x_i, \xi_i) = \sum_{i=1}^{n} f(\xi_i)[\alpha(x_i) - \alpha(x_{i-1})], \quad \xi \in [x_{i-1}, x_i], \tag{9.66}$$

is known as a Stieltjes sum.

If for each positive $\varepsilon > 0$, there is a $\delta = \delta(\varepsilon) > 0$ such that $|I - \sigma| < \varepsilon$ for every Stieltjes sum associated with any partition P of $[a,b]$ (i.e., any selection of points ξ_i) with $\mu(P) < \delta$, we say that the number I is the limit of the Stieltjes sum (9.66) as $\mu(P) \to 0$.

Definition 9.16. If the Stieltjes sum (9.66) has a finite limit I as $\mu(P) \to 0$, then the number I is called the *Stieltjes* (or *Riemann–Stieltjes*) *integral of* f with respect to α over $[a,b]$, and is denoted

$$I = \int_a^b f(x) d\alpha(x).$$

Sometimes, α is called the integrating function. It is easy to see that, in particular, if $\alpha(x) = x + c$ (c a constant), then it is the same as the Riemann integral.

We will give a condition for the existence of the Stieltjes integral. Firstly, we assume that the function α is strictly monotone. For $x_i \in P$, we put $\Delta\alpha(x_i) = \alpha(x_i) - \alpha(x_{i-1}) > 0$ $(i = 1, \ldots, n)$. It is understood that in the results already obtained for the Riemann integral, we must put $\Delta\alpha(x_i)$ in place of Δx_i.

By analogy, we introduce the upper and lower Darboux–Stieltjes sums,

$$\begin{aligned}
S = \sum_{i=1}^{n} M_i \Delta\alpha_i, \qquad & s = \sum_{i=1}^{n} m_i \Delta\alpha_i, \\
M_i = \sup_{x \in [x_{i-1}, x_i]} f(x), \qquad & m_i = \inf_{x \in [x_{i-1}, x_i]} f(x).
\end{aligned} \tag{9.67}$$

For every partition P, the sums (9.67) satisfy the inequality $s(P) \leqslant \sigma(x_i, \xi_i) \leqslant S(P)$. Darboux–Stieltjes sums enjoy the following properties:

(1) The upper Darboux–Stieltjes sums do not increase (and lower Darboux–Stieltjes sums do not decrease) upon refinement P^* of P.

(2) If P_1 and P_2 are any partitions of $[a, b]$, then

$$s(P_1) \leqslant S(P_2),$$

i.e., no lower Darboux–Stieltjes sum is ever greater than any upper Darboux–Stieltjes sum, regardless of the respective partitions used.

For Darboux–Stieltjes integrals I^* and I_*,

$$s(P) \leqslant I_* \leqslant I^* \leqslant S(P),$$

where

$$I^* = \inf_P S(P), \quad I_* = \sup_P s(P)$$

are called the *upper* and *lower Darboux–Stieltjes integrals*, respectively.

Furthermore, I^*, I_* are the limits of the upper and lower Darboux–Stieltjes sums, respectively, as $\mu(P) \to 0$.

The theorem is proved similarly to Theorem 9.7.

Remark 9.13. This theorem is true for $\alpha(x)$ of bounded variation. The proof of the necessary and sufficient conditions for the existence of the Stieltjes integral is simply a reiteration of the proof of Theorem 9.6, and is omitted.

Theorem 9.29. *A function f is Stieltjes integrable with respect to the strictly monotone function α on $[a, b]$ $(a < b)$ if and only if for every $\varepsilon > 0$ there is a partition P of $[a, b]$ such that $S(P) - s(P) < \varepsilon$.*

We now determine the basic class of functions which are Stieltjes integrable.

Theorem 9.30. *If f is continuous on $[a, b]$ and α is strictly monotone on $[a, b]$, then f is Stieltjes integrable with respect to α over $[a, b]$.*

Indeed, by Theorem 9.28, the function $\alpha(x)$ can be represented as

$$\alpha(x) = \beta(x) - \gamma(x)$$

where $\beta(x)$ and $\gamma(x)$ are strictly monotone and bounded functions. Thus, the Darboux–Stieltjes sum (9.66) can be rewritten:

$$\sigma(x_i, \xi_i) = \sum_{i=1}^{n} f(\xi_i) \Delta \alpha(x_i) = \sum_{i=1}^{n} f(\xi_i) \Delta \beta(x_i) - \sum_{i=1}^{n} f(\xi_i) \Delta \gamma(x_i) = \sigma_1(x_i, \xi_i) - \sigma_2(x_i, \xi_i),$$

where

$$\Delta\alpha(x_i) = \alpha(x_i) - \alpha(x_{i-1}), \quad \Delta\beta(x_i) = \beta(x_i) - \beta(x_{i-1}), \quad \Delta\gamma(x_i) = \gamma(x_i) - \gamma(x_{i-1})$$

Since $\beta(x)$ and $\gamma(x)$ are increasing functions, the Darboux–Stieltjes sums $\sigma_1(x_i, \xi_i)$ and $\sigma_2(x_i, \xi_i)$ have finite limits as $\mu(P) \to 0$. It follows that $\sigma(x_i, \xi_i)$ has a finite limit, and so the Stieltjes integral exists.

Theorem 9.31. *If a function f defined on $[a,b]$ is Riemann integrable and α satisfies a Lipschitz condition, then f is Stieltjes integrable with respect to α on $[a,b]$.*

☐ As was shown in Example 9.25, such a function α is of bounded variation. At first, suppose α is an increasing function satisfying the Lipschitz condition with the constant L, i.e., $|\alpha(x') - \alpha(x'')| \leqslant L|x' - x''|$, $x', x'' \in [a,b]$.
Then note that

$$S - s = \sum_{i=1}^{n} (M_i - m_i)\Delta\alpha(x_i) \leqslant L \sum_{i=1}^{n} (M_i - m_i)\Delta x_i. \tag{9.68}$$

Since f is Riemann integrable, the sum $\sum_{i=1}^{n}(M_i - m_i)\Delta x_i$ can be made as small as desired by choosing a suitable partition P. Consequently, for given $\varepsilon > 0$, the value $S - s < \varepsilon$, if $\mu(P)$ is sufficiently small. By Theorem 9.29, this means that the Stieltjes integral exists.

Next, we treat the general case, when α satisfies the Lipschitz condition but is not necessarily increasing. Here, too, we also have a representation,

$$\alpha(x) = Lx - [Lx - \alpha(x)] = \alpha_1(x) - \alpha_2(x), \quad \alpha_1(x) = Lx, \quad \alpha_2(x) = Lx - \alpha(x),$$

where α_1 and α_2 are both increasing and both satisfy the Lipschitz condition. Clearly, the integrability follows. ∎

Theorem 9.32. *If f and φ are both Riemann integrable on $[a,b]$, and we have*

$$\alpha(x) = A + \int_a^x \varphi(t)\,dt,$$

then f is Stieltjes integrable with respect to φ on $[a,b]$.

☐ Since φ is Riemann integrable, there is a constant L such that $|\varphi(t)| \leqslant L$. Thus,

$$|\alpha(x') - \alpha(x'')| = \left| \int_{x'}^{x''} \varphi(t)\,dt \right| \leqslant L|x' - x''|,$$

and α satisfies a Lipschitz condition. Now it suffices to use Theorem 9.31. ∎

Sometimes a Stieltjes integral can be expressed as a Rieman integral. For example, if the derivative $\alpha'(x)$ is bounded and Riemann integrable on $[a,b]$, then

$$\int_a^b f(x)\,d\alpha(x) = \int_a^b f(x)\alpha'(x)\,dx.$$

2. The Basic Properties of the Stieltjes Integral. These four properties follow from the definition of the Stieltjes integral:

(i) $\int_a^b d\alpha(x) = \alpha(b) - \alpha(a)$.

(ii) If f_1 and f_2 are Stieltjes integrable with respect to α on $[a,b]$ and f is Stieltjes integrable with respect to α_1 and α_2 on $[a,b]$, then

$$\int_a^b (k_1 f_1 + k_2 f_2) d\alpha = k_1 \int_a^b f_1 d\alpha + k_2 \int_a^b f_2 d\alpha,$$

$$\int_a^b f d(k_1 \alpha_1 + k_2 \alpha_2) = k_1 \int_a^b f d\alpha_1 + k_2 \int_a^b f d\alpha_2 \quad (k_1, \ k_2 \text{ are constants}).$$

(3) If $a < c < b$, then

$$\int_a^b f(x) \, d\alpha(x) = \int_a^c f(x) d\alpha(x) + \int_c^b f(x) d\alpha(x)$$

assuming that these three integrals exist.

Note that the existence of the two integrals on the right does not necessarily imply the existence of $\int_a^b f(x) d\alpha(x)$. For example, let

$$f(x) = \begin{cases} 0, & -1 \leqslant x \leqslant 0, \\ p, & 0 < x \leqslant 1, \ p \neq 0 \end{cases}$$

$$\alpha(x) = \begin{cases} 0, & -1 \leqslant x < 0, \\ q, & 0 \leqslant x \leqslant 1, \ q \neq 0 \end{cases}$$

the Darboux–Stieltjes sum for the integral $\int_{-1}^0 f(x) d\alpha(x)$ is zero, because $f(x) = 0$, $-1 \leqslant x \leqslant 0$ for all partitions of $[-1,1]$. Similarly, since $\Delta \alpha(x_i) = \alpha(x_i) - \alpha(x_{i-1}) = 0$ for all partitions of $[-1,1]$ the Darboux–Stieltjes sum for $\int_a^b f(x) d\alpha(x)$ is zero. It follows that both integrals exist and are equal to zero. On the other hand, it is easy to see that the integral $\int_{-1}^1 f(x) d\alpha(x)$ does not exist. Indeed, let $P = \{x_0, x_1, \ldots, x_n\}$ be a partition of $[-1,1]$ such that $0 \notin P$. Then the Darbou–Stieltjes sum $\sigma(x_i, \xi_i) = \sum_{i=1}^n f(\xi_i) \Delta \alpha(x_i)$ has only one term

$$f(\xi_k)[\alpha(x_k) - \alpha(x_{k-1})] = q f(x_k)$$

such that $0 \in [x_{k-1}, x_k]$. Therefore, we have $\sigma = 0$ if $\xi_k \leqslant 0$ and $\sigma = p \cdot q \neq 0$ if $\xi_k > 0$. This means that the limit of σ, as $\mu(P) \to 0$, does not exist.

In fact, this occurs because 0 is a point of discontinuity for both $f(x)$ and $\alpha(x)$.

(4) The mean value theorem holds for the Stieltjes integral.

Suppose f is bounded on $[a,b]$, that is, there are real numbers m, M such that $m \leqslant f(x) \leqslant M$ on $[a,b]$. Furthermore, let α be an increasing function on this closed interval. Then there is a constant k $(m \leqslant k \leqslant M)$ such that

$$\int_a^b f(x) d\alpha(x) = k[\alpha(b) - \alpha(a)].$$

If f is continuous on $[a,b]$, then there exists a number $\xi \in [a,b]$ such that $k = f(\xi)$.

9.12   The Lebesgue Integral

1. The Lebesgue Measure of Subsets of the Real Line.   Since an interval I is a set of points, we see that its length could be regarded as the value of a function whose value on I is $m(I)$. If we consider two dimensions, we can think of a rectangle as a generalized interval. By analogy in three dimensions we can consider volume as a generalized length. Thus, the idea of length is generalized to sets in any higher dimensional Euclidean spaceby the use of the measure concept. In this section we shall be interested in extending the notion of length to subsets of the real line other than intervals. Without loss of generality we assume that the set is a subset of $S = [0, 1]$.

Suppose E is a subset of $[0, 1]$. Its complement relative to S is denoted by $C_S E$ or more briefly CE (see Section 1.1).

Definition 9.17. A family $\{\alpha_j = (p_j, q_j) : j \in J\}$ of open intervals is said to be an *open covering* of a set E if every point of E belongs to some member of $\{\alpha_j : j \in J\}$, that is

$$E \subset \bigcup_{j \in J} \alpha_j.$$

If the set of positive numbers J is finite (denumerable), then we say that E has a *finite (denumerable) covering* $\{\alpha_j : j \in J\}$.

If $J_0 \subset J$ and $E \subset \bigcup_{j \in J} \alpha_j$ we call $\{\alpha_j : j \in J_0\}$ an open subcovering of E.

It is understood that a collection of open intervals can be constructed such that at least every point of E belongs to one of the intervals α_j $(j \in J)$. The sum of lengths of these intervals α_j $(j \in J)$ is denoted by $\sum_{j \in J} (q_j - p_j)$. Obviously, for any open covering $\{\alpha_j : j \in J\}$, this sum is positive, i.e., $\sum_{j \in J} (q_j - p_j) > 0$. Therefore, the set of values of the form $\sum_{j \in J} (q_j - p_j)$ for different coverings $\{\alpha_j : j \in J\}$ of E is bounded below, and so there is a greatest lower bound $\left(\inf \sum_{j \in J} (q_j - p_j) \right)$. This greatest lower bound, clearly depending only on E, is called the outer measure of E and is denoted by $m^* E$.

By definition of outer measure, for each $\varepsilon > 0$ there exists an open covering $\{\alpha_j : j \in J\}$ of E such that

$$m^* E \leqslant \sum_{j \in J} (q_j - p_j) \leqslant m^* E + \varepsilon. \tag{9.69}$$

The difference between the length of the closed interval $S = [0, 1]$ and the outer measure of CE is called the interior or inner measure of E and is denoted by $m_* E$:

$$m_* E = 1 - m^* CE.$$

We give some properties of outer and inner measures.

(1) The outer and inner measures of any set are nonnegative.

□   Indeed, as a greatest lower bound of the set of positive numbers $\sum_{j \in J}(q_j - p_j)$ the outer measure is nonnegative,i.e, $m^*E \geqslant 0$. On the other hand, since $CE \subset S$ the outer measure $m^*CE \leqslant 1$ and so $m_*E = 1 - m^*CE \geqslant 0$.   ∎

(2) For any set E, m^*E and m_*E always exist and

$$m^*E \geqslant m_*E$$

□   Indeed, write the right hand side of the inequality (9.69) for CE

$$\sum_{j \in J}(q'_j - p'_j) \leqslant m^*CE + \varepsilon \qquad (9.70)$$

where $\{(p'_j, q'_j) : j \in J\}$ is an open covering of CE. It follows from the inequalities (9.69) and (9.70) that

$$m^*E + m^*CE + 2\varepsilon > \sum_{j \in J}(q_j - p_j) + \sum_{j \in J}(q'_j - p'_j).$$

On the other hand since the sum of the right hand side of this inequality is greater than the length of S, we have

$$m^*E + m^*CE + 2\varepsilon > 1$$

or

$$m^*E > 1 - m^*CE - 2\varepsilon$$

Because ε is arbitrary, we have

$$m^*E \geqslant 1 - m^*CE = m_*E.$$

                                                                                        ∎

(3) If $E_1 \subset E$, then $m^*E_1 \leqslant m^*E$ and $m_*E_1 \leqslant m_*E$ (monotonicity).

□   Let A and B be the sets of sums of lengths of intervals belonging to all possible covering of the sets E_1 and E, respectively.

     Then,

$$m^*E_1 = \inf A, \quad m^*E = \inf B.$$

It is easy to see that $A \supseteq B$ and so $\inf A \leqslant \inf B$. Thus, from the last equalities, we have

$$m^*E_1 \leqslant m^*E.$$

On the other hand,

$$m^*CE_1 \geqslant m^*CE$$

$$1 - m^*CE_1 \leqslant 1 - m^*CE$$

$$m_*E_1 \leqslant m_*E.$$

■

Definition 9.18. If the outer and inner measures of a set E are equal, then we say that E is *Lebesgue measurable* (or, more briefly, *measurable*) and define the *Lebesgue measure of E* as $m^*E = m_*E = mE$.

It can be easily proved that the measurability of E implies the measurability of CE and conversely. Indeed, if E is measurable, then

$$mE = m_*E = 1 - m^*CE \tag{9.71}$$

On the other hand, $m_*CE = 1 - m^*E$ and so

$$mE = m^*E = 1 - m_*CE \tag{9.72}$$

Thus (9.71) and (9.72) imply $m^*CE = m_*CE$, that is, CE is measurable.

Example 9.28. Show that a finite set E is measurable. Suppose that the number of points of E is N. Let $\varepsilon > 0$ be sufficiently small. For each point take an interval with length ε/N containing this point. The collection of these intervals is an open covering of E with length ε. Therefore $m^*E = 0$. Then $0 \leqslant m_*E \leqslant m^*E = 0$ and so $m^*E = 0$. Consequently, we have $mE = m^*E = m_*E = 0$.

It follows that from Example 9.28 that $m(\varnothing) = 0$.

Example 9.29. Show that a denumerable set E is measurable and its measure is zero. Let the points of the set E be $a_1, a_2, a_3, \ldots$ Then we can enclose $a_1, a_2, a_3, \ldots$ in open intervals of lengths $\dfrac{\varepsilon}{2^i}$ $(i = 1, 2, 3, \ldots)$, respectively. Here $\varepsilon > 0$ is arbitrarily small. Then this collection of intervals is an open covering of E. The sum of the lengths of these intervals is not more than the sum of the geometric series

$$\frac{\varepsilon}{2} + \frac{\varepsilon}{2^2} + \frac{\varepsilon}{2^3} + \cdots + \frac{\varepsilon}{2^n} + \cdots = \frac{\varepsilon(1/2)}{1 - 1/2} = \varepsilon.$$

Thus $m^*E = 0$. It follows that $m_*E = 0$, too. Consequently, $mE = m^*E = m_*E = 0$.

Remember that, by Theorem 1.4, the set of rational numbers contained in $[0,1]$ is denumerable, and then its measure is zero.

Example 9.30. Show that the set of irrational numbers contained in $[0,1]$ is measurable and its measure is one. If E denote the set of irrational numbers, then CE is theset of rational numbers contained in $[0,1]$. As was seen in the previous example, its measure is zero. Then, by (9.71), we have

$$mE = 1 - mCE = 1.$$

Thus the set of irrational numbers contained in $[0,1]$ is measurable and its measure is 1.

Example 9.31. Show that an open set is measurable. Let E $(E \subset [0,1])$ be an arbitrary open set. This open set E is a finite or countable union of mutually disjoint intervals $\alpha_1, \alpha_2, \ldots$ (see, for example [18]). Indeed, let x be an arbitrary point of E. Obviously, there is at least one open interval containing x and contained in E. Let I_x be the union of all such open intervals. Then we prove, that I_x is itself an open interval. If we denote $a = \inf I_x$, $b = \sup I_x$, then $I_x \subset (a,b)$. Suppose now, that y is an arbitrary point of (a,b) distinct from x, where for example, assume, that $a < y < x$. Clearly, there is a point y_1 such that $a < y_1 < y$. Therefore, E contains an open interval containing the points y_1 and x. Then it follows, that $y \in I_x$. The case $y > x$ is treated by analogy. Besides, since $x \in I_x$, then $I_x \supset (a,b)$ and so $I_x = (a,b)$. Consequently, I_x is an open interval. Moreover, by construction the interval (a,b) is not contained in E and is not a subset of a lager interval contained in E. On the other hand, it is obvious, that two intervals I_x and I_{x_1} $(x \neq x_1)$ either coincide or else are disjoint (otherwise I_x and I_{x_1} would both be contained in a lager interval $I_x \cup I_{x_1} \subset E$). It is easy to see, that there are no more than countably many such pairwise intervals I_x. Indeed, choosing an arbitrary rational point in every I_x, we establish a one-to-one correspondence between the intervals and a subset of the rational numbers. At last, it is clear, that $E = \bigcup_x I_x$. Thus,

$$m(E) = \sum_{k=1}^{\infty} |\alpha_k|,$$ where $|\alpha_i|$ is the length of α_i $(i = 1, 2, \ldots)$ and the series is convergent.

And any closed set F is measurable. In this case, F is either $[0,1]$ or $[0,1] \setminus CF$. Since CF is measurable as an open set, then F is measurable and $mF = 1 - mCF$. Note that the meauseres of both closed $[a,b]$ $(a < b)$ and open intervals (a,b) are their lengths $b - a$.

A property which is true except for a set of measure zero is said to hold almost everywhere (a.e.). For example, the Dirichlet function

$$D(x) = \begin{cases} 1 & x \text{ rational,} \\ 0 & x \text{ irrational} \end{cases}$$

is zero a.e., because $D(x) = 1$ is true for a set of rational numbers of measure zero. Similarly, if $0 \leqslant x \leqslant 1$, then $\lim_{n \to \infty} x^n = 0$ a.e., because $\lim_{n \to \infty} x^n = 0$ only at $x = 1$, a set of measure zero.

Let us state, without proof, the fact that Lebesgue measure is countably additive, viz., any countable union of mutually disjoint measurable sets $E_1, E_2, \ldots$ is measurable and its measure is the sum of the measures of the sets $E_1, E_2, \ldots$:

$$m\left[\bigcup_{k=1}^{\infty} E_k\right] = mE_1 + mE_2 + \cdots \tag{9.73}$$

This result, so-called σ-additivity of the Lebesgue measure, though intuitively obvious is not easy to prove. The proof can be found, for example, in [12,18].

It follows from (9.73) that the intersection $\bigcap_{k=1}^{\infty} E_k$ is measurable. Indeed, the complement sets $CE_k, k = 1, 2, \ldots$ are measurable and so, by formula (9.73), $\bigcup_{k=1}^{\infty} CE_k$ is measurable.

Since $C\left[\bigcap_{k=1}^{\infty} E_k\right] = \bigcup_{k=1}^{\infty} CE_k$, then by Theorem 1.1, $\bigcap_{k=1}^{\infty} E_k$ is measurable, too.

Example 9.32. Let us prove that the Cantor set has measure zero. One of the interesting properties of the Cantor set is that although it is non-countable it has measure zero.

Consider the closed interval $[0, 1]$. Trisect the interval at points $1/3$, $2/3$ and remove the open interval $\left(\frac{1}{3}, \frac{2}{3}\right)$ (called the middle third). Thus we obtain the set E_1:

$$E_1 = \left[0, \frac{1}{3}\right] \cup \left[\frac{2}{3}, 1\right]$$

By trisecting the intervals $\left[0, \frac{1}{3}\right]$ and $\left[\frac{2}{3}, 1\right]$ and again removing the open middle thirds, we obtain E_2:

$$E_2 = \left[0, \frac{1}{9}\right] \cup \left[\frac{2}{9}, \frac{3}{9}\right] \cup \left[\frac{6}{9}, \frac{7}{9}\right] \cup \left[\frac{8}{9}, 1\right].$$

Continuing in this manner we obtain a sequence of closed sets $E_1, E_2, E_3, \ldots$ having the following properties:

(1) $E_1 \supset E_2 \supset E_3 \ldots$;

(2) For every n, the set E_n is the union of 2^n closed sets with lengths 3^{-n}.

The Cantor set, denoted by $P = \bigcap_{n=1}^{\infty} E_n$, is the intersection of $E_1, E_2, E_3, \ldots$

It is easy to see that the length of E_n is

$$m(E_n) = \left(\frac{2}{3}\right)^n \quad (n = 1, 2, 3, \ldots).$$

Thus, since $P \subset E_n$ for all n we have $m(P) = 0$.

It would seem that there is practically nothing left to the Cantor set P. However, it turns out that the set has cardinal number c, i.e., is equivalent to $[0, 1]$. It also has many other remarkable properties. In fact, there can be established a one to one and onto function $f : [0, 1] \to P$, and so the set P has cardinality c^4.

We next show that there is a non-measurable set in the interval $\left[-\dfrac{1}{2}, +\dfrac{1}{2} \right]$.

Example 9.33. Consider the interval $I = \left[-\dfrac{1}{2}, +\dfrac{1}{2} \right]$. Subdivide all points of I into classes, setting two points x and y into the same class if and only if their difference is a rational number. Thus, for each point $x \in I$, let the class $K(x)$ consist of all those points of the interval I which have the form $x + r$:

$$K(x) = \{ y \in I : x - y \in Q \} = \{ x + r \in I : r \in Q \}$$

where, as usual, Q denotes the set of rational numbers. In particular, $x \in K(x)$. We claim that if $K(x_1) \neq K(x_2)$ $(x_1, x_2 \in I)$, then $K(x_1) \cap K(x_2) = \varnothing$. In fact, assume the contrary, viz., that $z \in K(x_1) \cap K(x_2)$. Then $z = x_1 + r_1 = x_2 + r_2$, where r_1, r_2 are rational numbers: thus, $x_2 = x_1 + r_1 - r_2$. On the other hand, if $t \in K(x_2)$, then $t = x_2 + r = x_1 + (r_1 - r_2 + r) = x_1 + r'$ so that $t \in K(x_1)$ and $K(x_2) \subset K(x_1)$. We establish in the same way, that $K(x_1) \subset K(x_2)$ and so $K(x_1) = K(x_2)$, i.e., $K(x_1)$ and $K(x_2)$ are one and the same class, contrary to the hypothesis that they are distinct. Hence, the set of all classes obtained in this manner form a partition of $[-1, 1]$. We now choose one point from each class and designate by A the set of all points selected. The set A is non-measurable To prove this, we enumerate all rational points of the closed interval $[-1, 1]$:

$$r_0 = 0, \ r_1, \ r_2, \ r_3, \ldots$$

And denote by A_k the set obtained from the set A by the translation $x + r_k$: $A_k = \{ x + r_k : x \in A \}$ $(A_0 = A)$. All the sets A_k are congruent with one another and so

$$m^* A_k = m^* A = \alpha, \ \ m_* A_k = m_* A = \beta, \ \ k = 0, 1, 2, \ldots$$

Now we show that $\beta > 0$. To do this, we note that

$$\bigcup_{k=0}^{\infty} A_k \subset \left[-\frac{3}{2}, +\frac{3}{2} \right]. \tag{9.74}$$

In fact, if $x \in \left[-\dfrac{1}{2}, +\dfrac{1}{2} \right]$, then x lies in one of the classes of the partition formed above. If x_0 is the representative of this class in the set A, the difference $x - x_0$ is a rational number which

[4]K. Stromberg, *Real and Abstract Analysis*, Springer, 1969.

obviously belongs to the interval $[-1, +1]$. That is, $x - x_0 = r_k$ and $x \in A_k$ and $x \in \bigcup_{k=0}^{\infty} A_k$. This proves the first inclusion in (9.74). On the other hand, if $x \in A_k$, then $x = x_0 + r_k$, $|x_0| \leqslant \frac{1}{2}$, $r_k \leqslant 1$ and so $A_k \subset \left[-\frac{3}{2}, +\frac{3}{2}\right]$ for arbitrary k. Consequently, $\bigcup_{k=0}^{\infty} A_k \subset \left[-\frac{3}{2}, +\frac{3}{2}\right]$ and so the second inclusion in (9.74) is true. Then it follows from the first inclusion, that

$$1 = m^* I \leqslant m^* \left[\bigcup_{k=0}^{\infty} A_k\right] \leqslant \sum_{k=0}^{\infty} m^* A_k,$$

i.e.,

$$1 \leqslant \beta + \beta + \beta + \cdots.$$

Obviously, this means that $\beta > 0$. On the other hand, it is easy to show that $\alpha = 0$. To do this, we show first that $A_n \cap A_m = \varnothing$ for $n \neq m$. In fact, if $z \in A_n \cap A_m$, the points $x_n = z - r_n$, $x_m = z - r_m$ would be distinct points of the set A, i.e., representatives of two distinct classes, which is impossible, because their difference $x_n - x_m = r_m - r_n$ is a rational number. This prove that $A_n \cap A_m = \varnothing$. Then it follows from the second inclusion in (9.74) that

$$3 = m_* \left[-\frac{3}{2}, +\frac{3}{2}\right] \geqslant m_* \left[\bigcup_{k=0}^{\infty} A_k\right] \geqslant \sum_{k=0}^{\infty} m_* A_k,$$

whence

$$\alpha + \alpha + \alpha + \cdots \leqslant 3$$

and $\alpha = 0$. Consequently, we obtain $m_* A = \alpha < \beta = m^* A$, which proves the non-measurability of the set A.

If we had initially partitioned not the interval $\left[-\frac{1}{2}, +\frac{1}{2}\right]$ into classes, but rather an arbitrary measurable set E of positive measure, then repeating this argument literally, we would have arrived at a non-measurable set $A \subset E$. Therefore, we conclude that every set of positive measure contains a non-measurable subset.

2. Measurable Functions. Let E be a measurable set and f be a real valued function defined on E. The set of values $x \in E$ for which $f(x) > A$ we denote $E[f(x) > A]$, i.e.,

$$E[f(x) > A] = \{x \in E : f(x) > A\}.$$

Definition 9.19. We say that f is *Lebesgue measurable* (or, briefly, *measurable*) on E if for each real number A the set $E[f(x) > A]$ is measurable.

It follows at once from the definition that a function f defined on a set of measure zero is measurable. On the other hand it can be shown that if f is measurable on E, then it is measurable on all measurable subset B of E.

Indeed since $B[f > A] = B \cap E[f > A]$ and the intersection of sets on the right is measurable, then $E[f > A]$ is measurable.

Note that a function f can be defined as measurable on E if any one of the sets $E[f \geqslant A]$, $E[f \leqslant A]$, $E[f < A]$ is measurable for each A. Indeed, because $E[f \geqslant A] = \bigcap_{k=1}^{\infty} E\left[f > A - \frac{1}{k}\right]$ and $E_k \equiv E\left[f > A - \frac{1}{k}\right]$ is measurable, it follows that $E[f \geqslant A]$ is measurable.

On the other hand since $E[f \leqslant A] = E \setminus E[f > A]$, $E[f < A] = E \setminus E[f \geqslant A]$ it follows that the sets $E[f \leqslant A]$, $E[f < A]$ are measurable. Indeed it only remains to prove that for measurable two sets E_1 and E_2 their difference $E_1 \setminus E_2$ is measurable. But $E_1 \setminus E_2$ is measurable, because its complement $C(E_1 \setminus E_2) = CE_1 \cup CE_2$ as a union of two measurable sets, is again measurable.

Example 9.34. (a) Let $f(x) = c$ (c-constant) be a constant function defined on E.

Obviously

$$E[f > A] = \begin{cases} E, & \text{if } A < c, \\ \varnothing, & \text{if } A \geqslant c. \end{cases}$$

Then, the measurability of E and $\varnothing$ implies that $E[f > A]$, and so f, is measurable.

(b) Show that the Dirichlet function defined on $E = [a, b]$

$$D(x) = \begin{cases} 1, & x \in \mathbb{Q}, \\ 0, & x \in \mathbb{R} \setminus \mathbb{Q} \end{cases}$$

is measurable ($\mathbb{Q}$ is the set of rational numbers).

It is easy to see that

$$E[D > A] = \begin{cases} [a, b] \cap \mathbb{Q}, & 0 \leqslant A < 1, \\ [a, b], & A < 0, \\ \varnothing, & A \geqslant 1. \end{cases}$$

Then, since the sets $[a, b]$, $\mathbb{Q}$, and $\varnothing$ are measurable, $E[D > A]$, and so $D(x)$, is too.

Example 9.35. Show that $f(x) = [\![x]\!]$ ($[\![x]\!]$ is the integer part of x) defined on $E = [a, b]$ is measurable. Indeed,

$$E[f > A] = \begin{cases} [\![k]\!] + 1, b], & k < A \leqslant k+1 \leqslant f(b), \\ 0, & A > f(b) \end{cases}$$

Because $[\![k]\!] + 1, b]$ is measurable, and so $E[f > A]$ is measurable, then f is measurable.

Example 9.36. Let $E = [a,b]$. Show that the characteristic function of $B \subset [a,b]$

$$\eta_B(x) = \begin{cases} 1, & \text{if } x \in B, \\ 0, & \text{if } x \in E \setminus B \end{cases}$$

is measurable for measurable B. Obviously, we have that

$$E[\eta_B(x) > A] = \begin{cases} B, & 0 \leqslant A < 1, \\ [a,b], & A < 0, \\ \varnothing, & A \geqslant 1. \end{cases}$$

But $[a,b]$ and $\varnothing$ are measurable, and B is measurable by hypothesis. It follows that $\eta_B(x)$ is measurable.

Theorem 9.33. *Let f be a function defined on a closed set E. Then for the continuity of f on E, it is necessary and sufficient that for each real A the sets $E[f(x) \geqslant A]$ and $E[f(x) \leqslant A]$ are closed.*

☐ **Necessity.** Let f be continuous on E. Take an arbitrary A and set $F = E[f(x) \geqslant A]$. Moreover, let $x_n \in F$ be any sequence the limit of which is x_0. Since E is closed, it follows that $x_0 \in E$ (Definition 1.22). On the other hand, the continuity of f implies that $f(x_n) \to f(x_0)$. Clearly, $x_n \in F$ implies that for all n we have $f(x_n) \geqslant A$. By passing to the limit in the last inequality, we have $f(x_0) \geqslant A$, or $x_0 \in F$, i.e., $E[f(x) \geqslant A]$ is closed. The closedness of $E[f(x) \leqslant A]$ is proved similarly.

Sufficiency. Suppose now that for every real A, F is closed. We will prove that f is continuous. Let $x_n \in E$ be a sequence and suppose that $x_n \to x_0$ ($x_0 \in E$). For any given $\varepsilon > 0$, put

$$F_1 = E[f(x) \geqslant f(x_0) + \varepsilon], \quad F_2 = E[f(x) \leqslant f(x_0) - \varepsilon].$$

By hypothesis, F_1, F_2 are closed, and so $F = F_1 \cup F_2$ is closed (Theorem 1.14). Since $x_0 \notin F$, that is, x_0 is not a limit point of F, there is a $\delta > 0$ such that an open ball (interval) $B_\delta(x_0)$ centered at x_0 contains no point of F. But there is a positive integer N such that $x_n \in B_\delta(x_0)$ for all $n \geqslant N$. Hence $x_n \notin F$ ($n \geqslant N$) and

$$f(x_0) - \varepsilon < f(x_n) < f(x_0) + \varepsilon.$$

Consequently, $f(x_n) \to f(x_0)$, that is, f is continuous. ∎

Example 9.37. Let f be continuous on a closed set, E. Show that f is measurable on this set.

By Theorem 9.33, $E[f(x) \geqslant A]$ is closed for arbitrary A and is measurable (Example 9.31). Thus by Definition 9.19 f is measurable on E.

Theorem 9.34. *Let a function f be defined on a finite or countable union $E = \bigcup_k E_k$ of measurable sets E_k, $k = 1, 2, \ldots$. Then if f is measurable on each E_k, then it measurable on E.*

☐ For every A, clearly $E[f > A] = \bigcup_k E_k[f > A]$. By hypothesis, $E_k[f > A]$ is measurable for all k, and so $E[f > A]$ is measurable. ∎

Now, using Definition 9.19 and Theorem 9.31, we can prove the following theorem.

Theorem 9.35. *If f is measurable on a measurable set E, then the functions $f(x) + k$ (k-constant), $k f(x)$, $|f(x)|$, $[f(x)]^2$, $\dfrac{1}{f(x)}$ are measurable (in the last case, provided $f(x) \neq 0$ for all $x \in E$).*

Theorem 9.36. *If f and g are measurable on a measurable set E, then the set $E(f > g)$ is measurable.*

☐ Enumerate the rational points of R as $r_1, r_2, r_3, \ldots$. Then,

$$E(f > g) = \bigcup_{k=1}^{\infty} [E(f > r_k) \cap E(g < r_k)].$$

Each of the sets $E(f > r_k)$, $E(g < r_k)$ is measurable, and this countable union of measurable sets is measurable by (9.73). ∎

Theorem 9.37. *If f and g are measurable on E, then the functions $f \pm g$, $f \cdot g$ and f/g are measurable on E (in the last case, provided $g(x) \neq 0$, $x \in E$).*

(1) Because $E[f \pm g > A] = E[f > A \mp g]$ for each A, the sets $E[f \pm g > A]$ are measurable by Theorems 9.35 and 9.36, and so $f \pm g$ are measurable.

(2) Since $f(x) \cdot g(x) = \dfrac{1}{4} \left\{ [f(x) + g(x)]^2 - [f(x) - g(x)]^2 \right\}$, the functions $f(x) \pm g(x)$, and so $[f(x) + g(x)]^2$, are measurable (Theorem 9.35). Hence, $f(x) \cdot g(x)$ is measurable.

(3) Since $\dfrac{f(x)}{g(x)} = f(x) \cdot \dfrac{1}{g(x)}$, then by Theorem 9.35 and (2), f/g is measurable.

Remark 9.14. It is not hard to see that if a function f is measurable on E, then E itself is measurable. Indeed, obviously,

$$E = \bigcup_{n=1}^{\infty} E[f(x) > -n].$$

Then, since each $E[f(x) > -n]$ ($n = 1, 2, \ldots$) is measurable, their countable union E is measurable, too. So in the preceding theorems and definitions, the set E need not be assumed to be measurable.

3. The Lebesgue Integral and its Basic Properties.

In this chapter we have seen that there are bounded functions defined on closed intervals that are not Riemann integrable. The Riemann integral has certain defects which can be remedied by means of the Lebesgue integral. Although it is possible to define the Lebesgue integral in many ways, we shall give a procedure which parallels as closely as possible the definition for Riemann's integral.

Suppose that a function f is bounded and measurable on the interval E. Suppose that m and M are any two real numbers such that $m \leqslant f(x) \leqslant M$. Given n, we partition the interval $[m, M]$ into n subintervals by choosing points $\{y_i\}$, $i = 0, 1, \ldots, n$ such that

$$m = y_0 < y_1 < \cdots < y_n = M.$$

Thus our partition of $[m, M]$ is $P = \{y_0, y_1, \ldots, y_n\}$.

Note that these values are represented geometrically by points on the $0y$ axis.

Let E_i, $i = 1, 2, 3, \ldots, n$ be the set of all x in E such that $y_i \leqslant f(x) < y_{i+1}$, that is

$$E_i = \{x : y_i \leqslant f(x) < y_{i+1}\}, \quad i = 0, 1, \ldots, n-1.$$

Because f is measurable, these sets are measurable and disjoint.

Consider the upper and lower sums S and s, defined by

$$S(P) = \sum_{i=0}^{n-1} y_{i+1} mE_i,$$

$$s(P) = \sum_{i=0}^{n-1} y_i mE_i \tag{9.75}$$

respectively. Here mE_i is the measure of E_i ($i = 0, 1, 2, \ldots, n-1$). By varying the partition we obtain sets of values for $S(P)$ and $s(P)$ satisfying the inequality $s(P) \leqslant I_* \leqslant I* \leqslant S(P)$, where $I^* = \inf_P S(P)$, $I_* = \sup_P s(P)$.

Usually, the Lebesgue integral is defined as the common limit S and s given by (9.75), respectively, as the number of subdivision points becomes infinite in such a manner that $\alpha \to 0$, where $\alpha = \max\{(y_{i+1} - y_i) : i = 0, 1, \ldots, n-1\}$.

Definition 9.20. Assuming that f is bounded, we define the Lebesgue integral as the limit of $s(P) = \sum_{i=0}^{n-1} y_i mE_i$ as $\alpha \to 0$. We denote the Lebesgue integral of f on E by $(L) \int_E f(x)dx$ or $\int_E f(x)dx$. In particular, if $E = [a, b]$, then we denote the Lebesgue integral by

$$(L) \int_a^b f(x)dx, \text{ or simply } \int_a^b f(x)dx.$$

Theorem 9.38. *If f is bounded and measurable on E, then f is Lebesgue integrable on E.*

☐ We easily reduce the existence of the Lebesgue integral to the existence of the Stieltjes integral:

$$\lim_{\alpha \to 0} s(P) = \lim_{\alpha \to 0} \sum_{i=0}^{n-1} y_i mE_i$$

$$= \lim_{\alpha \to 0} \sum_{i=0}^{n-1} y_i mE(y_i \leqslant f(x) < y_{i+1})$$

$$= \lim_{\alpha \to 0} \sum_{i=0}^{n-1} y_i [mE(y_{i+1} > f) - mE(y_i > f)]$$

$$= \lim_{\alpha \to 0} \sum_{i=0}^{n-1} y_i [g(y_{i+1}) - g(y_i)] = \int_m^M y dg(y).$$

Here, g denotes the function defined as $g(y) = mE(y > f)$, and is monotone increasing, because $y_1 < y_2$ implies $mE(y_1 > f) < mE(y_2 > f)$. Obviously, $f(y) = y$ is continuous and so in the last equality, the existence of limit follows from the existence of the corresponding Stieltjes integral (Theorem 9.30).   ■

Below we give the basic properties of the Lebesgue integral. We assume, unless otherwise stated, that f is bounded and measurable and thus Lebesgue integrable, and that all the sets involved are measurable.

(1) If E has measure zero, i.e., $mE = 0$, then

$$\int_E f(x)dx = 0.$$

☐ In fact, each subset $E(y_i \leqslant f(x) < y_{i+1})$ has measure zero, and so any Lebesgue integral sum is zero. Thus, in the theory of the Lebesgue integral any set of measure zero can be neglected.   ■

(2) If $A \leqslant f \leqslant B$, then

$$AmE \leqslant \int_E f(x)dx \leqslant BmE.$$

Sometimes this is called the *mean value theorem for Lebesgue integrals*.

☐ The validity of this property follows from considering the integral $\int_E \varphi_i(x)dx$ ($i = 1, 2, 3$) for each of the following three cases, $\varphi_1(x) \equiv A$, $\varphi_2(x) \equiv f(x)$ and $\varphi_3(x) \equiv B$. Since $E(y_0 \leqslant A < y_1) = E$, $y_0 = A$ in case $\varphi_1(x) \equiv A$, we have

$$s(P) = \sum_{i=0}^{n-1} y_i mE_i = y_0 mE(y_0 \leqslant A < y_1) = y_0 mE.$$

Thus, $\int_E A\,dx = AmE$. Similarly, $\int_E B\,dx = BmE$. At the same time, if $\varphi_2(x) \equiv f(x)$ then the integral sum satisfies the inequality

$$AmE = \sum_{i=0}^{n-1} AmE(y_i \leqslant f(x) < y_{i+1})$$

$$\leqslant \sum_{i=0}^{n-1} y_i mE(y_i \leqslant f(x) < y_{i+1})$$

$$\leqslant \sum_{i=0}^{n-1} BmE(y_i \leqslant f(x) < y_{i+1}) = BmE.$$

By passing to the limit as $\alpha \to 0$, the desired inequality follows. In particular, if $f(x) \equiv C$ (C is constant), then $\int_E f(x)dx = CmE$. ∎

(3) If $E = E_1 \cup E_2$, where E_1 and E_2 are disjoint, i.e., $E_1 \cap E_2 = \varnothing$, then

$$\int_E f(x)dx = \int_{E_1} f(x)dx + \int_{E_2} f(x)dx.$$

The result is easily generalized to finite unions of mutually disjoint sets.

☐ The validity of the assertion follows from the relation

$$\int_E f(x)dx = \lim_{\alpha \to 0} s(P) = \lim_{\alpha \to 0} \sum_{i=0}^{n-1} y_i mE(y_i \leqslant f(x) < y_{i+1})$$

$$= \lim_{\alpha \to 0} \sum_{i=0}^{n-1} y_i mE_1(y_i \leqslant f(x) < y_{i+1}) + \lim_{\alpha \to 0} \sum_{i=0}^{n-1} y_i mE_2(y_i \leqslant f(x) < y_{i+1})$$

$$= \int_{E_1} f(x)dx + \int_{E_2} f(x)dx.$$

∎

(4) If $f(x) = g(x)$ almost everywhere on E, that is, $mE(f \neq g) = 0$, then

$$\int_E f(x)dx = \int_E g(x)dx.$$

☐ Let denote $mE(f \neq g) = E_0$. Then, by property (3),

$$\int_E f(x)dx = \int_{E \setminus E_0} f(x)dx + \int_{E_0} f(x)dx$$

$$\int_E g(x)dx = \int_{E \setminus E_0} g(x)dx + \int_{E_0} g(x)dx$$

Since $f(x) = g(x)$, $x \in E \setminus E_0$, on comparing these equalities, we deduce that the first integrals on the right hand sides are equal, and by property (1), the second integrals are zero. ∎

(5) If f is both Riemann[5] and Lebesgue integrable on $[a,b]$, then

$$(\text{L}) \int_E f(x)dx = (\text{R}) \int_E f(x)dx.$$

---

[5] Actually, the existence of its Riemann integral implies the existence of its Lebesgue integral.

☐ Let $\{x_i\}$ $(i = 0, 1, \ldots, n)$ be subdivision points of any partition of $[a, b]$, that is,

$$a = x_0 < x_1 < \cdots < x_n = b.$$

Put

$$M_i = \sup_{x \in [x_i, x_{i+1}]} f(x), \quad m_i = \inf_{x \in [x_i, x_{i+1}]} f(x), \quad i = 0, 1, \ldots, n-1.$$

By property (2), we have

$$m_i(x_{i+1} - x_i) \leqslant (L) \int_{x_i}^{x_{i+1}} f(x)dx \leqslant M_i(x_{i+1} - x_i).$$

Taking into account (2), summing these inequalities over i gives us

$$\sum_{i=0}^{n-1} m_i(x_{i+1} - x_i) \leqslant (L) \int_{x_i}^{x_{i+1}} f(x)dx \leqslant \sum_{i=0}^{n-1} M_i(x_{i+1} - x_i). \qquad (9.76)$$

The inequality (9.76) remains true if, in place of L, we write R (i.e., if we take the Riemann integral instead of the Lebesgue integral). Then, by passing to the limit as $\alpha = \max(x_{i+1} - x_i) \to 0$, we have

$$(L) \int_a^b f(x)dx = (R) \int_a^b f(x)dx.$$

∎

(6) Let f be a bounded and measurable function on E. If $f(x) \geqslant 0$ a.e. and $\int_E f(x)dx = 0$, then $f(x) = 0$ a.e. on E.

☐ Let us denote

$$E_+ = E[f(x) > 0], \quad E_- = E[f(x) < 0], \quad E_0 = E[f(x) = 0].$$

Since $m(E_-) = 0$ it follows that $\int_{E_-} f(x)dx = 0$. Besides $\int_{E_0} f(x)dx = 0$ and so

$$\int_{E_+} f(x)dx = \int_E f(x)dx = 0. \qquad (9.77)$$

It is clear that $E_+ = \bigcup_{k=1}^{\infty} P_k$, where $P_k = E[f(x) \geqslant 1/k]$. Then by property (2) and the fact that the integral over P_k cannot exceed the integral over E_+, we have

$$\int_{E_+} f(x)dx \geqslant \int_{P_k} f(x)dx \geqslant \frac{m(P_k)}{k}.$$

which contradicts the hypothesis that the integral over E is zero. ∎

(7) Let $\{f_n\}$ be a sequence of functions measurable on E such that $\lim_{n \to \infty} f_n(x) = f(x)$. Then if the sequence is uniformly bounded, i.e., if there exists a constant C such that $|f_n(x)| \leqslant C$ for all n, we have

$$\lim_{n \to \infty} \int_E f_n(x)dx = \int_E \lim_{n \to \infty} f(x)dx = \int_E f(x)dx.$$

This important theorem, which is not true for Riemann integrals, shows the superiority of the Lebesgue integral.

(8) If f is bounded and measurable on E, then $|f|$ is Lebesgue integrable on E. In addition, if $|f|$ is bounded and measurable on E, then f is Lebesgue integrable on E.

☐ It is easy to see that, if f is measurable on E, then $|f|$ is measurable on E. ∎

We have seen that for Riemann integrals the second part of property (8) is not true (see Property 9.8 and Remark 9.7).

Example 9.38. Show that the Dirichlet function $D(x)$ (Example 9.34) is Lebesgue integrable on $E = [0, 1]$.

Denote by R and I, respectively, the rational and irrational numbers contained in $E = [0, 1]$. Since $I \cap R = \varnothing$, it follows from property (3) that

$$\int_E D(x)dx = \int_I D(x)dx = \int_R D(x)dx.$$

Because $mR = 0$ (Example 9.29), we derive $\int_R D(x)dx = 0$ from property (1). On the other hand, $D(x) \equiv 0$, $x \in I$, and so, by property (2), we have $\int_I D(x)dx = 0$. Thus, the desired Lebesgue integral is zero. Recall that the Dirichlet function is not Riemann integrable (see Remark 9.3).

Example 9.39. (1) Show that the Riemann function

$$R(x) = \begin{cases} \dfrac{1}{n}, & \text{if } x = \dfrac{m}{n} \text{ noncancellable fraction,} \\ 0, & \text{if } x \text{ irrational} \end{cases}$$

is Lebesgue integrable on any interval $E = [a, b]$. Indeed, following the same procedure as in Example 9.38 yields

$$\int_E R(x)dx = \int_R R(x)dx + \int_I R(x)dx = \int_R \frac{1}{n}dx + 0 = \frac{mR}{n} = 0 \quad (n \neq 0).$$

(2) Show that the function

$$f(x) = \begin{cases} A, & x \in P, \\ 0, & x \in [0, 1] \setminus P \end{cases}$$

is Lebesgue integrable, where P is the Cantor set (Example 9.7), for A, B any real numbers.

Indeed, since $mP = 0$ (Example 9.32) we can write

$$\int_0^1 f(x)dx = \int_P f(x)dx + \int_{[0,1] \setminus P} f(x)dx$$

$$= A \int_P dx + B \int_{[0,1] \setminus P} dx$$

$$= 0 + B \left[\int_0^1 dx - \int_P dx \right] = B.$$

4. The Space $L_2[a,b]$ of Square-Integrable Functions. This space is the set of all real measurable functions f on $[a,b]$ whose squares are Lebesgue integrable on $[a,b]$. The usual, pointwise, addition and scalar multiplication can be defined in this space. Besides, if $f, g \in L_2[a,b]$, then their product $f \cdot g$ is Lebesgue integrable. This fact follows from the inequality

$$|f(x)g(x)| \leqslant \frac{1}{2}\left[f^2(x) + g^2(x)\right]$$

and the basic properties of the Lebesgue integral. In particular, if $g(x) \equiv 1$, then it follows that any element $f \in L_2[a,b]$ is Lebesgue integrable. Inner products in $L_2[a,b]$ are defined by the formula

$$\langle f, g \rangle = \int_a^b f(x)g(x)dx, \quad f, g \in L_2[a,b].$$

The modulus $\|f\|$ of a function $f \in L_2[a,b]$ is defined as follows, and called the norm of $f \in L_2[a,b]$:

$$\|f\| = \sqrt{\langle f, f \rangle} = \sqrt{\int_a^b f^2(x)dx}.$$

Note that if $\|f - g\| = 0$, then $f(x) = g(x)$ almost everywhere on $[a,b]$. Hence, it is convenient to make this into an equivalence relation. We say that f and g are equivalent if $f(x) = g(x)$ almost everywhere.

In $L_2[a,b]$ (as well as in any other Euclidean space), for every pair of functions $f, g \in L_2[a,b]$, we have the Cauchy–Bunyakowsky's inequality:

$$\langle f, g \rangle^2 \leqslant \langle f, f \rangle \langle g, g \rangle,$$

or

$$\left(\int_a^b f(x)g(x)dx\right)^2 \leqslant \left(\int_a^b f^2(x)dx\right)\left(\int_a^b g^2(x)dx\right).$$

Remark 9.15. One of the other usual ways to define the Lebesgue integral is based on the concept of a step-function. Let $P = \{x_0, \ldots, x_n\}$ be a partition of $[a,b]$. Then a function φ defined as $\varphi(x) = m$, $x \in (x_i, x_{i+1})$ $(i = 0, 1, \ldots, n-1)$ (M_i being constants) is called a *step-function*.

Denote by $L^+[a,b]$ the class of functions which can be obtained as the almost everywhere limit of a nondecreasing sequence of step-functions $\{\varphi_n(x)\}$ such that the numerical sequence $\left\{\int_a^b \varphi_n(x)dx\right\}$ is bounded. Then a function f is Lebesgue integrable on $[a,b]$, if it may be represented as the difference $f = f_1 - f_2$ of two such functions, $f_1, f_2 \in L^+[a,b]$. The Lebesgue integral of f on $[a,b]$ is defined by

$$\int_a^b f(x)dx = \int_a^b f_1(x)dx - \int_a^b f_2(x)dx.$$

Until now we defined the Lebesgue integral and obtained its properties for bounded measurable functions. Now we define the Lebesgue integral for unbounded measurable functions.

First, consider the case where f is unbounded and measurable but nonnegative. If N is a natural number we shall use the notation

$$[f(x)]_N = \begin{cases} f(x) & \text{for all } x \in E \text{ such that } f(x) \leqslant N, \\ N & \text{for all } x \in E \text{ such that } f(x) > N. \end{cases}$$

Thus for every positive integer N, the function $[f(x)]_N$ is bounded and measurable, and so, Lebesgue integrable. We define the Lebesgue integral of unbounded butnonnegative f on E as

$$\int_E f(x)dx = \lim_{N \to \infty} \int_E [f(x)]_N dx.$$

This limit is either a nonnegative number or is infinite. If the limit is infinite, then f is not Lebesgue integrable on E. If the limit is a nonnegative number, then we say that f is Lebesgue integrable (or, briefly, integrable) on E and that its integral is equal to this number.

Now, consider the case $f(x) \leqslant 0$. In this case we define the Lebesgue integral of f as follows

$$\int_E f(x)dx = -\int_E |f(x)|dx,$$

where the integral on the right hand side is defined as above since $|f(x)| \geqslant 0$.

In the general case, we take

$$f^+(x) = \begin{cases} f(x), & \text{for all } x \in E \text{ such that } f(x) \geqslant 0, \\ 0, & \text{for all } x \in E \text{ such that } f(x) < 0 \end{cases}$$

$$f^-(x) = \begin{cases} 0, & \text{for all } x \in E \text{ such that } f(x) \geqslant 0, \\ -f(x), & \text{for all } x \in E \text{ such that } f(x) < 0. \end{cases}$$

Obviously, these functions are both nonnegative and

$$f(x) = f^+(x) - f^-(x).$$

Consequently, this leads us to define the integral of f on E as the difference of the integrals of these nonnegative functions, that is,

$$\int_E f(x)dx = \int_E f^+(x)dx - \int_E f^-(x)dx.$$

If the two integrals on the right exist, then we say that f is Lebesgue integrable on E.

5. Complex-valued functions. In Remark 3.7 we discussed the convergence of complex sequences and series. Now we define the integral of complex-valued functions.

Let f_1, f_2 be real-valued functions defined on $A \subseteq R$. Then the function f defined a $f(x) = f_1(x) + if_2(x)$, is called a *complex-valued function*, and $f_1(x)$ and $f_2(x)$ are called its *real* and *imaginary parts*. A function f is differentiable on A if at every x in A,

$$f'(x) = f_1'(x) + if_2'(x).$$

It is easy to see that the sum, product, and quotient of differentiable complex-valued functions is differentiable.

Example 9.40. Find the derivative of the function defined by $f(x) = e^{\alpha x}$, $\alpha = a + ib$. Since $e^{x+iy} = e^x(\cos y + i\sin y)$ (Section 5.7), we can write

$$f'(x) = \frac{d}{dx}\left[e^{(a+ib)x}\right] = \frac{d}{dx}\left[e^{ax}(\cos bx + i\sin bx)\right].$$

Denoting

$$f_1(x) = e^{ax}\cos bx, \quad f_2(x) = e^{ax}\sin bx$$

we derive that

$$f_1'(x) = e^{ax}(a\cos bx - b\sin bx), \quad f_2'(x) = e^{ax}(a\sin bx + b\cos bx).$$

Thus

$$\frac{d}{dx}(e^{\alpha x}) = \alpha e^{\alpha x}. \tag{9.78}$$

An antiderivative of the function f is a function $F(x) = F_1(x) + iF_2(x)$ such that $F'(x) = f(x)$ for all $x \in A$, where $F_1(x)$, $F_2(x)$ are antiderivatives of the functions $f_1(x)$, $f_2(x)$, respectively. Obviously, if $F(x)$ is an antiderivative, then every antiderivative of f is of the form $F(x) + C$ (C being some constant). The expression $\int f(x)dx$, with no limits on the integral symbol, is called the indefinite integral of the function f, in contrast with the definite integral, which has upper and lower limits, and we set

$$\int f(x)dx = \int f_1(x)dx + i\int f_2(x)dx.$$

Clearly, for these indefinite integrals, properties (1)–(4) of Section 8.2 are true.

Example 9.41. Evaluate $I = \int e^{\alpha x}dx$.

Taking into account (9.78), we have

$$I = \frac{1}{\alpha}e^{\alpha x}x + C\,(\alpha \neq 0)$$

Example 9.42. Evaluate $I_k(\alpha) = \int x^k e^{\alpha x} dx$ $(\alpha \neq 0)$, where k is a natural number and α is a complex number. Applying integration by parts, we derive

$$I_k(\alpha) = \frac{1}{\alpha} \int x^k de^{\alpha x} = \frac{1}{\alpha} x^k e^{\alpha x} - \frac{k}{\alpha} \int x^{k-1} e^{\alpha x} dx = \frac{1}{\alpha} x^k e^{\alpha x} - \frac{k}{\alpha} I_{k-1}(\alpha),$$

$$I_k(\alpha) = \frac{1}{\alpha} x^k e^{\alpha x} - \frac{k}{\alpha} I_{k-1}(\alpha) \quad (k \geqslant 1).$$

Example 9.43. Evaluate $I = \int e^{\alpha x} \cos bx \, dx$, where α is complex and b is real.

By using the Euler formulas(Section 5.7), we can write

$$I = \int e^{\alpha x} \left[\frac{e^{ibx} + e^{-ibx}}{2} \right] dx = \frac{1}{2} \int e^{(a+ib)x} dx + \frac{1}{2} \int e^{(a-ib)x} dx$$

$$= \frac{e^{ax}}{2} \left[\frac{e^{ibx}}{a+ib} + \frac{e^{-ibx}}{a-ib} \right] + C = e^{ax} \frac{a \cos bx + b \sin bx}{a^2 + b^2} + C.$$

Let F be an antiderivative of a function f defined on $[a,b]$. Then by the Fundamental Theorem of Calculus, the definite integral $\int_a^b f(x) dx$ is

$$\int_a^b f(x) dx = F(b) - F(a)$$

If the functions f_1 and f_2 are Riemann integrable on $[a,b]$, then

$$\int_a^b f(x) dx = \int_a^b f_1(x) dx + i \int_a^b f_2(x) dx$$

is called the *Riemann integral* of f on $[a,b]$.

Note that the basic properties of Newton's and Riemann's integrals for real-valued functions given in Sections 9.1 and 9.4 can be proved for complex-valued functions. In particular, if a function is integrable in both Newton's and Riemann's senses, then the two different definitions yield the same results for the values of the definite integrals.

9.13 Problems

In each of problems (1)–(3), for an arbitrary given n, subdivide the indicated intervals into n equal subintervals and compute the upper and lower Riemann sums.

(1) $f(x) = x^3, -2 \leqslant x \leqslant 3.$ *Answer:* $S = 4 + \dfrac{175}{2n} + \dfrac{125}{4n^2}; s = 4 - \dfrac{175}{2n} + \dfrac{125}{4n^2}.$

(2) $f(x) = \sqrt{x}, 0 \leqslant x \leqslant 1.$ *Answer:* $S = \dfrac{1}{n} \sum_{i=1}^{n} \sqrt{\dfrac{i}{n}}; s = \dfrac{1}{n} \sum_{i=1}^{n-1} \sqrt{\dfrac{i}{n}}.$

(3) $f(x) = 2^x, 0 \leqslant x \leqslant 10.$ *Answer:* $S = \dfrac{10230 \cdot 2^{\frac{10}{n}}}{n \left(2^{\frac{10}{n}} - 1 \right)}; s = \dfrac{10230}{n \left(2^{\frac{10}{n}} - 1 \right)}.$

In problems (4)–(8), by passing to the limit in the Riemann sums, σ, evaluate the definite integrals.

(4) $\displaystyle\int_a^b x^2 dx.$ *Answer:* $\dfrac{b^3 - a^3}{3}.$

Suggestion: Subdivide the interval $[a,b]$ into n equal subintervals by choosing points $x_i = aq^i$ $(i = 0, 1, \ldots, n)$, $q = \sqrt[n]{\dfrac{b}{a}}.$

(5) $\displaystyle\int_a^b \dfrac{dx}{x}, 0 < a < b.$ *Answer:* $\ln\dfrac{b}{a}.$

(6) $\displaystyle\int_a^b \sqrt{x}\, dx.$ *Answer:* $\dfrac{2}{3}\left(b^{\frac{3}{2}} - a^{\frac{3}{2}}\right).$

(7) $\displaystyle\int_a^b \sin x\, dx.$ *Answer:* $\cos a - \cos b.$

Suggestion: Use the formula

$$\sin a + \sin(a+h) + \cdots + \sin[a+(n-1)h] = \frac{\cos(a-h) - \cos(a+nh)}{2\sin h}.$$

(8) $\displaystyle\int_a^b \cos x\, dx.$ *Answer:* $\sin b - \sin a.$

(9) Let f be bounded and concave downward on $[a,b]$. Prove that

$$(b-a)\frac{f(a)+f(b)}{2} \leqslant \int_a^b f(x)dx \leqslant (b-a)f\left(\frac{a+b}{2}\right).$$

(10) Let f be a Riemann integrable function on $[a,b]$ and let

$$f_n(x) = \sup_{x\in[x_i,x_{i+1}]} f(x); \quad x_i = a + \frac{i}{n}(b-a) \ (i = 0, 1, \ldots, n; \ n = 1, 2, \ldots), \ x \in [x_i, x_{i+1}];$$

Prove that

$$\lim_{n\to\infty} \int_a^b f_n(x)dx = \int_a^b f(x)dx.$$

(11) Let the functions f and g be integrable. Investigate whether their composition $f \circ g$ is integrable. Consider the case where g is the Riemann function (Example 9.38) and

$$f(x) = \begin{cases} 0, & \text{if } x = 0, \\ 1, & \text{if } x \neq 0. \end{cases}$$

(12) Let f be Riemann integrable on $[a,b]$. Prove that

$$\int_a^b f^2(x)dx = 0$$

if and only if $f(x) = 0$ at all points of continuity of $[a,b]$.

In problems (13)–(16), find the limits by using the definite integral.

(13) $\lim\limits_{n\to\infty}\left(\dfrac{1}{n^2}+\dfrac{2}{n^2}+\cdots+\dfrac{n-1}{n^2}\right).$ *Answer:* $\dfrac{\pi}{4}$.

(14) $\lim\limits_{n\to\infty}\left(\dfrac{n}{n^2+1^2}+\dfrac{n}{n^2+2^2}+\cdots+\dfrac{n}{n^2+n^2}\right).$ *Answer:* $\dfrac{\pi}{4}$.

(15) $\lim\limits_{n\to\infty}\dfrac{1}{n}\left(\sin\dfrac{\pi}{n}+\sin\dfrac{2\pi}{n}+\cdots+\sin\dfrac{(n-1)\pi}{n}\right).$ *Answer:* $\dfrac{2}{\pi}$.

(16) $\lim\limits_{n\to\infty}\dfrac{\sqrt[n]{n!}}{n}.$ *Answer:* $\dfrac{1}{e}$.

Use integration by parts to compute the integrals in problems (17)–(22):

(17) $\displaystyle\int_0^\pi x\sin x\,dx.$ *Answer:* π.

(18) $\displaystyle\int_0^{\ln 2} xe^{-x}\,dx.$ *Answer:* $\dfrac{1}{2}\ln\dfrac{e}{2}$.

(19) $\displaystyle\int_0^1 \arccos x\,dx.$ *Answer:* 1.

(20) $\displaystyle\int_0^{2\pi} x^2\cos x\,dx.$ *Answer:* 4π.

(21) $\displaystyle\int_0^{\ln 2} \sqrt{e^x-1}\,dx.$ *Answer:* $2-\dfrac{\pi}{2}$.

(22) $\displaystyle\int_0^1 \dfrac{\arcsin\sqrt{x}}{\sqrt{x(1-x)}}\,dx.$ *Answer:* $\dfrac{\pi^2}{4}$.

In each of problems (23)–(30), apply the Fundamental Theorem of Calculus to find the integrals.

(23) $\displaystyle\int_{\sqrt[3]{a}}^x x^2\,dx.$ *Answer:* $\dfrac{x^3-a}{3}$.

(24) $\displaystyle\int_1^e \dfrac{dx}{x}.$ *Answer:* 1.

(25) $\displaystyle\int_0^{\frac{\sqrt{2}}{2}} \dfrac{dx}{\sqrt{1-x^2}}.$ *Answer:* $\dfrac{\pi}{4}$.

(26) $\displaystyle\int_0^{\frac{\pi}{2}} \cos^2 x\,dx.$ *Answer:* $\dfrac{\pi}{4}$.

(27) $\displaystyle\int_0^{\frac{\pi}{3}} \tan x\,dx.$ *Answer:* $\ln 2$.

(28) $\displaystyle\int_{\sinh 1}^{\sinh 2} \dfrac{dx}{\sqrt{1+x^2}}.$ *Answer:* 1.

(29) $\int_0^{\frac{\pi}{2}} \sin^n x\, dx.$ *Answer:* $\dfrac{(2k-1)!!}{(2k)!!} \dfrac{\pi}{2}, n = 2k; \quad \dfrac{(2k)!!}{(2k+1)!!}, n = 2k+1,$ where

$$m!! = \begin{cases} 1\cdot 3\cdot 5\cdots m, & m = 2k-1, \\ 2\cdot 4\cdot 6\cdots m, & m = 2k. \end{cases}$$

(30) $\int_0^1 x^m (\ln x)^n dx.$ *Answer:* $\dfrac{(-1)^n n!}{(m+1)^{n+1}}.$

(31)   Use integration by parts to compute the Euler integral $B(m,n)$, where m, n are positive integers

$$B(m,n) = \int_0^1 x^{m-1}(1-x)^{n-1} dx. \qquad \textit{Answer:} \qquad \dfrac{(m-1)!(n-1)!}{(m+n-1)!}.$$

(32)   The Legendre polynomial is defined as follows:
$$P_n(x) = \frac{1}{2^n n!} \frac{d^n}{dx^n}\left[(x^2 - 1)^n\right] \quad (n = 0, 1, 2, \ldots).$$
Prove that

$$\int_{-1}^1 P_m(x) P_n(x)\, dx = \begin{cases} 0, & m \neq n, \\ \dfrac{2}{2n+1}, & m = n. \end{cases}$$

In problems(33)–(35), use the mean value theorem to determine which integral is greater in value:

(33) $\int_0^1 e^{-x} dx$ or $\int_0^1 e^{-x^2} dx.$ *Answer:*   second.

(34) $\int_\pi^{2\pi} e^{-x^2} \cos^2 x\, dx$ or $\int_0^\pi e^{-x^2} \cos^2 x\, dx.$ *Answer:*   first.

(35) $\int_0^{\frac{\pi}{2}} \sin^2 x\, dx$ or $\int_0^{\frac{\pi}{2}} \sin^{10} x\, dx.$ *Answer:*   second.

(36)   Prove that
$$\lim_{n\to\infty} \int_n^{n+p} \frac{\sin x}{x}\, dx = 0 \quad (p > 0).$$
In problems (37)–(41) evaluate the improper integrals:

(37) $\int_1^\infty \dfrac{dx}{x^5}.$ *Answer:* $\dfrac{1}{4}.$

(38) $\int_0^\infty e^{-x} dx.$ *Answer:*   1.

(39) $\int_0^1 \ln x\, dx.$ *Answer:* $-1.$

(40) $\int_0^\infty \dfrac{dx}{9+x^2}.$ *Answer:* $\dfrac{\pi}{6}.$

(41) $\int_{-\infty}^{+\infty} \dfrac{dx}{(x^2+x+1)^2}.$ *Answer:* $\dfrac{4\pi}{3\sqrt{3}}.$

Determine the convergence of the improper integrals in problems (42)–(46).

(42) $\displaystyle\int_0^\infty x\sin x\,dx.$ *Answer:*   divergent.

(43) $\displaystyle\int_{-1}^1 \frac{dx}{x^4}.$ *Answer:*   divergent.

(44) $\displaystyle\int_0^{+\infty} \frac{x^2\,dx}{x^4-x^2+1}.$ *Answer:*   convergent.

(45) $\displaystyle\int_0^{+\infty} \frac{\sin^2 x}{x}\,dx.$ *Answer:*   divergent.

(46) $\displaystyle\int_0^{+\infty} \frac{\ln(1+x)}{x^n}\,dx.$ *Answer:*   convergent for $1 < n < 2$.

(47) Prove that if f is monotone on the interval $(0,b)$ and $\displaystyle\int_0^b x^\alpha f(x)\,dx$ exists, then
$$\lim_{x\to 0+0} x^{\alpha+1} f(x) = 0.$$
In problems (48)–(52), evaluate the integrals in the Cauchy principal value sense.

(48) p.v. $\displaystyle\int_0^{+\infty} \frac{dx}{x^2+3x+2}.$ *Answer:* $\ln\dfrac{1}{2}$.

(49) p.v.   *Answer:*   0.

(50) p.v. $\displaystyle\int_1^2 \frac{dx}{x\ln x}.$ *Answer:*   0.

(51) p.v. $\displaystyle\int_{-\infty}^{+\infty} \sin x\,dx.$ *Answer:*   0.

(52) p.v. $\displaystyle\int_{-\infty}^{+\infty} \arctan x\,dx.$ *Answer:*   0.

(53) Prove that for $t \geqslant 0$, the integral p.v. $\displaystyle\int_0^t \frac{dx}{\ln x}$ exists.

(54) Calculate the rectangular approximation to the integral $\displaystyle\int_0^{2\pi} x\sin x\,dx$, with $n = 12$. *Answer:* -6.2832.

(55) Calculate the trapezoidal approximation to the integral $\displaystyle\int_0^{\frac{\pi}{2}} \sqrt{1-\frac{1}{4}\sin^2 x}\,dx$ with $n = 6$.   *Answer:*   1.4675.

Calculate Simpson's approximation to the integrals (56)–(57).

(56) $\displaystyle\int_0^{\frac{\pi}{2}} \frac{\sin x}{x}\,dx$ $(n = 10)$. *Answer:*   1.3709.

(57) $\displaystyle\int_0^1 \frac{x\,dx}{\ln(1+x)}$ $(n = 6)$. *Answer:*   0.2288.

(58)   Let α be monotone on $[a,b]$ and continuous at $x_0 \in [a,b]$. Prove that the function f defined as

$$f(x) = \begin{cases} 0, & \text{if } x \neq x_0, \\ 1, & \text{if } x = x_0 \end{cases}$$

is Stieltjes integrable with respect to α and $\int_a^b f(x)d\alpha(x) = 0$.

(59)   Let f be of bounded variation over $[0,2\pi]$ and suppose $f(2\pi) = f(0)$. Prove that the absolute values of the integrals

$$\int_0^{2\pi} f(x)\cos nx\,dx, \quad \int_0^{2\pi} f(x)\sin nx\,dx,$$

are bounded with the value $\dfrac{V_0^{2\pi}}{n}$ ($V_0^{2\pi}$ is a total variation of f). *Suggestion:* Show that

$$n\int_0^{2\pi} f(x)\cos nx\,dx = \int_0^{2\pi} f(x)d(\sin nx) = \int_0^{2\pi} \sin nx\,df(x).$$

(60)   Show that, $\int_0^3 xd(\llbracket x \rrbracket - x) = \dfrac{3}{2}$, where $x - 1 < \llbracket x \rrbracket \leqslant x$.

(61)   Suppose $f \geqslant 0$ and $\int_E f(x)dx = 0$. Show that $f(x) = 0$ almost everywhere on E.

Suggestion: Let $E_n = \left\{ x \in E : f(x) > \dfrac{1}{n} \right\}$. Then $m\left[\bigcup_n E_n\right] = 0$ if and only if $m(E_n) = 0$ ($n = 0, 1, 2, \ldots$).

(62)   Prove that if $\int_A f(x)dx = 0$ for all measurable subsets A of E, then $f(x) = 0$ almost everywhere on E.

(63)   Show that

$$\int_0^{2\pi} e^{inx}e^{-imx}dx = \begin{cases} 0, & m \neq n, \\ 2\pi, & m = n \end{cases}$$

where n and m are integers.

Suggestion: Use Euler's formula $e^{ix} = \cos x + i\sin x$.

Chapter 10

Applications of the Definite Integral

First we define the concepts of length, area, and volume: these geometric concepts are given analytic definitions using the concept of the definite integral which we have developed. We compute the arc length, the surface area, and the volume for some special cases, such as geometric figures produced by revolving one-dimensional curves in space. Then, the centroids of plane regions and curves, and the moments of curves and of solids are calculated. It is shown that the definite integral can be applied to the law of conservation of mechanical energy, to Einstein's theory of relativity, and to calculate the escape velocity from the earth. At the end of the chapter, we consider examples connected with atomic energy, critical mass and atomic reactors.

10.1 Arc length

1. Arc Length in Cartesian Coordinates. Let L be the graph of a continuous function f defined on the closed interval $[a,b]$. Let $P = \{x_0, \dots, x_n\}$ be any partition of $[a,b]$,

$$a = x_0 < x_1 < \cdots < x_n = b.$$

We denote by $P_i = (x_i, f(x_i))$ $(i = 0, 1, \dots, n)$ the point on the arc L corresponding to the i-th subdivision point x_i, and inscribe in L a polygonal arc $l = l(P)$, that is, the union of the line segments $P_{i-1}P_i$, $i = 1, \dots, n$. If we denote the length of the polygonal arc l corresponding to the partition P, by $|S| = |S(P)|$, then we have

$$|S| = \sum_{i=1}^{n} \sqrt{(x_i - x_{i-1})^2 + (f(x_i) - f(x_{i-1}))^2}.$$

Definition 10.1. If the set of positive real numbers $\{|S|\}$ corresponding to the lengths of polygonal arcs $y = p$ "inscribed in" L for all possible partitions P of $[a,b]$ is bounded, then we say that the length S of L exists, and $S = \sup\{|S|\}$.

Remark 10.1. It can be shown that there exists a curve which does not have a length.[1]

---

[1] See, for example, B.A. Ilyin and E.G. Poznjak, *Introduction to Mathematical Analysis*, M, 1982, pp. 382–386.

E. Mahmudov, *Single Variable Differential and Integral Calculus*,
DOI: 10.2991/978-94-91216-86-2_10, © Atlantis Press 2013

Remark 10.2. Suppose that L is the curve in the plane given parametrically by the pair of functions $x = \varphi(t)$, $y = \psi(t)$, $t \in [\alpha, \beta]$. Then in the length formula for $|l|$ corresponding to the partition $T = \{t_0, \ldots, t_n\}$ of $[\alpha, \beta]$, the terms under the radical sign may be taken to be $[\varphi(t_i) - \varphi(t_{i-1})]^2 + [\psi(t_i) - \psi(t_{i-1})]^2$. It should be noted that in this case, the number of times that different values of the parameter t are assigned to the same point (x, y), should be finite.[2]

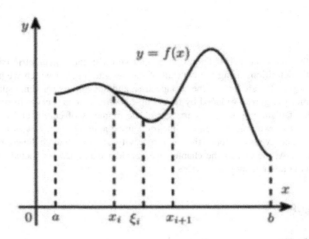

Fig. 10.1   A polygonal arc inscribed in curve $y = f(x)$ is a union of segments with length ΔS_i.

Now, by Δs_i, we denote the length of the line segments with end points $(x_i, f(x_i))$ and $(x_{i+1}, f(x_{i+1}))$ $(i = 0, 1, \ldots, n-1)$. Then, evidently,

$$S = \lim_{\max \Delta s_i \to 0} \sum_{i=0}^{n-1} \Delta s_i. \tag{10.1}$$

If the derivative $f'(x)$ of $f(x)$ is continuous on $[a, b]$, then the limit (10.1) exists and so the arc L has the length S. Put $\Delta x_i = x_{i+1} - x_i$ and $\Delta y_i = f(x_{i+1}) - f(x_i)$. Then

$$\Delta s_i = \sqrt{(\Delta x_i)^2 + (\Delta y_i)^2} = \sqrt{1 + \left(\frac{\Delta y_i}{\Delta x_i}\right)^2} \Delta x_i.$$

By Theorem 5.4, $f(x_{i+1}) - f(x_i) = f'(x_0)(x_{i+1} - x_i)$, $x_0 \in (x_i, x_{i+1})$ and $\dfrac{\Delta y_i}{\Delta x_i} = f'(x_0)$. Thus, $\Delta s_i = \sqrt{1 + [f'(\xi_i)]^2} \Delta x_i$, and the length $|S|$ of the polygonal arc l is

$$|S| = \sum_{i=0}^{n-1} \sqrt{1 + [f'(\xi_i)]^2} \Delta x_i. \tag{10.2}$$

---

[2]See, for example, L. Bieberbach, *Introduction to Differential and Integral Calculus*, vol. 1, p. 156.

By hypothesis, the function $\sqrt{1+[f'(\xi_i)]^2}$ is continuous. It follows that the limit S of the sum (10.2) exists,

$$S = \lim_{\mu(P)\to 0} \sum_{i=1}^{n-1} \sqrt{1+[f'(\xi_i)]^2}\,\Delta x_i = \int_a^b \sqrt{1+[f'(x)]^2}\,dx$$

or, more briefly,

$$S = \int_a^b \sqrt{1+[f'(x)]^2}\,dx. \tag{10.3}$$

Remark 10.3. If the arc L has parametric equations $x = \varphi(t)$, $y = \psi(t)$, $t \in [\alpha,\beta]$, then $\dfrac{dy}{dx} = \dfrac{\psi'(t)}{\varphi'(t)}$ ($\varphi'(t) \neq 0, t \in [\alpha,\beta]$). Hence, by substituting $x = \varphi(t)$ in (10.3), we have

$$S = \int_\alpha^\beta \sqrt{[\varphi'(t)]^2 + [\psi'(t)]^2}\,dt \quad (a = \varphi(\alpha),\ b = \varphi(\beta)). \tag{10.4}$$

Remark 10.4. If in (10.3), for the upper bound b of the integral we write instead the independent variable x, then

$$S(x) = \int_a^x \sqrt{1+[f'(x)]^2}\,dx$$

and so

$$\frac{dS}{dx} = \sqrt{1+[f'(x)]^2}.$$

2. Arc Length in Polar Coordinates.

Suppose we have an arc given by the equation $\rho = f(\theta)$ in polar coordinates with radial coordinate ρ and angular coordinate θ. Since the basic relations connecting Cartesian and polar coordinates are $x = \rho\cos\theta$, $y = \rho\sin\theta$, its parametric equations can also be written

$$x = f(\theta)\cos\theta, \quad y = f(\theta)\sin\theta.$$

It is not hard to see that $\left(\dfrac{dx}{d\theta}\right)^2 + \left(\dfrac{dy}{d\theta}\right)^2 = \rho'^2 + \rho^2$. Therefore, using (10.4), the length of the arc from $\theta = \alpha$ to $\theta = \beta$ is

$$S = \int_\alpha^\beta \sqrt{\rho'^2 + \rho^2}\,d\theta. \tag{10.5}$$

Example 10.1. Compute the length of the circle $x^2 + y^2 = R^2$. Because $y = \sqrt{x^2 - R^2}$, $x \in [0,R]$ the derivative is $\dfrac{dy}{dx} = -\dfrac{x}{R^2 - x^2}$.
Thus the length, S, of the circle is

$$S = 4\int_0^R \sqrt{1+[f'(x)]^2}\,dx = 4\int_0^R \frac{R}{R^2 - x^2}\,dx = 4R\arcsin\frac{x}{R}\Big|_0^R = 2\pi R.$$

Example 10.2. Find the length of the cardioid $\rho = 3(1+\sin\theta)$ (Figure 10.2)

The cardioid is symmetric about the straight line $\theta = \dfrac{\pi}{2}$.

Thus, we assume that $-\dfrac{\pi}{2} \leqslant \theta \leqslant \dfrac{\pi}{2}$.

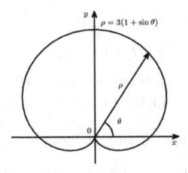

Fig. 10.2   The cardioid curve of Example 10.2

We easily derive that

$$\rho^2 + \rho'^2 = 3^2(1+\sin\theta)^2 + 3^2\cos^2\theta = 2\cdot 3^2(1+\sin\theta) = 18\left(\sin\frac{\theta}{2}+\cos\frac{\theta}{2}\right)^2.$$

Thus,

$$S = 2\int_{-\frac{\pi}{2}}^{\frac{\pi}{2}} \sqrt{18}\left(\cos\frac{\theta}{2}+\sin\frac{\theta}{2}\right)d\theta = 6\sqrt{2}\left(2\sin\frac{\theta}{2}-2\cos\frac{\theta}{2}\right)\Big|_{-\frac{\pi}{2}}^{\frac{\pi}{2}} = 24\sqrt{2}\sin\frac{\pi}{4} = 24.$$

Example 10.3. Find the length of the astroid $x = 4\cos^3 t$, $y = 4\sin^3 t$ given in parametric equations. (The astroid in Cartesian coordinates is $x^{2/3}+y^{2/3}=1$.)

Because the curve is symmetric about both x- and y-axes, we compute one-fourth of the length of the astroid by assuming that $t \in \left[0, \dfrac{\pi}{2}\right]$.

Obviously,

$$\frac{dx}{dt} = -12\cos^2 t\sin t, \qquad \frac{dy}{dt} = 12\sin^2 t\cos t.$$

Thus,

$$S = 4\int_0^{\frac{\pi}{2}} \sqrt{144\cos^4 t\sin^2 t + 144\sin^4 t\cos^2 t}\, dt = 48\int_0^{\frac{\pi}{2}} \sqrt{\cos^2 t\sin^2 t}\, dt$$

$$= 48\int_0^{\frac{\pi}{2}} \sin t\cos t\, dt = 48\,\frac{\sin^2 t}{2}\Big|_0^{\frac{\pi}{2}} = 24,$$

or,

$$S = 24.$$

Example 10.4. Find the length of the logarithmic spiral $r = ae^{b\varphi}$ between the points (φ_0, r_0) and (φ_1, r_1).

By Formula 10.5, we have

$$S = \int_{\varphi_0}^{\varphi_1} \sqrt{a^2 e^{2b\varphi} + a^2 b^2 e^{2b\varphi}} = a\sqrt{1+b^2} \int_{\varphi_0}^{\varphi_1} e^{b\varphi} \, d\varphi$$

$$= \frac{a}{b}\sqrt{1+b^2}\left(e^{b\varphi_1} - e^{b\varphi_0}\right) = \frac{\sqrt{1+b^2}}{b}(r_1 - r_0),$$

$$S = \frac{\sqrt{1+b^2}}{b}(r_1 - r_0).$$

10.2 Area and the Definite Integral

1. Area in Cartesian Coordinates. Using the area formula for a triangle, we can find the area of an arbitrary polygonal figure. The reason is that any polygon can be divided into non-overlapping triangles. This approach to area goes back several thousand years, to the ancient civilizations of Egypt and Sumeria.

Let P be a bounded polygonal region in the plane. The area of P is denoted by $m(P)$. Recall that the area of a polygon is the nonnegative number with the properties:

(1) (*Additivity*) For arbitrary polygons P_1 and P_2 such that $P_1 \cap P_2 = \varnothing$

$$m(P_1 \cup P_2) = m(P_1) + m(P_2).$$

(2) (*Invariance*) If the polygons P_1 and P_2 are equal (that is, there is a one to one and onto $f : P_1 \to P_2$ such that $\rho(x,y) = \rho(f(x), f(y))$, $x, y \in P_1$ with metric ρ), then

$$m(P_1) = m(P_2).$$

(3) (*Monotonicity*) If $P_1 \subseteq P_2$, then $m(P_1) \leqslant m(P_2)$.

To investigate the area of a curvilinear bounded region f in the plane, we might inscribe in it and also circumscribe about it all possible polygons P and Q, that is $P \subseteq F \subseteq Q$. Obviously, the set of numbers $\{m(Q)\}$ is bounded below (for instance, by zero). It follows that there is the least upper bound

$$m_* = m_*(F) = \sup_{P \subset F} m(P).$$

Similarly, there exists the greatest lower bound

$$m^* = m^*(F) = \inf_{Q \supset F} m(Q).$$

By the monotonicity property (3), we have

$$m_*(F) \leqslant m^*(F).$$

Definition 10.2. If the values $m_*(F)$, $m^*(F)$ turn out to be equal, $m_*(F) = m^*(F)$, then the area of the set f we started with must be given by $m = m(F) = m_* = m^*$.

Let f be a continuous, positive valued function on $[a,b]$, and suppose we want to compute the area f under the graph of f from $x = a$ to $x = b$.

By Theorem 9.5, the function f is Riemann integrable on $[a,b]$, so for any $\varepsilon > 0$ there is a partition of $[a,b]$ such that $S - s < \varepsilon$, where S and s are upper and lower Riemann sums, respectively.

Suppose now that P is the inscribed rectangular polygon and Q is the circumscribed rectangular polygon for f, that is $P \subseteq F \subseteq Q$. Because $S = m(Q)$ and $s = m(P)$ it follows that $m(Q) - m(P) < \varepsilon$ and $s \leqslant m(F) \leqslant S$. Thus the area of f exists and is the Riemann integral $\int_a^b f(x)dx$ (see Sections 9.2 and 9.3). Consequently, denoting $A = m(F)$, we have

$$A = \int_a^b f(x)dx. \tag{10.6}$$

If $f(x) \leqslant 0$ on $[a,b]$, then the integral (10.6) is a nonpositive number, $-A$. Since the area is nonnegative, we define

$$A = \int_a^b |f(x)|dx.$$

Let f_1, f_2 be continuous with $f_1(x) \geqslant f_2(x)$ for x in $[a,b]$. Then the area A of the region bounded by the curves $y = f_1(x)$ and $y = f_2(x)$ and the vertical lines $x = a$, $x = b$ is

$$A = \int_a^b [f_1(x) - f_2(x)]dx. \tag{10.7}$$

More generally, consider a continuous function f with a graph that crosses the x-axis at finitely many points x_i, $i = 1, \ldots, k$ between a and b. Then we write

$$\int_a^b f(x)dx = \int_a^{x_1} f(x)dx + \int_{x_1}^{x_2} f(x)dx + \cdots + \int_{x_k}^b f(x)dx.$$

Hence, we see that $\int_a^b f(x)dx$ is equal to the area under $y = f(x)$ above the x-axis, minus the area over $y = f(x)$ below the x-axis.

Finally, let f_1, f_2 be continuous, with $f_1(y) \geqslant f_2(y)$ for y in $[c,d]$. Then the area A of the region bounded by the curves $x = f_1(y)$ and $x = f_2(y)$ and the vertical lines $y = c$ and $y = d$ is

$$A = \int_c^d [f_1(y) - f_2(y)]dy.$$

Suppose we have parametric equations for a curve,

$$x = \varphi(t), \quad y = \psi(t), \quad t \in [\alpha, \beta], \quad (\varphi(\alpha) = a, \ \varphi(\beta) = b). \tag{10.8}$$

In order to calculate the area under the graph from α to β, assume that the equations (10.8) define a function $y = f(x)$ on $[a,b]$. Then, by substituting $x = \varphi(t)$ in formula (10.6) so that $dx = \varphi'(t)dt$, $y = f[\varphi(t)] = \psi(t)$, we obtain

$$A = \int_\alpha^\beta \psi(t)\varphi'(t)dt. \tag{10.9}$$

Example 10.5. Find the area A of the region bounded by the curves $y = x^\alpha$ and $x = y^\alpha$, $\alpha \geqslant 1$. The points of intersection of these curves are $(0,0)$ and $(1,1)$. Because $x^{1/\alpha} \geqslant x^\alpha$ for all x in $[0,1]$, it follows from (10.7) that

$$A = \int_0^1 (x^{1/\alpha} - x^\alpha)dx = \left(\frac{\alpha x^{\frac{\alpha+1}{\alpha}}}{\alpha+1} - \frac{x^{\alpha+1}}{\alpha+1} \right)\Big|_0^1 = \frac{\alpha-1}{\alpha+1}.$$

Example 10.6. Find the area A of the region bounded by the curves $y = \sin x$, $x \in [0, 2\pi]$, and the x-axis. We have,

$$A = \int_0^{2\pi} |\sin x|dx = \int_0^\pi \sin x\, dx + \left| \int_\pi^{2\pi} \sin x\, dx \right|$$
$$= -(\cos\pi - \cos 0) + |-(\cos 2\pi - \cos\pi)| = 2 + |-2| = 4,$$

that is, $A = 4$.

Example 10.7. Find the area A of the region bounded by the x-axis and the cycloid, $x = 2(t - \sin t)$, $y = 2(1 - \cos t)$, $t \in [0, 2\pi]$.

By (10.9),

$$A = \int_0^{2\pi} 2(1-\cos t)2(1-\cos t)dt = 4\int_0^{2\pi}(1-\cos t)^2dt$$
$$= 4\left[\int_0^{2\pi} dt - 2\int_0^{2\pi}\cos t\, dt + \int_0^{2\pi}\cos^2 t\, dt \right]$$
$$= 4(2\pi - 0 + \pi) = 12\pi,$$

That is, $A = 12\pi$.

2. Area in Polar Coordinates.

Suppose that the region R is bounded by the two radial lines $\theta = \alpha$ and $\theta = \beta$ and by the curve $\rho = f(\theta)$, $\theta \in [\alpha,\beta]$. To approximate the area A of R, we begin with a partition $T = \{\theta_0, \theta_1, \ldots, \theta_n\}$ of the interval $[\alpha,\beta]$,

$$\alpha = \theta_0 < \theta_1 < \cdots < \theta_n = \beta.$$

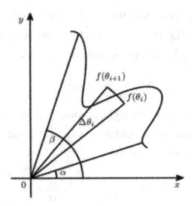

Fig. 10.3   The area formula from Reimann sums in polar coordinates.

Let $\xi_i \in [\theta_i, \theta_{i+1}]$ $(i = 0, 1, \ldots, n - 1)$ be some selection of points for T, and $\Delta\theta_i = \theta_{i+1} - \theta_i$. It is not hard to see that the area of the sector bounded by the lines $\theta = \theta_i$, $\theta = \theta_{i+1}$ and the curve $\rho = f(\theta)$, is approximately equal to the area of the circular sector $\frac{1}{2}[f(\xi_i)]^2\Delta\theta_i$ with radius $f(\xi_i)$ and bounded by the same angle (Figure 10.3).

We add the areas of these sectors for $i = 0, 1, \ldots, n - 1$, and thereby find that

$$A \approx \frac{1}{2}\sum_{i=0}^{n-1}[f(\xi_i)]^2\Delta\theta_i.$$

The right-hand sum is a Riemann sum for the integral $\frac{1}{2}\int_\alpha^\beta f^2(\theta)d\theta$. We conclude that the area A of the region R bounded by lines $\theta = \alpha$ and $\theta = \beta$ and by the curve $\rho = f(\theta)$, $\theta \in [\alpha, \beta]$, is the limit of the sum as the mesh $\mu(T) \to 0$:

$$\lim_{\mu(T)\to 0}\frac{1}{2}\sum_{i=0}^{n-1}[f(\xi_i)]^2\Delta\theta_i = \frac{1}{2}\int_\alpha^\beta f^2(\theta)d\theta,$$

or

$$A = \frac{1}{2}\int_\alpha^\beta \rho^2 d\theta. \tag{10.10}$$

Now consider two curves $\rho_1 = f_1(\theta)$ and $\rho_2 = f_2(\theta)$ $(f_1(\theta) \geqslant f_2(\theta))$ for $\theta \in [\alpha, \beta]$. Then the area of the region bounded by these two curves and the rays $\theta = \alpha$ and $\theta = \beta$ may be subtracting the area bounded by the inner curve from that bounded by the outer curve. That is, the area between the curves is given by

$$A = \frac{1}{2}\int_\alpha^\beta [\rho_1^2 - \rho_2^2]d\theta.$$

Example 10.8. (1) Find the area of the region bounded by the three-leaved rose,

$$\rho = a\cos 3\theta.$$

As is seen in Figure 10.4, for $\theta \in \left[0, \dfrac{\pi}{6}\right]$ we have the one-sixth part of the desired area. So, by (10.10), we get

$$A = 6 \cdot \frac{1}{2} \cdot \int_0^{\frac{\pi}{6}} (a\cos 3\theta)^2 d\theta = 3a^2 \int_0^{\frac{\pi}{6}} \frac{1+\cos 6\theta}{2} d\theta = 3a^2 \left[\frac{\pi}{12} + \frac{\sin 6\theta}{12}\right]\Big|_0^{\frac{\pi}{6}} = \frac{\pi a^2}{4},$$

that is,

$$A = \frac{\pi a^2}{4}.$$

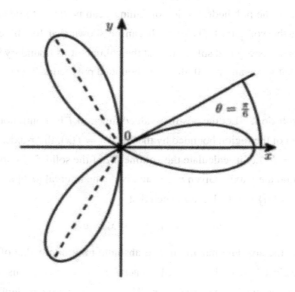

Fig. 10.4   The area of the three-leaved rose.

(2) Find the area of the region bounded by the lemniscate $\rho = 4\sqrt{\cos 2\theta}$.

It is easy to see, if by A we denote the area of the lemniscate, that for $\theta \in \left[0, \dfrac{\pi}{4}\right]$,

$$\frac{1}{4}A = \frac{1}{2}\int_0^{\frac{\pi}{4}} \rho^2 d\theta,$$

and so,

$$\frac{1}{4}A = \frac{1}{2} \cdot 4^2 \int_0^{\frac{\pi}{4}} \cos 2\theta d\theta = 8 \frac{\sin 2\theta}{2}\Big|_0^{\frac{\pi}{4}} = 8 \cdot \frac{1}{2} = 4.$$

Thus,

$$A = 16.$$

10.3   Volume and the Definite Integral

In this section we show how to use integrals to calculate the volumes of certain solids f in three-dimensional space. We begin with the idea that volume has properties, (1)–(3) (additivity, invariance, and monotonicity), analogous to the properties of area listed in Section 10.2. Denote by Q a bounded polyhedral solid containing f, and by P, a similar solid that is contained in f. Taking $m^* = m^*(F) = \operatorname*{int}_{Q \supset F} m(Q)$ and $m_* = m_*(F) = \sup_{P \subset F} m(P)$, we see that $m_*(F) \leqslant m^*(F)$. If $m_*(F) = m^*(F)$, we say that f has a volume, and that $m = m(F) = m_* = m^*$ is the volume of f.

If the volume of the polyhedral solid containing f can be taken to be arbitrarily small, then we say that the volume of f is zero. It can be shown that for the existence of the volume of F, it is necessary and sufficient that the volume of its boundary is equal to zero. In other words, that given any $\varepsilon > 0$, there are bounded polyhedra P, Q ($P \subseteq F \subseteq Q$) such that $m(Q) - m(P) < \varepsilon$.

1. Solids of Revolution.   Let the positive-valued function f be continuous on $[a, b]$. Denote by R the area of the region bounded by the curve $y = f(x)$, the x-axis, and the vertical lines $x = a$ and $x = b$. Let us calculate the volume, V, of the solid f obtained by revolving the region R around the x-axis. Given n, we subdivide the interval $[a, b]$ into n subintervals by choosing points $\{x_i\}$ ($i = 0, 1, \ldots, n$) such that

$$a = x_0 < x_1 < \cdots < x_n = b.$$

Let M_i and m_i be the absolute maximum and absolute minimum value of the continuous function f on the closed interval $[x_{i-1}, x_i]$. For each i ($i = 0, 1, \ldots, n$), consider the rectangle with base $[x_{i-1}, x_i]$, height M_i, and area $M\Delta x_i$. The union of these n rectangles contains the region R under the graph: it is the circumscribed rectangular polygon associated with the partition of $[a, b]$. Similarly, the rectangle with base $[x_{i-1}, x_i]$ and height m_i lies within the region R. The union of these rectangles is the inscribed rectangular polygon associated with our partition of $[a, b]$ (see Figure 9.2). The solids obtained by revolving these circumscribed and inscribed rectangular polygons around the x-axis, we denote by Q and P, respectively. Obviously, $P \leqslant F \leqslant Q$ and the sums

$$m(Q) = \pi \sum_{i=1}^{n} M_i^2 \Delta x_i, \quad m(P) = \pi \sum_{i=1}^{n} m_i^2 \Delta x_i$$

are upper and lower Riemann sums for $\pi f^2(x)$, respectively. Since $\pi f^2(x)$ is Riemann integrable, given arbitrarily small positive ε, the difference between these sums can be made less than ε, by a suitable choice of mesh size. Hence, the solid f has a volume. Thus, by passing to the limit as the mesh (μ) tends to zero, we obtain

$$V = \pi \int_a^b f^2(x)dx. \tag{10.11}$$

Suppose that $f_1(x) \geqslant f_2(x)$, and that the solid f is generated by revolving the region between the curves $y = f_1(x)$, $y = f_2(x)$, and the vertical lines $x = a, x = b$, about the x-axis. Then the volume of f is

$$V = \pi \int_a^b [f_1^2(x) - f_2^2(x)]dx.$$

Similarly, if the solid f is generated by revolving the region between the curves $x = f_1(y)$, $x = f_2(y)$ and vertical lines $y = c, y = d$ about the y-axis, then

$$V = \pi \int_c^d [f_1^2(y) - f_2^2(y)]dy.$$

2. Volumes by the Method of Cross Sections.   This method computes the volume of a solid that is described in terms of its cross sections by planes perpendicular to either the x-axis or y-axis. We first suppose that the solid f with volume V lies opposite the interval $[a,b]$ on the x-axis. Let $A = A(x)$ denote the area of the cross section of f by the perpendicular plane that meets the x-axis at the point x of $[a,b]$. Assume that this cross-sectional area function $A(x)$ is continuous on $[a,b]$. To set up an integral formula for V, we begin with a partition $a = x_0 < x_1 < \cdots < x_n = b$ of $[a,b]$ into subintervals with lengths $\Delta x_i = x_{i+1} - x_i$. Let F_i denote the slab (or slice) of the solid. Moreover, suppose that the cross sections are obtained as a result of intersecting the solid f that lies opposite the i-th subinterval $[x_i, x_{i+1}]$. We denote the volume of this i-th slab of f by ΔV_i, so that $V = \sum_{i=0}^{n-1} \Delta V_i$. To approximate ΔV_i, we choose an arbitrary point ξ_i $(i = 0, 1, \ldots, n-1)$ in $[x_i, x_{i+1}]$, and consider the cylinder with height $\Delta x_i = x_{i+1} - x_i$ and base $A(\xi_i)$. Obviously, if the mesh of the partition is small, then $A(\xi_i)\Delta x_i$ is a good approximation to ΔV_i. Then we find that

$$V = \sum_{i=0}^{n-1} \Delta V_i \approx \sum_{i=0}^{n-1} A(\xi_i)\Delta x_i.$$

We recognize the approximating sum here as a Riemann sum that approaches $\int_a^b A(x)dx$ as the mesh $\mu \to 0$. Thus, we have $V = \lim_{\mu \to 0} \sum_{i=0}^{n-1} A(\xi_i)\Delta x_i$, or, more conveniently,

$$V = \int_a^b A(x)dx. \tag{10.12}$$

Formula 10.12 is sometimes called Cavalieri's[3] principle.

Similarly, if the solid F lies opposite the interval $[c,d]$ on the y-axis and has a continuous

---

[3]B.F. Cavalieri (1598-1647), Italian matematician.

cross-sectional area function $A(y)$, then its volume is

$$V = \int_c^d A(y)dy.$$

Example 10.9. Find the volume of the solid that is generated by rotating the plane region bounded by the curve $y = \sin x$, $x \in [0, \pi]$ about the x-axis.

By Formula 10.11, we can write

$$V = \pi \int_0^\pi \sin^2 x\, dx = \pi \int_0^\pi \frac{1 - \cos 2x}{2}\, dx = \frac{\pi^2}{2},$$

that is,

$$V = \frac{\pi^2}{2}.$$

Example 10.10. Find the volume of the ellipsoid $\dfrac{x^2}{a^2} + \dfrac{y^2}{b^2} + \dfrac{z^2}{c^2} = 1$.

Clearly, the ellipsoid is symmetric about each of the three coordinate planes, and for any x, $-a < x < a$, this equation can be reduced to the form

$$\frac{y^2}{b^2} + \frac{z^2}{c^2} = 1 - \frac{x^2}{a^2}$$

or

$$\frac{b^2}{\left[b\sqrt{1 - \dfrac{x^2}{a^2}}\right]^2} + \frac{z^2}{\left[c\sqrt{1 - \dfrac{x^2}{a^2}}\right]^2} = 1,$$

which is the equation of an ellipse with semiaxes

$$b_1 = b\sqrt{1 - \frac{x^2}{a^2}}, \quad c_1 = c\sqrt{1 - \frac{x^2}{a^2}}.$$

Parametric equations of this ellipse are $x = b_1 \cos t$, $y = c_1 \sin t$. Then, by (10.9), we compute that

$$A = 2\int_\pi^0 (c_1 \sin t)(-b_1 \sin t\, dt) = -2b_1 c_1 \int_\pi^0 \sin^2 t\, dt = 2b_1 c_1 \int_0^\pi \frac{1 - \cos 2t}{2}\, dt = \pi b_1 \cdot c_1.$$

Hence, the area of the ellipse is $\pi b_1 \cdot c_1$. Thus, the cross-sectional area of the ellipsoid is

$$A = A(x) = \pi bc \left(1 - \frac{x^2}{a^2}\right)$$

Then, by using (10.12), we deduce that the required volume is

$$V = \pi bc \int_{-a}^a \left(1 - \frac{x^2}{a^2}\right) dx = \pi bc \left(x - \frac{x^3}{3a^2}\right)\Big|_{-a}^a = \frac{4}{3}\pi abc.$$

In particular, in the case where $a = b = c = R$, we find the familiar formula for the volume of the sphere with radius R:

$$V = \frac{4}{3}\pi R^3.$$

3. Volume by the Method of Cylindrical Shells.

The method of cylindrical shells is a second way of computing the volumes of solids of revolution. A cylindrical shell is a region bounded by two concentric circular cylinders of the same height.

Suppose that it is required to find the volume V of the solid generated by revolving the region under $y = f(x)$ from $x = a$ to $x = b$ around the y-axis. We assume that f is nonnegative on $[a,b]$ and that $0 \leqslant a < b$. Suppose that we have a partition $\{x_i\}$ $(i = 0, 1, \ldots, n)$ of $[a,b]$ such that

$$a = x_0 < x_1 < \cdots < x_n = b.$$

Let x_i^* denote the midpoint of the subinterval $[x_i, x_{i+1}]$, and consider the rectangle with base $[x_i, x_{i+1}]$ and height $f(x_i^*)$. When this rectangle is revolved about the y-axis, it sweeps out a cylindrical shell of average radius x_i^*, thickness Δx_i, and height $f(x_i^*)$. This cylindrical shell approximates the solid with volume ΔV_i that is obtained by revolving the region under $y = f(x)$ and over $[x_i, x_{i+1}]$.

It follows that

$$\Delta V_i \approx \pi \left(x_i^* + \frac{\Delta x_i}{2} \right)^2 f(x_i^*) - \pi \left(x_i^* - \frac{\Delta x_i}{2} \right)^2 f(x_i^*) = 2\pi x_i^* f(x_i^*) \Delta x_i.$$

Thus, we obtain the approximation

$$V = \sum_{i=0}^{n-1} \Delta V_i \approx \sum_{i=0}^{n-1} 2\pi x_i^* f(x_i^*) \Delta x_i.$$

This is a Riemann sum that approaches $\int_a^b 2\pi x f(x) dx$ as the mesh μ of the partition $\{x_0, \ldots, x_n\}$ tends to zero.

Thus, we have

$$V = \lim_{\mu \to 0} \sum_{i=0}^{n-1} 2\pi x_i^* f(x_i^*) \Delta x_i$$

or

$$V = \int_a^b 2\pi x f(x) dx. \qquad (10.13)$$

The method of cylindrical shells is also useful for computing the volumes of solids of revolution of regions bounded by curves $x = f(y)$ from $y = c$ to $y = d$ about the x-axis,

$$V = \int_c^d 2\pi y f(y) dy.$$

Suppose that it is required to find the volume V of the solid generated by revolving the region between the curves $y = f_1(x)$, $y = f_2(x)$ $(f_2(x) \geqslant f_1(x))$ over the interval $[a,b]$, where $0 \leqslant a < b$, around the line $x = c$. As before, we have

$$V = 2\pi \int_a^b (x - c)[f_2(x) - f_1(x)] dx.$$

Example 10.11. Use the method of cylindrical shells to find the volume of the solid generated by revolving the region bounded by the curves $y^2 = x$ and $y = x^2$ about the y-axis.

According to the last formula in this example, $c = 0$ and $f_1(x) = x^2$, $f_2(x) = \sqrt{x}$, and so

$$V = 2\pi \int_0^1 x\left(\sqrt{x} - x^2\right) dx = 2\pi \int_0^1 (x^{3/2} - x^3)dx$$
$$= 2\pi \left(\frac{2}{5}x^{5/2} - \frac{x^4}{4}\right)\Big|_0^1 = 2\pi \left(\frac{2}{5} - \frac{1}{4}\right) = \frac{3\pi}{10},$$

so that,

$$V = \frac{3\pi}{10}.$$

10.4   Surface Area

A surface of revolution is one obtained by revolving a curve about an axis lying in its plane. First we inscribe a polygonal arc in the curve to be revolved. Then we consider, as an approximation to the area of the surface, the area of the surface generated by revolving the polygonal arc.

Suppose that we have a surface, with area A_x, from revolving about the x-axis the smooth arc $y = f(x)$ on $[a,b]$, where f is a nonnegative function. (Figure 10.5)

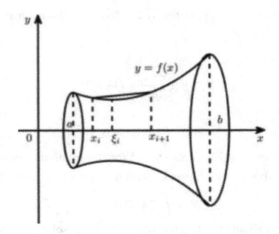

Fig. 10.5   The surface area of revolution of curve $y = f(x)$.

Suppose $P = \{x_0, x_1, \ldots, x_n\}$ is a partition of $[a,b]$. Moreover, as in Section 10.1, denote by Δs_i $(i = 0, 1, \ldots, n-1)$ the length of the line segments with end points $M_i M_{i+1}$. Because $\Delta x_i = x_{i+1} - x_i$ and $\Delta y_i = f(x_{i+1}) - f(x_i)$, the line segment $M_i M_{i+1}$ has length

$$\Delta s_i = \sqrt{(\Delta x_i)^2 + (\Delta y_i)^2} = \sqrt{1 + \left(\frac{\Delta y_i}{\Delta x_i}\right)^2} \Delta x_i = \sqrt{1 + [f'(\xi_i)]^2} \Delta x$$

for some point x_0 in the i-th subinterval $[x_i, x_{i+1}]$. The conical frustum obtained by revolving the segment $M_i M_{i+1}$ around the x-axis has slant height Δs_i and average radius

$$\frac{1}{2}[f(x_i) + f(x_{i+1})].$$

Then it is known from elementary geometry that the area of this conical frustum is

$$\Delta A_i = 2\pi \frac{f(x_i) + f(x_{i+1})}{2} \Delta s_i \quad (i = 0, 1, \ldots, n-1).$$

Therefore,

$$\Delta A_i = \pi[f(x_i) + f(x_{i+1})] \sqrt{1 + [f'(\xi_i)]^2} \Delta x_i.$$

Then, by revolving around the x-axis the polygonal arc inscribed in the smooth curve $y = f(x)$, we have a surface with the area

$$A_x \approx \sum_{i=0}^{n-1} \Delta A_i = \pi \sum_{i=0}^{n-1} [f(x_i) + f(x_{i+1})] \sqrt{1 + [f'(\xi_i)]^2} \Delta x_i.$$

On the other hand, by passing to the limit as the mesh tends to zero, we have

$$\lim_{\mu(P) \to 0} \pi \sum_{i=0}^{n-1} [f(x_i) + f(x_{i+1})] \sqrt{1 + [f'(\xi_i)]^2} \Delta x_i$$

$$= \lim_{\mu(P) \to 0} \pi \sum_{i=0}^{n-1} f(x_i) \sqrt{1 + [f'(\xi_i)]^2} \Delta x_i + \lim_{\mu(P) \to 0} \pi \sum_{i=0}^{n-1} f(x_{i+1}) \sqrt{1 + [f'(\xi_i)]^2} \Delta x_i$$

$$= 2\pi \int_a^b f(x) \sqrt{1 + [f'(x)]^2} \, dx.$$

Finally, we have the formula for the area of a surface of revolution around x-axis,

$$A_{ox} = 2\pi \int_a^b f(x) \sqrt{1 + [f'(x)]^2} \, dx \tag{10.14}$$

Similarly, if our smooth arc is given by $x = g(y)$, $y \in [c,d]$, then the area of the surface generated by revolving it around the y-axis is

$$A_{oy} = 2\pi \int_c^d g(y) \sqrt{1 + [g'(y)]^2} \, dy \tag{10.15}$$

Taking into account Formulas 10.14 and 10.15, it is not hard to see that if our smooth arc is given in polar coordinates by $\rho = f(\theta)$, $\theta \in [\alpha, \beta]$, then the areas A_0 and $A_{\pi/2}$ of the surfaces (see (10.5)) generated by revolving around $\theta = 0$ and $\theta = \dfrac{\pi}{2}$ are

$$A_\theta = 2\pi \int_\alpha^\beta \rho \sin\theta \sqrt{\rho^2 + \rho'^2}\, d\theta,$$

$$A_{\frac{\pi}{2}} = 2\pi \int_\alpha^\beta \rho \cos\theta \sqrt{\rho^2 + \rho'^2}\, d\theta$$

respectively.

Example 10.12. By considering a sphere of radius R as a surface of revolution, derive the formula $A = 4\pi R^2$ for its surface area.

Since the equation of the circle with radius R is $x^2 + y^2 = R^2$, we define the area $A = A_x$ of the sphere generated by revolving the semicircle

$$y = f(x) = \sqrt{R^2 - x^2}, \quad x \in [-R, R]$$

around the x-axis. Since

$$\sqrt{1 + [f'(x)]^2} = \frac{R}{R^2 - x^2},$$

we obtain

$$A = 2\pi \int_{-R}^R \sqrt{R^2 - x^2} \frac{R}{R^2 - x^2}\, dx = 4\pi \int_0^R R\, dx = 4\pi R^2.$$

Example 10.13. Find the area of the surface obtained by revolving the cycloid, $x = 2(t - \sin t)$, $y = 2(1 - \cos t)$, $t \in [0, 2\pi]$ (see Example 10.7), around the x-axis.

Obviously,

$$\sqrt{\left(\frac{dx}{dt}\right)^2 + \left(\frac{dy}{dt}\right)^2} = \sqrt{[2(1 - \cos t)]^2 + [2\sin t]^2} = 4\sin\frac{t}{2}$$

and so

$$A = 2\pi \int_0^{2\pi} 2(1 - \cos t)\sqrt{\left(\frac{dx}{dt}\right)^2 + \left(\frac{dy}{dt}\right)^2}\, dt = 32\pi \int_0^{2\pi} \sin^3\frac{t}{2}\, dt$$

$$= 32\pi \int_0^{2\pi} \left(1 - \cos^2\frac{t}{2}\right) \sin\frac{t}{2}\, dt = 32\pi \left(\frac{2}{3}\cos^3\frac{t}{2} - 2\cos\frac{t}{2}\right)\Bigg|_0^{2\pi} = \frac{256\pi}{3},$$

so,

$$A = \frac{256\pi}{3}.$$

10.5 Centroids of Plane Regions and Curves

Consider arbitrary masses $m_1, m_2, \ldots, m_n$ at points $x_1, x_2, \ldots, x_n$ of the x-axis. Then the point

$$\bar{x} = \frac{1}{m} \sum_{i=1}^{i=n} m_i x_i, \quad m = m_1 + m_2 + \cdots + m_n$$

is called the *center of mass* and the sum $\sum_{i=1}^{i=n} m_i x_i$ is called the *moment about the origin*.

Now, suppose that we have a system of n particles with masses $m_1, m_2, \ldots, m_n$ located in the plane at the points $(x_1, y_1), (x_2, y_2), \ldots, (x_n, y_n)$.

Similarly to the one-dimensional case above, we define the moment M_x and M_y of this system about the x-axis and about the y-axis, respectively, by the equations

$$M_x = \sum_{i=1}^{n} m_i y_i \text{ and } M_y = \sum_{i=1}^{n} m_i x_i.$$

The center of mass of this system of n particles is the point $(\bar{x}, \bar{y})$, with coordinates $\bar{x}$ and $\bar{y}$ defined by

$$\bar{x} = \frac{M_y}{m}, \quad \bar{y} = \frac{M_x}{m}, \quad m = m_1 + m_2 + \cdots + m_n. \tag{10.16}$$

Consequently, $(\bar{x}, \bar{y})$ is the point where a single particle of mass m would have the same moments, about the two coordinate axes, as does the system. In general, $(\bar{x}, \bar{y})$ is called the *centroid of the plane region*.

1. Centroids of Curves. Let $y = f(x)$, $x \in [a, b]$, be a smooth plane curve of constant linear density (the mass per unit length) α. Let us divide the curve into n parts, of lengths $\Delta s_i = s_{i+1} - s_i$ $(i = 0, \ldots, n-1)$. The mass of the part of length Δs_i is $\Delta m_i = \alpha \Delta s_i$. Pick, within each part, an arbitrary point with abscissa x_0. Then, to any part of the curve of length Δs_i and mass $\alpha \Delta s_i$, we associate a particle $(x_0, f(\xi_i))$. Thus, by (10.16), the coordinates of the centroid of the curve are approximated by

$$\bar{x} \approx \frac{\sum_{i=0}^{n-1} \xi_i \alpha \Delta s_i}{\sum_{i=0}^{n-1} \alpha \Delta s_i}, \quad \bar{y} \approx \frac{\sum_{i=0}^{n-1} f(\xi_i) \alpha \Delta s_i}{\sum_{i=0}^{n-1} \alpha \Delta s_i}.$$

If a function f is smooth on $[a, b]$, then by passing to the limit as $\max \Delta x_i \to 0$, we can define the coordinates of the centroid:

$$\bar{x} = \frac{\int_a^b x \, ds}{\int_a^b ds} = \frac{\int_a^b x \sqrt{1 + f'^2(x)} \, dx}{\int_a^b \sqrt{1 + f'^2(x)} \, dx},$$

$$\bar{y} = \frac{\int_a^b f(x) \, ds}{\int_a^b ds} = \frac{\int_a^b f(x) \sqrt{1 + f'^2(x)} \, dx}{\int_a^b \sqrt{1 + f'^2(x)} \, dx}. \tag{10.17}$$

Example 10.14. Find the centroid of the upper half of the circle $x^2 + y^2 = r^2$.

Note that $\bar{x} = 0$ by symmetry. Let us compute $\bar{y}$. Since $ds = \dfrac{r}{r^2 - x^2} dx$ (Example 10.1), by formula (10.17) we have

$$\bar{y} = \frac{\int_{-r}^{r} \sqrt{r^2 - x^2} \, \dfrac{r}{r^2 - x^2} dx}{\pi r} = \frac{2r^2}{\pi r} = \frac{2r}{\pi}.$$

2. Centroids of Plane Regions. Suppose that we have a plane region of constant density β bounded by the curves $y = f_1(x)$, $y = f_2(x)$ and vertical lines $x = a$ and $x = b$. We shall define the centroid $(\bar{x}, \bar{y})$ of the given region. Let $P = \{x_0, \ldots, x_n\}$ be a partition of $[a, b]$, $x = x_i$ ($x_0 = a$, $x_n = b$), and let x_0 be the midpoint of the i-th subinterval $[x_i, x_{i+1}]$. The mass of each rectangle is the product of β and the area of the rectangle with base $[x_i, x_{i+1}]$ and height $f_2(\xi_i) - f_1(\xi_i)$ $\left(\xi_i = \dfrac{x_i + x_{i+1}}{2} \right)$. Thus,

$$\Delta m_i \approx \beta[f_2(\xi_i) - f_1(\xi_i)]\Delta x_i, \quad i = 0, 1, \ldots, n - 1.$$

On the other hand, it is clear that the centroid of the region between the curves $y = f_1(x)$, $y = f_2(x)$; $x = x_i$, $x = x_{i+1}$ and the centroid of the rectangle described above are approximately the same, that is,

$$\bar{x}_i \approx \xi_i, \quad \bar{y}_i \approx \frac{f_2(\xi_i) + f_1(\xi_i)}{2}, \quad i = 0, \ldots, n - 1.$$

Then, by formula (10.16), we can write

$$\bar{x} \approx \frac{\sum\limits_{i=0}^{n-1} \xi_i \beta[f_2(\xi_i) - f_1(\xi_i)]\Delta x_i}{\sum\limits_{i=0}^{n-1} \beta[f_2(\xi_i) - f_1(\xi_i)]\Delta x_i},$$

$$\bar{y} \approx \frac{\dfrac{1}{2}\sum\limits_{i=0}^{n-1} [f_2(\xi_i) + f_1(\xi_i)]\beta[f_2(\xi_i) - f_1(\xi_i)]\Delta x_i}{\sum\limits_{i=0}^{n-1} \beta[f_2(\xi_i) - f_1(\xi_i)]\Delta x_i}.$$

By taking the limit as the mesh of the partition P tends to zero, i.e., $\mu(P) \to 0$, we have

$$\bar{x} = \frac{1}{A} \int_a^b x[f_2(x) - f_1(x)]dx, \quad \bar{y} = \frac{1}{2A} \int_a^b \left[f_2^2(x) - f_1^2(x) \right] dx$$

where

$$A = \int_a^b [f_2(x) - f_1(x)]dx.$$

Example 10.15. Find the centroid of the plane region bounded by the parabola $y^2 = 5x$ and the vertical line $x = 5$.

By the symmetry of this region about the x-axis, $\bar{y} = 0$. We compute $\bar{x}$. Setting $f_2(x) = \sqrt{5x}$, $f_1(x) = -\sqrt{5x}$, we find that

$$\bar{x} = \frac{2\int_0^5 x\sqrt{5x}\,dx}{2\int_0^5 \sqrt{5x}\,dx} = \frac{2\sqrt{5}\frac{2}{5}x^{5/2}\Big|_0^5}{2\sqrt{5}\frac{2}{3}x^{3/2}\Big|_0^5} = \frac{\frac{4}{5}5^3}{\frac{4}{3}5^2} = 3.$$

Consider again a finite system of particles with masses $m_1, m_2, \ldots, m_n$ and coordinates, $(x_1,y_1), (x_2,y_2), \ldots, (x_n,y_n)$, respectively. Then an important physical concept, the moment of inertia I_0 with respect to the origin, is defined this way:

$$I_0 = \sum_{i=1}^{n} r_i^2 m_i, \quad r_i^2 = \sqrt{x_i^2 + y_i^2}. \tag{10.18}$$

Let $y = f(x)$ be a continuous mass curve defined on $[a,b]$ of constant linear density α. Subdivide the curve into n parts: $\Delta s_0, \Delta s_2, \ldots, \Delta s_{n-1}$, of masses $\Delta m_i = \alpha\Delta s_i$ ($i = 0, 1, \ldots, n-1$), where $\Delta s_i = \sqrt{\Delta x_i^2 + \Delta y_i^2}$. Choose an arbitrary point with abscissa ξ_i in each of these parts of the mass curve. Then to any part of the curve of length Δs_i and of mass $\alpha\Delta s_i$ we associate a particle $(\xi_i, f(\xi_i))$. Thus, by (10.18), we have the approximation for the moment of inertia with respect to the origin,

$$I_0 \approx \sum_{i=0}^{n-1} (\xi_i^2 + \eta_i^2)\alpha\Delta s_i, \quad \eta_i = f(\xi_i). \tag{10.19}$$

If the derivative f' is continuous, then the sum on the right of (10.19) has a limit as $\max \Delta s_i \to 0$. This limit is the definite integral, expressing the moment of inertia with respect to the origin,

$$I_0 = \alpha \int_a^b [x^2 + f^2(x)] \sqrt{1 + f'^2(x)}\,dx. \tag{10.20}$$

(i) The Moment of Inertia of a Circle.

Suppose we have a mass circle of constant density α and with radius R. Then the mass of the circle is $m = 2\pi R \cdot \alpha$ and so

$$I_0 = mR^2 = \alpha 2\pi R \cdot R^2 = \alpha 2\pi R^3.$$

(ii) The Moment of Inertia of a Uniform Disk.

Suppose we have a uniform disk of constant density β. Consider a ring with inner and outer radii r_i and $r_i + \Delta r_i$, respectively. (Figure 10.6) Approximately, the mass of this ring is $\Delta m_i \approx \beta 2\pi r_i \Delta r_i$. Then the moment of inertia with respect to the origin of the ring is

$$\Delta I_0^i \approx \beta 2\pi r_i \Delta r_i \cdot r_i^2 = \beta 2\pi r_i^3 \cdot \Delta r_i.$$

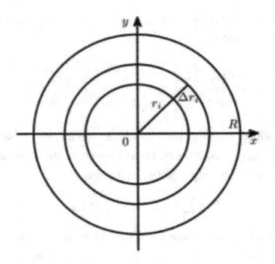

Fig. 10.6   The interior r_i and outer $r_i + \Delta r_i$ radii of a disk.

Then the moment of inertia of disk with respect to the origin is approximately

$$I_0 = \sum_{i=0}^{n-1} \Delta I_0^i \approx \sum_{i=0}^{n-1} \beta 2\pi r_i^3 \Delta r_i.$$

This sum is a Riemann sum for the definite integral of the function $\beta 2\pi r^3$ on $[0,R]$, corresponding to the partition $P = \{r_0, r_1, \ldots, r_n\}$ ($r_0 = 0$, $r_n = R$). If the mesh of P tends to zero, we have the following formula for the moment of inertia:

$$I_0 = \beta 2\pi \int_O^R r^3 dr = \pi \beta \frac{R^4}{2}.$$

If the mass of the disk is m, then $\beta = \dfrac{m}{\pi R^2}$, and so

$$I_0 = \frac{mR^2}{2}. \tag{10.21}$$

10.6   Additional Applications: Potential and Kinetic Energy, the Earth's Gravitational Field and Escape Velocity, Nuclear Reaction

The concept of work is introduced in Physics to measure the cumulative effect of a force in moving a body from one point to another. The work done by the force on the body is defined as the product of the force with the distance through which it acts. Thus, if a body moves a distance $s_1 - s_0$ from $x = s_0$ to $x = s_1$ in the positive direction along a coordinate line while subjected to a constant force F, the work done by the force is:

$$W = F(s_1 - s_0).$$

Suppose given a force function $F = F(x)$, defined at each point x of the line segment $[s_0, s_1]$. We take a partition $P = \{x_0, \ldots, x_n\}$ of $[s_0, s_1]$

$$s_0 = x_0 < x_1 < \cdots < x_n = s_1$$

and subdivide the interval $[s_0, s_1]$ into n subintervals $[x_i, x_{i+1}]$ $(i = 0, \ldots, n-1)$ with lengths $\Delta x_0, \Delta x_1, \ldots, \Delta x_{n-1}$. Select a point ξ_i in each $[x_i, x_{i+1}]$. Note that if the i-th subinterval is short and f is continuous, then the force will not vary much over this subinterval. Thus, the work done over the i-th subinterval is approximately $F(\xi_i)\Delta x_i$. Then the work done over the entire interval $[s_0, s_1]$ is approximately $\sum_{i=0}^{n-1} F(\xi_i)\Delta x_i$. If we now increase the number of subintervals in such a way that $\mu(P) \to 0$, the limit is just the definite integral,

$$W = \int_{s_0}^{s_1} F(x)dx. \tag{10.22}$$

Consequently, if a body moves in the positive direction over the interval $[s_0, s_1]$ while subjected to a variable force $F = F(x)$ in the direction of motion, then the work done by the force is given by (10.22). Obviously, if the direction of the force and the direction of motion are opposite and, more generally, if the upper bound of the integral (10.12) is not fixed, then

$$W(s) = -\int_{s_0}^{s} F(x)dx. \tag{10.23}$$

Example 10.16. Hooke's law states that (under appropriate conditions) a spring stretched x units beyond its equilibrium position pulls back with a force $F = kx$, where k is the spring constant. For example, suppose that a spring exerts a force of $1\,\mathrm{kg}$ for each $1\,\mathrm{cm}$ stretched beyond its natural length. How much work is required to stretch the spring to a length of $10\,\mathrm{cm}$?

At first we express the x in meters and the force f in kg. Then if $x = 0.01$, the force is $F = 1$. It follows that $1 = k \cdot (0.01)$. Thus the spring constant $k = 100$, and so $F = 100x$. Therefore, by (10.22), we have

$$W = \int_0^{0.1} 100x\,dx = 100\frac{x^2}{2}\Big|_0^{0.1} = 0,5\,\mathrm{kg\,m}.$$

1. The Conservation of Mechanical Energy. The value of $W(s)$ defined by formula (10.23) is called the *potential energy* of a body at the position s. If the force f is the weight of a body of mass m, that is $F = -mg$ (g is the gravitational acceleration), its potential energy at height S is

$$W(s) = -\int_0^s (-mg)dx = mgs. \tag{10.24}$$

It is a law of Physics that the kinetic energy K of a moving body of mass m and velocity v is

$$K = \frac{1}{2}mv^2 \tag{10.25}$$

By the conservation of mechanical energy during linear motion under the force $F = F(x)$, the sum of the kinetic energy and potential energy of a body is constant,

$$E = K + W,$$

where E is the total energy.

Consider the derivatives of the functions K and W with respect to t. By the chain rule for the derivative of a composite function, we have

$$\frac{dK}{dt} = \frac{1}{2}m\frac{d(v^2)}{dt} = \frac{1}{2}m\frac{d(v^2)}{dv}\frac{dv}{dt} = mv\frac{d^2s}{dt^2},$$

or

$$\frac{dK}{dt} = mav, \quad a = \frac{d^2s}{dt^2}.$$

On the other hand, because

$$W(s(t)) = -\int_{s_0}^{s(t)} F(x)dx$$

by the Fundamental Theorem of Calculus, we conclude $\dfrac{dW}{ds} = -F(s)$. Applying the chain rule again, we see that $\dfrac{dW}{dt} = \dfrac{dW}{ds}\cdot\dfrac{ds}{dt} = -Fv.$
Consequently,

$$\frac{dE}{dt} = \frac{dW}{dt} + \frac{dK}{dt} = mva - Fv = (ma - F).$$

Moreover, by Newton's Second Law of Motion, $F = ma$, and so, $\dfrac{dE}{dt} = 0$. Hence, for all t, the sum $K + W$ is constant.

We apply the conservation law of mechanical energy to a body falling under the influence of gravity. By formula (10.24), it follows that $E = \frac{1}{2}mv^2 + mgs$. If the height of the body starts out as s_0, and if we suppose its initial velocity is zero, then $E = mgs_0$. At the moment when it reaches the ground ($s = 0$), its velocity and $\frac{1}{2}mv_1^2 = mgs_0$. Then we find that $v_1 = \sqrt{2gv_0}$.

Conversely, if the body is thrown upward with velocity v_0, then the potential energy $W = 0$ and so $E = \frac{1}{2}mv_0^2$. At the end of its motion (the potential energy is zero) the kinetic energy of the body will be $\frac{1}{2}mv_1^2$, for v_1 the velocity at the time of reaching the earth's surface. Hence $\frac{1}{2}mv_1^2 = \frac{1}{2}mv_0^2$. Since $v_0 > 0$ and $v_1 < 0$ it follows that $v_1 = -v_0$.

2. Energy and Mass in the Theory of Relativity. In Section 4.5 we introduced the Special Theory of Relativity of Einstein, including the equation

$$m_0 \frac{d}{dt} \frac{v}{\sqrt{1 - \left(\frac{v^2}{c^2}\right)}} = F.$$

where c is the velocity of light and m_0 is the rest-mass of a body. We have

$$\frac{m_0 a}{\left(\sqrt{1 - \left(\frac{v^2}{c^2}\right)}\right)^3} = F, \quad a = \frac{dv}{dt}. \tag{10.26}$$

In relativistic physics, the kinetic energy of a body with velocity v is

$$K = \frac{m_0 c^2}{\sqrt{1 - \left(\frac{v^2}{c^2}\right)}}. \tag{10.27}$$

This is based on the fact that during the motion of the object the total energy $E = K + W$ is constant.

Indeed, by (10.26), we can write

$$\frac{dE}{dt} = \frac{dK}{dv}\frac{dv}{dt} + \frac{dW}{ds}\frac{ds}{dt}$$

$$= \frac{m_0 c^2 \left(-\frac{1}{2}\right)\left(-\frac{2v}{c^2}\right)}{\sqrt{\left(1 - \frac{v^2}{c^2}\right)^3}}\frac{dv}{dt} + (-F)v = \frac{m_0 v}{\sqrt{\left(1 - \frac{v^2}{c^2}\right)^3}}a - Fv = 0.$$

The quantity

$$m = \frac{m_0}{\sqrt{1 - \left(\frac{v^2}{c^2}\right)}} \tag{10.28}$$

is called the *mass of the moving object*. From (10.27) and (10.28), we have

$$K = mc^2. \tag{10.29}$$

In words, this last formula says that increasing (decreasing) the energy of an object implies also increasing (decreasing) its mass m.

3. Escape Velocity Let M denote the mass of the earth and R its radius: $R \approx 6370\,$km. Suppose we have launched an object with mass m as shown in Figure 10.7, at distance s from the earth's center O. The only force acting on the mass m is the earth's gravity, which, by Newton's law of Gravitation, yields

$$F = -\frac{\gamma M m}{s^2},\tag{10.30}$$

where γ is a constant of Nature. Assume that the distance between the launched object and the earth's surface is x, so that $s = R + x$.

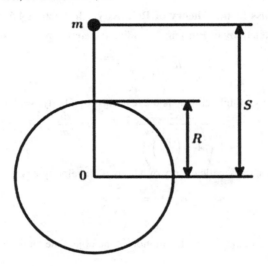

Fig. 10.7 Escape velocity.

Then,

$$F = -\frac{\gamma M m}{(R+x)^2}.$$

Now, if the value $\dfrac{x}{R}$ is sufficiently small, then, approximately,

$$\frac{1}{(R+x)^2} \approx \frac{1}{R^2}.$$

Indeed, for $x > 0$, we can write

$$\frac{1}{R^2} - \frac{1}{(R+x)^2} = \frac{2Rx+x^2}{(R+x)^2 R^2} < \frac{2x}{R^3} + \frac{x^2}{R^4}.\tag{10.31}$$

For instance, if $x < 6,3\,$km, then the difference in (10.31) is less than 10^{-19} cm^2. It follows that for objects close to the earth's surface, we can take $F = -\dfrac{\gamma M m}{R^2}$.

Let us denote

$$\frac{\gamma M}{R^2} = g \tag{10.32}$$

Then, from Newton's Third Law of Motion and from (10.32), we see that $-mg = ma$ (a is the acceleration), or, $a = -g$.

But for sufficiently large s, it follows from (10.30) that

$$ma = -\frac{\gamma Mm}{s^2}$$

or

$$\frac{d^2s}{dt^2} = -\frac{\gamma M}{s^2}. \tag{10.33}$$

Then, by (10.23) and (10.30), we have

$$W = -\int_r^s \frac{-\gamma Mm}{x^2}\,dx = \gamma Mm\left(\frac{1}{R} - \frac{1}{s}\right).$$

With the use of this formula we now compute the "escape velocity" from the earth, i.e., the minimum initial velocity v_0^{ev} with which an object (say spaceship) must be launched straight upward from the earth's surface so that it will continue forever to move away from the earth.

First assume that the needed height for our object is $s = A$, where A is a constant.

At the beginning of the motion, $K = \frac{1}{2}mv_0^2$ and $W = 0$. At the distance $s = A$, the velocity is $v = 0$. Hence, we deduce that

$$K = 0 \quad \text{and} \quad W = \gamma Mm\left(\frac{1}{R} - \frac{1}{A}\right).$$

By the law of conservation of mechanical energy, we have

$$\frac{1}{2}mv_0^2 = \gamma Mm\left(\frac{1}{R} - \frac{1}{A}\right)$$

which tells us that

$$v_0 = \sqrt{2\gamma M\left(\frac{1}{R} - \frac{1}{A}\right)}.$$

By substituting $\gamma M = gR^2$ (see 10.32) into this formula, we obtain

$$v_0 = \sqrt{2g\left(R - \frac{R^2}{A}\right)}.$$

This, in turn, becomes

$$v_0^{ev} = \sqrt{2gR}$$

after passing to the limit as A approaches infinity.

This is just the required "escape velocity." By substituting $R \approx 6370\,\text{km}$ and $g = 980\,\text{cm·s}^{-2}$ into the formula $v_0^{ev} = \sqrt{2gR}$, we have

$$v_0^k = 1.17 \cdot 10^6\,\text{cm} \cdot \text{s}^{-1}.$$

4. Nuclear Reactions. The mathematical models of real situations frequently involve equations containing derivatives of unknown functions. An equation such as

$$y'(x) - \alpha y(x) = \varphi(x) \tag{10.34}$$

is called a *first order linear nonhomogeneous differential equation*, where α is some constant, φ is a given function, and y is the unknown function sought. Firstly, in the case where $\varphi(x) = 0$ (called, the *homogeneous case*), we can find a solution $y = y(x)$ of (10.34) which satisfies the initial-value condition $y(0) = A$ (A a constant). It is easy to see that $y(x) = Ae^{\alpha x}$ is the unique solution of this problem. Indeed, the derivative of $y_1(x) = y(x)e^{-\alpha x}$

$$y_1'(x) = y'(x)e^{-\alpha x} - \alpha y(x)e^{-\alpha x} = e^{-\alpha x}[y'(x) - \alpha y(x)] = 0$$

and $y_1(0) = y(0) = A$, that is $y(x) = Ae^{\alpha x}$.

Next, we find the solution of the original equation (10.34), allowing $\varphi(x) \neq 0$. We search for a solution of the form $y(x) = c(x)e^{\alpha x}$ where the auxiliary function $c(x)$ is to be determined. If we substitute $y(x) = c(x)e^{\alpha x}$ into (10.34), we obtain

$$y'(x) - \alpha y(x) = c'(x)e^{\alpha x} + c(x)\alpha e^{\alpha x} - \alpha c(x)e^{\alpha x} = \varphi(x)$$

and so,

$$c'(x) = e^{-\alpha x}\varphi(x).$$

By the Fundamental Theorem of Calculus,

$$c(x) = \int_0^x e^{-\alpha t}\varphi(t)dt + A \quad (c(0) = A)$$

and so we find that

$$y(x) = Ae^{\alpha x} + e^{\alpha x}\int_0^x e^{-\alpha t}\varphi(t)dt. \tag{10.35}$$

Thus we have proved the following theorem.

Theorem 10.1. *The solution of the first order linear differential equation* (10.34) *with initial-value condition* $y(0) = A$ *has the form* (10.35) *and this solution is unique.*

(i) Atomic explosion

A nuclear reaction can be produced by using, for example, the uranium isotope (U^{235}). It occurs when some free neutrons (a kind of uncharged particle) come into collision with these nuclei. A large quantity of atomic energy is liberated as a result of the collision.

We will derive the differential equation of a model of this reaction. Suppose we have a radioactive substance with constant density in the form of a sphere of radius r. If at time t the number of neutrons is N, then during a short time-span h, as a result of the nuclear reaction, there will arise new neutrons, the number of which is proportional to N and h. The number of new neutrons produced will then be approximately αNh, α being a positive constant depending on the properties of the radioactive substance.

On the other hand, during the same time-span h, the number of neutrons leaving the sphere will be βNh, the number β depending on the surface area of the sphere. It follows that the number β is proportional to

$$\frac{4\pi r^2}{\frac{4}{3}\pi r^3} = \frac{3}{r}$$

or, in short, $\beta = \dfrac{\beta_0}{r}$, where β_0 depends only on the radioactive substance and its density. Then, by instant $t + h$, the number of total neutrons has become

$$N(t+h) \approx N(t) + \alpha N(t)h - \beta N(t)h.$$

Thus,

$$\frac{N(t+h) - N(t)}{h} \approx (\alpha - \beta)N(t).$$

By passing to the limit as $h \to 0$, we obtain a differential equation

$$\frac{dN}{dt} = (\alpha - \beta)N.$$

It follows from Theorem 10.1 that the solution of this equation that also satisfies the initial-value condition $N(0) = N_0$ (N_0 a positive integer) is

$$N = N_0 e^{(\alpha-\beta)t}. \tag{10.36}$$

Clearly, if the unit of time t is measured in seconds, then the dimensions of α, β are $1/s$ and so β_0 has the dimensions of a velocity.

The number $r_{cr.}$ corresponding to $\alpha = \beta = \beta_0/r$ is called the *critical radius*. The mass of the sphere of radioactive substance with radius $r_{cr.}$, is called the *critical mass*, that is, $m_{cr.} = \dfrac{4}{3}\pi r_{cr.}^3 \rho$, where ρ is the density of the radioactive substance. If $r > r_{cr.}$, then $\alpha - \beta > 0$, and by (10.36) the number of neutrons N is increasing exponentially: and for a short time "atomic explosion" will take place.

(ii) The Atomic Reactor

Suppose that a source of neutrons is introduced into the sphere considered above, producing q neutrons per unit time. If the number of existing neutrons at time t is N, then for small $h > 0$, the number of neutrons at time $t + h$ will be

$$N(t+h) \approx N(t) + \alpha N(t)h - \beta N(t)h + qh.$$

Consequently,

$$\frac{N(t+h) - N(t)}{h} \approx (\alpha - \beta)N(t) + q.$$

As the limit when $h \to 0$, we obtain

$$\frac{dN}{dt} = (\alpha - \beta)N + q. \tag{10.37}$$

By Theorem 10.1, the solution of this differential equation is

$$N = N_0 e^{(\alpha-\beta)t} + e^{(\alpha-\beta)t} \int_0^t e^{-(\alpha-\beta)\eta} q \, d\eta.$$

The integral existing on the right hand side is

$$\int_0^t e^{-(\alpha-\beta)\eta} q \, d\eta = -\frac{q e^{-(\alpha-\beta)\eta}}{\alpha - \beta}\bigg|_0^t = -\frac{q}{\beta - \alpha}\left[e^{-(\alpha-\beta)t} - 1\right].$$

Thus,

$$N = N_0 e^{(\alpha-\beta)t} + \frac{q}{\beta - \alpha}\left[1 - e^{(\alpha-\beta)t}\right]. \tag{10.38}$$

Assume that $\alpha - \beta \neq 0$ or, equivalently, $r \neq r_{cr}$. If $\alpha - \beta = 0$, then it follows from (10.37) that $N = N_0 + qt$. If $\alpha > \beta$, then according to (10.38), the number of neutrons is increasing exponentially, and so an atomic explosion occurs. In the case $\alpha < \beta$ the expression $e^{(\alpha-\beta)t}$ approaches zero, and so, by (10.38), we conclude

$$N \approx \frac{q}{\beta - \alpha}. \tag{10.39}$$

Consequently, the number of neutrons is approximately constant and our atomic reactor is guaranteed to produce a constant energy stream. In short, this is the working principle of the atomic reactor.

From (10.39) we also derive that for small q the number of neutrons can be increased, when the difference $\beta - \alpha$ is sufficiently small. On the other hand, it can be understood that a mass close to the critical mass is dangerous without a strict control of the reactor's work. Here we should mention that in 1986, there was some instability in the working of the nuclear power station of Chernobyl, in the Ukraine, resulting in an explosion in the reactor.

10.7 Problems

In problems (1)–(8), find the arc length:

(1) $\quad x = \dfrac{1}{4}y^2 - \dfrac{1}{2}\ln y \; (1 \leqslant y \leqslant e).$ $\qquad$ *Answer:* $\quad \dfrac{e^2 + 1}{4}.$

(2) $\quad y = \ln \cos x \; \left(0 \leqslant x \leqslant a < \dfrac{\pi}{2}\right).$ $\qquad$ *Answer:* $\quad \ln \tan\left(\dfrac{\pi}{4} + \dfrac{a}{2}\right).$

(3) $\quad x^{\frac{2}{3}} + y^{\frac{2}{3}} = a^{\frac{2}{3}}$ (astroid). $\qquad$ *Answer:* $\quad 6a.$

(4) $\quad x = \dfrac{c^2}{a}\cos^3 t, \; y = \dfrac{c^2}{b}\sin^3 t, \; c^2 = a^2 - b^2.$ $\qquad$ *Answer:* $\quad \dfrac{4(a^2 - b^2)}{ab}.$

(5) $\quad x = a(\sinh t - t), \; y = a(\cosh t - 1) \; (0 \leqslant t \leqslant T).$

$\qquad\qquad$ *Answer:* $\quad 2\left(\cosh\dfrac{T}{2}\sqrt{\cosh T} - 1\right) - \sqrt{2}\ln\dfrac{\sqrt{2}\cosh\dfrac{T}{2} + \sqrt{\cosh T}}{1 + \sqrt{2}}.$

(6) $\quad \rho = a\theta$ (Archimedes' spiral), $0 \leqslant \theta \leqslant 2\pi.$

(7) $\quad \rho = \dfrac{p}{1 + \cos\theta} \; \left(|\theta| \leqslant \dfrac{\pi}{2}\right).$ $\qquad$ *Answer:* $\quad p\left[\sqrt{2} + \ln\left(1 + \sqrt{2}\right)\right].$

(8) $\quad \rho = a\tanh\dfrac{\theta}{2} \; (0 \leqslant \theta \leqslant 2\pi).$ $\qquad$ *Answer:* $\quad a(2\pi - \tanh\pi).$

In problems (9)–(11) find the area of the bounded regions given by the curve in Cartesian coordinates.

(9) $\quad y = 2x - x^2, \; x + y = 0.$ $\qquad$ *Answer:* $\quad 4\dfrac{1}{2}.$

(10) $\quad y = 2^x, \; y = 2, \; x = 0.$ $\qquad$ *Answer:* $\quad 2 - \dfrac{1}{\ln 2}.$

(11) $\quad Ax^2 + 2Bxy + Cy^2 = 1 \; (A > 0, \; AC - B^2 > 0).$ $\quad$ *Answer:* $\quad \dfrac{\pi}{\sqrt{AC - B^2}}.$

In problems (12)–(15) find the area of the bounded regions given by the curve in parametric equations.

(12) $\quad x = 2t - t^2, \; y = 2t^2 - t^3.$ $\qquad$ *Answer:* $\quad \dfrac{8}{15}.$

(13) $\quad x = a(\cos t + t\sin t), \; y = a(\sin t - t\cos t) \; (0 \leqslant t \leqslant 2\pi).$ $\quad$ *Answer:* $\quad \dfrac{a^2}{3}(4\pi^3 + 3\pi).$

(14) $\quad x = a(2\cos t - \cos 2t), \; y = a(2\sin t - \sin 2t).$ $\qquad$ *Answer:* $\quad 6\pi a^2.$

(15) $\quad x = a\cos t, \; y = \dfrac{a\sin^2 t}{2 + \sin t}.$ $\qquad$ *Answer:* $\quad \pi a^2\left(\dfrac{16}{\sqrt{3}} - 9\right).$

In problems (16)–(18) find the area of the bounded regions given by the curve in polar coordinates

(16) $\quad \rho^2 = a^2\cos 2\theta.$ $\qquad$ *Answer:* $\quad a^2.$

(17) $\quad \rho = \dfrac{p}{1 + \varepsilon\cos\varphi} \; (0 < \varepsilon < 1).$ $\qquad$ *Answer:* $\quad \dfrac{p^2}{6}(3 + 4\sqrt{2}).$

(18) $\quad \rho = \dfrac{1}{\theta}, \; \rho = \dfrac{1}{\sin\theta} \; \left(0 < \theta \leqslant \dfrac{\pi}{2}\right).$ $\qquad$ *Answer:* $\quad \dfrac{1}{\pi}.$

In problems (19)–(22), find the volumes bounded by the surfaces.

(19) $\dfrac{x^2}{a^2} + \dfrac{y^2}{b^2} = 1, z = \dfrac{c}{a}x, z = 0.$ *Answer:* $\dfrac{2}{3}abc.$

(20) $\dfrac{x^2}{a^2} + \dfrac{y^2}{b^2} - \dfrac{z^2}{c^2} = 1, z = \pm c.$ *Answer:* $\dfrac{8\pi abc}{3}.$

(21) $x^2 + z^2 = a^2, y^2 + z^2 = a^2.$ *Answer:* $\dfrac{16}{3}a^2.$

(22) $x + y + z^2 = 1, x = 0, y = 0, z = 0.$ *Answer:* $\dfrac{4}{15}.$

(23)   If the area of the cross section of the solid region in the plane perpendicular to the x-axis is $s(x) = Ax^2 + Bx + C, a \leqslant x \leqslant b$, prove that its volume is

$$V = \frac{H}{6}\left[s(a) + 4s\left(\frac{a+b}{2}\right) + s(b)\right], \quad H = b - a.$$

In each of problems (24)–(29) find the volume of a solid of a revolution.

(24) $y = 2x - x^2, y = 0$ (a) about the x-axis,   *Answer:* $\dfrac{16\pi}{15},$

                                          (b) about the y-axis.   *Answer:* $\dfrac{8\pi}{3}.$

(25) $y = e^{-x}, y = 0 \ (0 \leqslant x < +\infty)$ (a) about the x-axis   *Answer:* $\dfrac{\pi}{2},$

                                          (b) about the y-axis.   *Answer:* $2\pi.$

(26) $x^2 - xy + y^2 = a^2$, about the x-axis.   *Answer:* $\dfrac{8\pi a^3}{3}.$

(27)   Prove that the volume of the region bounded by the curve in polar coordinates $0 \leqslant \alpha \leqslant \theta \leqslant \beta \leqslant \pi, 0 < \rho \leqslant \rho(\theta)$ (ρ and θ are polar coordinates) revolved about the polar axis is

$$V = \frac{2\pi}{3}\int_{\alpha}^{\beta}\rho^3(\theta)\sin\theta d\theta.$$

(28)   Find the volume of the region bounded by the curve $\theta = \pi\rho^2, \theta = \pi$, in polar coordinates, revolved about the polar axis. *Answer:* $\dfrac{2}{3}\pi.$

(29)   Find the volume of the region bounded by the curve $a \leqslant \rho \leqslant a\sqrt{2}\sin 2\theta$, in polar coordinates, revolved about the polar axis. *Answer:* $\dfrac{\pi^2 a^2}{2\sqrt{2}}.$

In (30)–(33), find surface area of the volume of revolution described.

(30) $y = \tan x \left(0 \leqslant x \leqslant \dfrac{\pi}{4}\right)$ about the x-axis.

Answer: $\pi \left[(\sqrt{5} - \sqrt{2}) + \ln \dfrac{(\sqrt{2}+1)(\sqrt{5}-1)}{2}\right]$.

(31) $y^2 = 2px \ (0 \leqslant x \leqslant x_0)$,

(a) about the x-axis, Answer: (a) $\dfrac{2\pi}{3}\left[(2x_0 + p)\sqrt{2px_0 + p^2} - p^2\right]$.

(b) about the y-axis. Answer: (b) $\dfrac{\pi}{4}\left[(p + 4x_0)\sqrt{2x_0(p + 2x_0)} - p^2 \ln \dfrac{\sqrt{2x_0} + \sqrt{p + 2x_0}}{\sqrt{p}}\right]$.

(32) $\rho = a(1 + \cos\theta)$, about the polar axis. Answer: $\dfrac{32}{5}\pi a^2$.

(33) $\rho^2 = a^2 \cos 2\theta)$, $\theta = \dfrac{\pi}{2}$, about the polar axis. Answer: $2\pi a^2 \sqrt{2}$.

(34)   Find the volume and surface area of the region bounded by the parabola $y^2 = 2px$ and straight line $x = \dfrac{p}{2}$ revolved about the line $y = p$. Answer: $V = \dfrac{4\pi}{3}p^2$, $A = 2\pi p^2 \left[(2 + \sqrt{2}) + \ln(1 + \sqrt{2})\right]$.

(35)   Find the centroid of the region bounded by the curve $x^2 + 4y - 16 = 0$ and x-axis.
Answer: $\left(0; \dfrac{8}{5}\right)$.

(36)   Find the centroid of the triangle with vertices $(0,0)$, $(0,3)$, and $(3,0)$. Answer:
$\left(\dfrac{9}{2}, \dfrac{9}{2}\right)$.

(37)   Find the centroid of the region bounded by the curve $y = \sin x \ (0 \leqslant x \leqslant \pi)$ and the line $y = 0$. Answer: $\left(\dfrac{\pi}{2}, \dfrac{\pi}{8}\right)$

(38)   How much work must be done (a) in order to lift a solid of mass m from the surface of the earth of radius R to a distance h above the earth's surface?
(b) if $h \to \infty$? Answer: (a) $W_h = mg\dfrac{Rh}{R+h}$, (b) $W_\infty = mgR$.

(39)   Find the centroid of the region $\dfrac{x^2}{a^2} + \dfrac{y^2}{b^2} \leqslant 1$, $0 \leqslant x \leqslant a$, $0 \leqslant y \leqslant b$. Answer:
$\left(\dfrac{4a}{3\pi}, \dfrac{4b}{3\pi}\right)$.

(40)   Find the centroid of the region bounded by the cycloid $x = a(t - \sin t)$, $y = a(1 - \cos t)$ $(0 \leqslant t \leqslant 2\pi)$, and the x-axis. Answer: $\left(\pi a, \dfrac{5}{6}a\right)$.

(41)   The radius and mass of Jupiter are $70{,}96 \cdot 10^6$ m and $1{,}88 \cdot 10^{27}$ kg, respectively. Find the escape velocity of a solid on the surface of the Jupiter.

(42)   For a given radioactive substance, $\alpha = 2 \cdot 10^8 \mathrm{s}^{-1}$ and $\beta_0 = 4 \cdot 10^{10}\,\mathrm{cm}\cdot\mathrm{s}^{-1}$. Find the critical mass, $r_{cr.}$.

Bibliography

[1] V.M. Alekseev, V.M. Tikhomirov, S.V. Fomin, *Optimal Control*, Moscow Consultants Bureau, New York, 1987.

[2] Alan Jeffrey, Hui Dai, *Handbook of Mathematical Formulas and Integrals*, Elsevier, 2008.

[3] Alex Poznyak, *Advanced Mathematical Tools for Control Engineers*, Vol. 1, Elsevier, 2008.

[4] L. Biberbach, *Introduction to Differential and Integral Calculus*, vol. 1, Leipzig and Berlin, B.G. Teubner, 1927.

[5] J.M. Borwein, *Pi and the AGM*, Wiley-Interscience, New York, NY, 1987.

[6] Charlambos Aliprantis, Owen Burkinshaw, *Problems in Real Analysis*, Academic Press, Elsevier, 1998.

[7] B.P. Demidovich, *Collection of Problems and Exercises on Mathematical Analysis*, 10th ed., Nauka, Moscow, 1990 (Russian).

[8] C.H. Edwards and David E. Penney, *Calculus and Analytic Geometry*, Prentice Hall, Englewood Cliffs, New Jersey, 1982.

[9] C.H. Jr. Edwards, *The Historical Development of the Calculus*, New York, Academic Press, 1999.

[10] G.M. Fihtengoltz, *Course on Differential and Integral Calculus*, Vol. I–III, Publisher for Mathematics – Physics Literature, Moscow, 1962 (Russian).

[11] J.A. Fridy, *Introductory Analysis: The Theory of Calculus*, Academic Press; 2nd edition, 2000.

[12] B. Gelbaum and J. Olmstad, *Counterexamples in Analysis*, Mir, Moscow, 1964 (Russian).

[13] Gerald B. Folland, *Real Analysis: Modern Techniques and Their Applications* (*Pure and Applied Mathematics*), New York, Academic Press, 1999.

[14] P.R. Halmos, *Measure Theory*, Vol. 18 of Graduate Texts in Mathematics, Springer-Verlag, New York, 1974.

[15] Howard Anton, *Calculus with Analytic Geometry*, 3rd ed., New York, John Wiley & Sons, 1988.

[16] John A. Fridy, *Introductory Analysis: The Theory of Calculus*, Academic Press; 2 edition, 2000.

[17] N.R. Howes, *Modern Analysis and Topology, Universitext*, Springer, New York, 1995.

[18] V.A. Ilyin and E.G. Poznyak, *Fundamentals of Mathematical Analysis*, Mir, Moscow, 1982.

[19] John Fridy, *Introductory Analysis*, Academic Press, Elsevier, 2000.

[20] A.N. Kolmogorov, S.V. Fomin, *Introductory Real Analysis*, Dover Publications, Inc, New York, 1970.

[21] C. Kubrusly, *Measure Theory*, Academic Press, 2006.

[22] A.G. Kurosh, *Course of Advanced Algebra*, Nauka, Moscow, 1975.

[23] N.N. Luzin, The Theory of Real Variable Functions, Uchpedgiz, 1948.

[24] LUH, W., Mathematik Für Naturwissenschaftler I, Akademische Verlagesellschaft, Wieschbaden, 1982.

[25] E.N. Mahmudov, *Mathematical Analysis and Its Applications*, Papatya, Istanbul, 2002.

E. Mahmudov, *Single Variable Differential and Integral Calculus*,
DOI: 10.2991/978-94-91216-86-2, © Atlantis Press 2013

[26]   M. Mureşan, *A Concrete Approach to Classical Analysis*, Springer-Verlag, 2000.

[27]   I.P. Natanson, *Theory of Functions of a Real Variable*, Frederick Ungar Publishing Co., New York, Vol. I (1955), Vol. II (1960).

[28]   S.M. Nikolsky, *A Course of Mathematical Analysis*, Mir, Moscow, 1981.

[29]   Omar Hijab, *Introduction to Calculus and Classical Analysis*, 2nd ed., Springer-Verlag, 2007.

[30]   Patrick M. Fitzpatrick, *Advanced Calculus: A Course in Mathematical Analysis*, New York, Hafner, 1995.

[31]   Peter Henrici, *Elements of Numerical Analysis*, John Wiley & Sons, New York, 1965.

[32]   Rangachary Kannan, Carole King Kruger, *Advanced Analysis on the Real Line*, New York, John Wiley & Sons, 1966.

[33]   Roger Godement, *Analysis II: Differential and Integral Calculus, Fourier Series, Holomorphic Functions*, Springer-Verlag, 2005.

[34]   W. Rudin, *Principles of Mathematical Analysis*, McGraw-Hill, New York, 1976.

[35]   Taylor, Angus E., W. Robert Mann, *Advanced Calculus*, 3rd ed., New York, John Wiley, 1983.

[36]   Thomas Dence, Joseph Dence, *Advanced Calculus, A Transition to Calculus*, Academic Press, Elsevier, 2009.

Index

A

Absolute convergence 287

Absolute value 12, 18, 48, 51, 101, 132, 165, 182, 332

Acceleration 109, 117, 118, 203, 356, 359

Addition law for limits 34, 73

Additivity 313, 339, 344

Algebraic functions 69, 221, 246

Alternating harmonic series 51, 55, 163

Alternating series 50, 51, 60

Alternating series test 51

Angle 18, 19, 25, 88, 89, 109, 127, 131, 132, 138, 199, 203, 219, 244, 342

Antiderivative 221–223, 225, 227, 257, 258, 260–262, 264, 265, 278, 281, 282, 285, 294, 326, 327

Antidifferentiation 221, 222, 228

Approximation 9–11, 112, 113, 118, 143, 160, 169, 171, 177, 181, 182, 294, 295–297, 299, 332, 345, 347, 348, 353

Arc length 132, 335, 337, 362

Arctangent function 164

Arcsine function 88–90, 121, 122, 142, 143, 251

Area 89, 202, 204, 205, 207, 219, 263, 264, 266, 267, 295–298, 335, 339–346, 348–350, 352, 363–365

Argument 19, 45

Astroid 338, 363

Average radius 347, 349

Atomic reactor 335, 361, 362

Asymptote 185, 211–215, 220

B

Base of logarithm 86

Bending downward and upward 208, 219

Bernstein polynomial 182, 183

Bijection (surjection and injection) function 4, 5

Binomial formula 19, 37, 158

Binomial series 158

Bolzano-Weierstrass theorem 41, 80, 98

Bounded monotone sequence property 37

Bounded region 339, 363

Bounded sequence 30, 34, 42, 80, 100

C

Cardioid 338

Cardinality 1, 5, 8, 15, 16, 28, 29, 257, 314

Cartesian product 21

Cauchy theorem 48, 53, 56, 58, 71, 93, 150, 286, 292

Cauchy sequence 31, 42–44, 48, 71, 72, 94, 95, 99, 100

Cauchy-Schwarz-Bunyakowsky inequality 23

Cavalieri 346

Cantor set 257, 313, 314, 324

Chebyshev polynomial 182

Center of mass 351

Centroid 335, 351, 352, 365

Change of logarithm base 86

Characteristic function 317

Chain rule 113, 225, 356

Circle 28, 32, 127, 131–137, 144, 165, 166, 219, 337, 350, 352, 353

Classification of conics 137, 138

Comparison test for series 54

Compact set 26, 27, 29, 45

Complex numbers 17–21, 30, 62, 165, 171, 175

E. Mahmudov, *Single Variable Differential and Integral Calculus*,
DOI: 10.2991/978-94-91216-86-2, © Atlantis Press 2013

Complex-valued function 67, 101, 326, 327
Composition of functions 113, 126, 225, 328
Continuity 67, 75, 77–82, 84, 85, 89, 97, 107, 109, 110, 147, 149, 165, 209, 262, 280, 293, 317, 329
Concave downward 209, 210, 215, 219, 220, 328
Conditionally convergent series 31, 48, 51, 56
Cone 137, 205, 219
Conjugate of complex numbers 18
Conjugate pairs of roots 175, 176, 231, 232
Continuous function 16, 67, 75, 78, 80, 81, 84, 88, 95, 97, 99, 100, 104, 147, 163, 194, 276, 278, 281, 302, 335, 340, 344
Convergence of improper integral 286–288, 291, 292, 331
Convergence of series 155, 157, 158, 161–166, 170
Cosine function 88, 89, 121, 122, 160, 161, 164, 166, 170, 245, 255
Curvature 107, 129, 131–135
Cycloid 127, 134, 136, 341, 350, 365
Cylinder 202, 205, 206, 345, 347

D

Decimal expansion 8–10
Decreasing function 82, 194
Definite integral 257, 259, 266, 281, 326–329, 339, 344, 353–355
Deleted neighborhood 189
De Moivre's formulas 20
Denumerable set 6, 7, 311
Dependent variable 67
Derivative 107–125, 127, 133, 134, 136, 139–147, 149, 151, 152, 155–157, 164, 165, 167, 168, 174, 178, 181, 185–187, 194–197, 200, 202, 203, 205–211, 215, 217, 221–225, 227, 229, 237, 266, 280, 281, 298, 303, 304, 307, 326, 336, 337, 353, 356, 360
Differential 107, 109, 110–112, 114, 117, 123, 125–127, 143, 165, 185, 228, 245, 360–362
Differentiable function 110, 11, 114, 126, 132, 140, 145, 154, 163, 178, 181, 194, 208, 228
Discontinuity 67, 75–77, 82–84, 104, 123, 149, 150, 215, 262, 266, 277, 278, 293, 303, 308

Distance 21, 25, 30, 31, 44, 128, 134, 198, 199, 203, 219, 258, 354, 355, 358, 359, 365
Divergence of improper integral 285–288
Divergence of series 31, 44, 46–48, 51, 53, 59, 61
Domain of function 3, 67, 68, 72, 85–88, 101, 102, 127, 165, 194, 215

E

Evolvent 107, 129, 131, 135, 136, 144
Eccentricity 138
Elementary function 85, 118, 145, 155, 221, 224, 281, 283, 294
Ellipse 28, 130, 137, 138, 144, 219, 347
Ellipsoid 346
Energy 335, 354, 356, 357, 359, 360, 362
Elliptic integral 225, 281
Extended mean value theorem 147, 150, 151, 261, 276, 278
Escape velocity 335, 354, 358, 359, 365

F

Factorial 37, 63, 68
Fermat's theorem 145
Fermat's principle of least time 198
Fibonacci sequence 31, 39
First order differential equation 360, 361
Force 117, 118, 128, 227, 354–356, 358
Functions 1, 16, 21, 25, 67–74, 77, 78, 81–83, 85–95, 97–102, 104, 107, 113, 114, 116–121, 124–127, 129, 139–141, 145, 151, 155, 159, 160, 163, 165, 166, 168, 170, 185–189, 192, 194, 200, 211, 219–225, 228–230, 234, 236, 237, 242, 243, 245–248, 254, 255, 257, 258, 260, 261, 273–276, 278, 280–283, 287, 289, 290, 294, 299, 300, 302, 303, 305–307, 316–319, 323-328, 336, 356, 360
Fundamental theorem of algebra 172
Fundamental theorem of calculus 257, 280, 281, 293, 294, 327, 329, 356, 360

G

Galilei's mechanics 109
Generalized length 309
General mean value theorem 278
Gravitation law 117, 118, 354, 356, 358
Gravitational acceleration 356
Greatest lower bound 12, 77, 269, 309, 310, 339

H

Half-angle Formula   244
Harmonic series   31, 46, 51, 52, 55, 59, 158, 163, 289, 302
Higher order derivatives   123, 127, 143, 164, 185, 197
Homogeneous differential equation   360
Hyperbola   87, 137–139
Hyperbolic functions   69, 86–88, 120, 166
Horizontal asymptotes   214, 215

I

Identities   86
Imaginary part of complex number   18, 231, 326
Improper integrals   259, 285, 291, 292, 330, 331
Increasing function   82, 83–85, 300, 303, 307, 309
Increment   107–113, 116, 139, 148
Inertia   353, 354
Infinite product   61, 62
Inflection point test   210, 211
Inscribed rectangular polygon   340, 344
Integrable function   276, 278, 280, 281, 324, 328
Integral   278, 280, 281–291, 297, 299, 305–309, 319, 335, 339, 344
Integral table   223, 224
Integral test   287, 288
Integrand   221
Integration by parts   228, 229, 246, 252, 259, 281, 282, 290, 327, 329, 330
Interior point   26
Intermediate value theorem   78
Inverse functions   82, 88, 113, 120, 121, 221
Inverse hyperbolic functions   120
Inverse trigonometric functions   121
Irreducible quadratic polynomial   235, 236

J

Jupiter   365

K

Kepler's problem   vi
Kinetic energy   354, 356, 357

L

Lagrange's form of remainder   153, 155, 157, 159, 160

Law for multiplication   11, 17
Laws of logarithms   86, 119
Least upper bound   12, 77, 79, 82, 207, 269, 299
Lebesgue measure   309, 311, 313
Lebesgue integral   257, 309, 319, 320, 322–325
Leibniz's series   159
Length of arc   132, 335, 337, 362
Lemniscate   343
L'Hopital's rule   185, 187, 188, 216, 217
Limit comparison test   54
Limit point   26, 27, 39, 41, 42, 44, 69, 70, 72, 75, 80, 94, 317
Linear approximation   112, 113, 143
Local maxima and minima   194
Logarithm   69, 85, 86, 119, 191, 193, 339
Logarithmic spiral   339
Lower bound   12, 77, 269, 281, 309, 310

M

Maclaurin series   145, 155–160, 185, 192, 283
Maclaurin's Formula   153, 160
Mass of the earth   358
Mass of the disk   354
Mass per unit length   351
Maximum of functions   145, 146, 194–206, 215, 217–219, 344
Maximum value   146, 194, 198, 203
Mean value theorem   145, 147, 150, 151, 168, 257, 260, 261, 276, 278, 279, 308, 320, 330
Measure zero   257, 278, 313, 316, 320
Mesh of partition   262–265, 270, 277, 278, 295, 342, 345, 347, 349, 352, 354
Minimum of function   145, 146, 194–197, 199–202, 211, 215, 217–219, 280, 344, 359
Method of cylindrical shells   347, 348
Method of substitution   221, 225, 226, 239–249, 251, 252, 255, 290, 296
Moment of inertia   353, 354
Monotone sequence   37, 63
Multiplication   11, 17, 69, 324

N

Natural exponential function   56, 159
Natural logarithm function   86
Necessary conditions for relative extrema   145, 194, 195

Neighborhood 41, 42, 91, 114, 115, 152, 155, 185, 188, 189, 194–197, 199, 209–211, 281
Nonincreasing sequence 9, 37, 38, 50, 58
Nested interval property 14, 15
Newton's second law of motion 117, 356
Normal line 129–131, 134–136, 144
Numerical integration 257, 294, 297, 298

O

Odd function 215
One-sided limits 70, 74, 76, 82, 108, 123, 146, 291
One-to-one function 83
Open interval 16, 26–29, 70, 74, 83, 85, 145–147, 150, 162, 164, 186, 202, 208, 278, 281, 291, 309, 311–313
Oscillation 271, 274–278
Ostrogradsky's method 237, 254

P

Parabola 129, 133, 136–139, 204, 219, 297, 298, 352, 365
Parabolic approximation 297
Parametric equations 107, 127, 130, 135, 144, 337, 338, 341, 346, 363
Partial sums of series 45–47, 49, 50, 58, 61, 92, 101, 165, 288
Partition 257, 262–278, 294, 295, 297, 299–308, 314, 315, 319, 322, 325, 335, 336, 340, 341, 344, 345, 347, 349, 352, 354, 355
Period 31, 88, 102
Perpendicular 129, 136, 345, 364
Point of discontinuity 75, 83, 84, 149, 215, 262, 266, 267, 277, 293, 303, 308
Point of inflection 210
Polar axis 138, 364, 365
Pole 137
Polygon 32, 266, 267, 335, 336, 339, 340, 344, 348, 349
Polynomials 68, 69, 73, 167, 171, 173–178, 181–183, 221, 230, 233, 237, 238, 242
Positive-term series 46, 49, 50, 54, 56, 57
Potential energy 356, 357
Power station 362
Power function 85, 119
Product of functions 229
Product of series 31, 59, 60
Proper rational fraction 230

Q

Quadratic polynomials 221, 238
Quadratic factors 232
Quotient of two functions 72, 73, 116, 185, 187, 188, 230, 326
Quotient of two polynomials 230
Quotient law for limits 72, 73

R

Radian measure 19, 113
Radius of convergence 162–165, 170
Rational function 68, 73, 214, 221, 230, 234–236, 242, 243, 246–248, 254
Rational number 7–11, 13, 14, 16, 26–29, 32, 44, 83, 85, 97–99, 102, 246, 278, 312, 314–316
Ratio test 51–53, 158, 163, 165
Reactor 335, 361, 362
Real numbers 1, 8–13, 15, 18, 21–24, 26, 29, 30, 32, 95, 47, 62, 67, 69, 75, 137, 202, 231, 278, 309, 319, 324, 335
Rearrangement 31, 55, 56
Rectangular approximation 294, 295, 299, 332
Reciprocal 11, 129
Refinement of partition 262, 268, 269, 271, 272, 306
Remainder of series 145, 152–154, 157, 159, 160, 171, 192, 197, 208, 211, 282
Repeated root 73, 174–176, 230, 231
Revolution of regions 347, 349
Revolution around x-axis 349
Riemann sum 263, 265–269, 272, 274, 294, 295, 327, 328, 340, 342, 344, 345, 347, 354
Right-hand limit 70, 76
Rolle's theorem 145, 148, 150, 167, 178
Roots of polynomials 174
Rose 343

S

Scalar(inner) multiplication 22, 324
Secant line 139, 167
Second derivative 123, 168, 178, 196, 202, 203, 206, 208–210, 215, 217, 295
Selection for partition 263, 265–269, 272, 273, 275, 297, 305, 342
Series 31, 32, 34, 45–62, 64–67, 92–96, 101, 105, 110, 145, 151, 155–166, 168–170,

185, 192, 257, 280, 282–284, 288, 289, 291, 302, 312, 326
Sketching graphs   211
Slope   109, 115, 129, 132, 134, 136, 139, 208
Solid of revolution   344–348, 364
Shell   347, 348
Space   22–26, 29, 41, 42, 44, 45, 100, 302, 303, 309, 324, 335, 344
Spiral   339, 363
Stieltjes integral   257, 305–308, 320
Substitution   90, 165, 175, 191, 221, 225–227, 234, 239, 241–249, 251, 252, 255, 290, 296
Surface area   202, 335, 348, 350, 361, 364, 365
Surface of revolution   348–350

T

Table of derivatives and integrals   121, 122, 223
Tangent function   86–88
Tangent line   115, 129–132, 134–136, 144, 147, 167, 208–210
Trigonometric functions   88, 90, 121, 160, 221, 245, 255
Taylor series   145, 155, 156, 164, 170, 185
Taylor's formula   145, 151–155, 197, 208, 211, 217, 296
Taylor's theorem   152
Termwise differentiation and integration   164, 170, 280
Test for concavity   208–210
Test for relative extrema   195, 197, 199
Transcendental function   69, 221, 237, 280
Trapezoidal approximation   296, 297, 299, 332
Trigonometric substitution   247

U

Uniform continuity   81, 97

Uniform convergence   92, 93, 100, 101
Union of sets   2, 3, 7, 26–29, 266, 312–314, 316, 318, 319, 321, 335, 336, 344
Upper and lower Riemann sums   266, 327, 340, 344
Upper bound   12, 13, 68, 77, 79, 82, 267, 269, 299, 337, 339, 355
Uranium isotopes   360

V

Vacuum   118, 227
Value of function   3, 67, 68, 72, 75, 77, 79, 82, 84, 85, 87, 101, 111, 117, 145, 146, 151, 154, 155, 165, 171, 173, 176, 178–181, 183, 194, 196–201, 203, 211, 215, 217, 218, 260, 280, 288, 309, 340
Value of the definite integral   260, 280, 284, 293, 299
Vector space   22–24
Velocity   108, 109, 117, 118, 128, 203, 227, 258, 335, 354, 356, 357–359, 361, 365
Vertically asymptote   212–214, 215
Vertical tangent line   208
Volumes by the Method of Cross Sections   345
Volume by cylindrical shell method   347, 348
Volume of solid of revolution   344, 345

X

x-axis   80, 129, 203, 263, 340, 341, 344–351, 353, 364, 365
x-coordinate   22, 135, 139, 351

Y

y-axis   79, 208, 210, 215, 345–349, 351, 364, 365
y-coordinate   138, 139, 351, 353

Z

Zeno's paradox   31
Zeros of polynomials   174

Printed in the United States
By Bookmasters